ECO-FRIENDLY TEXTILE DYEING AND FINISHING

Edited by **Melih Günay**

Eco-Friendly Textile Dyeing and Finishing

http://dx.doi.org/10.5772/3436
Edited by Melih Günay

CBS, Edition **2016**

Contributors

Bojana Voncina, George Z. Kyzas, Nikolaos Lazaridis, Dimitrios Bikiaris, Margaritis Kostoglou, Marija Gorjanc, Marija Gorenšek, Petar Jovančić, Miran Mozetič, Chaoxia Wang, Yunjie Yin, Riza Atav, Shahidi, Farah Chequer, Gisele Oliveira, Maria Valnice Zanoni, Danielle Palma De Oliveira, Juliano Cardoso, Elisa Ferraz, Heba Farouk Mansour, Dimitris Kekos, Styliani Kalantzi, Diomi Mamma

Published by InTech

Notice

Publishing Process Manager Sandra Bakic

Typesetting InTech DTP Team

Cover InTech Design Team

Additional hard copies can be obtained from orders@intechopen.com

Eco-Friendly Textile Dyeing and Finishing , Edited by Melih Günay
p. cm.
ISBN 978-953-51-0892-4

INTECH

open science | open minds

Contents

Preface

Initially the sole purpose of dyeing was to color textile substrates for fancy fabric appearances. Although, this was an impressive achievement at the time, the competitive challenges began to drive the development of highly functional fibers and substrates through advanced dyeing and finishing processes for higher added value in applications of; membrane filtration, coatings, composites, microelectronic devices, thin-film technology, super absorbency, antimicrobial materials, biocides and insecticides, flame reterdancy, improved reactivity and numerous others.

Polymeric fibers that are mechanically strong, chemically stable, and easy to process often have inert surfaces which makes them not suitable for these advanced applications. Consequently, there has been significant number of studies that focuses on enhancing the chemical, biological, physical, optical and dyeability properties of fibers without negatively effecting their mechanical and most desired properties. Among the techniques, perhaps the plasma treatment is one of the most investigated. Also, Cyclodextrins which can act as hosts and form inclusion compounds with various small molecules to provide certain desired attributes may be applied to textile substrates as reagent during the finishing processes. The majority of these studies often involve a) the embedding of novel nanoparticles for adding unique features to textiles, b) uniformly maximizing the loading capacity of textile substrates to improve nanoparticle adsorption for optimal surface property.

While for the purpose of coloration only natural dyes were used initially. Due to limitations in coloration and with the invent of synthetic fibers, natural dyes are mostly replaced with dyes themselves are either chemically hazardous or require auxiliary chemicals that are not good for the environment. At the beginning, we were not as concerned of the damage caused by dyeing to the environment. However over time as we come to understand that our being healthy and well being also depends on our environment, we have been increasingly paying attention to reduce our footprint on our ecosystem. In particular for dyeing of textiles, the efforts primarily focuses on reducing the water consumption, using of natural dyes or less harmful dyes and chemicals, right-first-time dyeing, the development of an effective degumming process based on enzymes as active agents , dyeing and energy optimization and development of advanced waste water treatment processes. In recent years, many attempts have also been made to improve various aspects of dyeing by the introduction and advancement of new technologies that used ultrasound, ultraviolet, ozone, plasma, microwave, gamma irradiation, laser, supercritical carbondioxide.

Consequently, as market forces demand unique and sophisticated products from the textile industry, recent research has been focused on the development of technologies for functional textiles some of which implemented advanced finishing and dyeing techniques. Meantime, the local governments and regulators also require the textile industry to become more environment friendly in their operations. Hence, this book aims to present the cutting edge research in both areas to advance the knowledge in this field.

Dr. Melih Günay
HueMetrix Inc. NC, USA

Advances in Dyeing Chemistry and Processes

Chapter 1

Multifunctional Textiles – Modification by Plasma, Dyeing and Nanoparticles

Marija Gorjanc, Marija Gorenšek, Petar Jovančić and
Miran Mozetič

Additional information is available at the end of the chapter

http://dx.doi.org/10.5772/53376

1. Introduction

The textile industry in developed countries is confronting the world's marketing conditions and competitive challenges which are driving towards the development of advanced, highly functional textiles and textiles with higher added value. The conventional textile finishing techniques are wet chemical modifications where water and rather hazardous chemicals are used in large quantities and wastewaters need to be processed before discharging effluent, whereas the most problematic factor are ecological impacts to the environment and effects to human health. The increasing environmental concerns and demands for an environmentally friendly processing of textiles leads to the development of new technologies, the use of plasma being one of the suitable methods [1]. Plasma technology is an environmentally friendly technology and a step towards creating solid surfaces with new and improved properties that cannot be achieved by conventional processes [2]. Plasma is the fourth state of matter. It is a gas with a certain portion of ionized as well as other reactive particles, e.g. ions, electrons, photons, radicals and metastable excited particles. Several types of plasma are known; however, only non-equilibrium or cold plasma is used for the modification of physical and chemical properties of solid materials such as textiles. Chemically reactive particles produced at a low gas temperature are a unique property of cold plasma; hence, there is minimal thermal degradation of a textile substrate during the plasma processing [3]. Cold plasma is a partially ionized gas with the main characteristic of a very high temperature of free electrons (typically of the order of 10,000 K, often about 50,000 K) and a low kinetic temperature of all other species. The average energy of the excited molecules is usually far from the values calculated from the thermal equilibrium at room temperature. The rotational temperature, for instance, is often close to 1000 K, while the vibrational temperature can be as

high as 10,000 K, although the kinetic (translation) temperature is close to room temperature. Furthermore, the dissociation fraction is often several percent, which is orders of magnitude larger than that calculated from thermal equilibrium at room temperature. This also applies to the ionization fraction; although this is often much lower than the dissociation fraction. Plasma with such characteristics readily interacts with solid surfaces, causing reactions that would otherwise occur only at elevated temperature of the solid material. For this reason, non-equilibrium plasma represents an extremely powerful medium for modification of the surface properties of solid materials. A medium of particular interest is weakly ionized highly dissociated oxidative plasma that can be sustained in high frequency discharges in oxygen, air, carbon dioxide, water vapor, and mixtures of these gases with a noble gas. Such plasma has been successfully utilized in an extremely wide range of applications from nanoscience to fusion reactors. Plasma is used for synthesis of nanostructures with interesting properties, removal of thin films of organic impurities, selective etching of composites, sterilization, passivation of metal, ashing of biological materials, etching of photoresists, functionalization of polymers, and conditioning of tokamaks with carbon walls [4-9].

The choice of discharge parameters is determined by the requirements of each particular application. For selective plasma etching, for instance, extremely aggressive plasma is needed; thus, it is created with powerful generators at a moderate pressure (where the O density is the highest) in pure oxygen or in a mixture of oxygen and argon. For treatment of delicate organic materials, on the other hand, weak plasma performs better, since aggressive plasma would destroy organic material in a fraction of a second. Therefore, extremely delicate organic materials are rather treated in an afterglow or in plasma created at low pressure and with a low-power generator. Water vapor is sometimes used instead of oxygen. The advantages of using plasma are ecological and economical. Moreover, the textiles subjected to the treatment are modified without an alteration of the bulk properties. Unlike wet chemical processes, which penetrate deep into the fibers, plasma produces no more than a surface reaction, the properties given to the material being limited to the surface layer of a few nanometers [10]. The modification of textile substrates using plasma enables different effects on the textile surfaces from the surface activation to a thin film deposition via plasma polymerization. In the first stage of the treatment, plasma reacts with the substrate surface where active species and new functional groups are created, which can completely change the reactivity of the substrate [11]. The changes in the surface morphology of fibers can be induced by plasma etching process where the nano- or micro-roughness of fibers is formed [12]. The nanostructured textile surfaces have a higher specific surface area, which leads to new or improved properties of the treated surface, i.e. increased surface activity, hydrophilic or hydrophobic properties, and increased absorption capacity towards different materials, i.e. nanoparticles and nano-composites [13- 18].

When researching the deposition of nanoparticles onto different substrates, it is important to understand the adhesion of particles, which is dependent on the interaction mechanism with a material. The mechanism of nanoparticle adhesion has not been completely explained yet, since there are many different opinions among the theorists on the subject. Thus, it is generally considered that attractive forces and chemical bonds play an important role in the adhesion of particles [19]. The physical or mechanical adhesion of nanoparticles mostly occurs due to van der Walls or electrostatic forces, while the chemical adhesion of particles is a consequence of ionic, covalent, metallic and hydrogen bonds [20]. Moreover, the nanoparticles can penetrate into certain parts of the substrate, such as pores, holes and crevices, and they lock mechanically to the substrate. This adhesion mechanism, which is called mechanical interlocking, has been solved from the perspective of surface roughness effects [21]. Since plasma causes etching of the fibers and leads to an increase of the surface roughness higher adhesion properties towards metal or ceramic particles onto substrates can be achieved [22-30]. The adhesion of TiN (titanium nitride) onto PP (polypropylene) and PC (polycarbonate) was increased after a modification of substrates with argon low-pressure plasma [31]. The roughness after plasma treatment increased from 15 nm to 17 nm for PP and from 12 nm to 30 nm for PC. Consequently, contact angles decreased from 95° to 59° for PC and from 87° to 35° for PP. The surface modification of polyethylene terephthalate (PET) polymer was created by oxygen and nitrogen plasma at different treatment times [32]. The surface of PET polymer was modified in order to achieve improved attachment of fucoidan, which is a bioactive coating with antithrombogenic properties. The attachment of fucoidan was improved by oxygen plasma treatment, especially due to the surface roughening. The adhesion work, the surface energy and the surface polarity of PA6 (polyamide-6) fibers were improved by dielectric barrier discharge (DBD) treatment in helium at atmospheric pressure. Furthermore, a new structure was observed at the nanoscale, with an increased roughness and a larger surface area, favoring the adsorption [33]. The self-cleaning and UV protective properties of PET fibers were drastically improved after a modification of PET fibers with oxygen plasma and loading of TiO_2 prepared by an aqueous sol-gel process [34]. Cotton also showed self-cleaning properties after RF plasma and TiO_2 treatment [35]. TiO_2 on textile substrates is also used for a biomedical application to improve antimicrobial effectiveness of the fabric [36]. By using oxygen radiofrequency plasma at a higher power input, the roughness of fibers increased and likewise the adhesion of TiO_2 onto treated fabric. Treatment of PA and PET with corona plasma increased the adhesion of colloidal silver which affected the antifungal protection of the fabrics [37]. The quantity of silver on plasma-treated fabric was three times higher than on untreated fabric.

Preparation of metal nanoparticles also enables the development of new biocides. Due to their large surface area and ability to detain moisture, the textile materials are an excellent environment for a microorganism growth. Microorganisms can cause milder, aesthetic unpleasantness to serious health related problems. Textile materials with an antimicrobial effectiveness are used for medical, military and technical textiles, textiles for sports and leisure and bedding. At nanotechnology researches in textiles, different forms of silver were used, such as metal silver nanoparticles, silver chloride (AgCl) and composite particles of silver and titanium dioxide ($Ag-TiO_2$) [3, 24, 25, 27, 38-47]. In the case of antimicrobial efficiency, the surface coating of nanosilver on titanium dioxide maximizes the number of particles per unit area in comparison with the use of an equal mass fraction of pure silver [48,49]. Different methods have been used for the deposition and loading of silver nanoparticles onto synthetic and natural textiles, i.e. sonochemical coating, sol-gel process, dip-coating, pad-batch and exhaustion method, the use of nanoporous structure of cellulose fibers as a nanoreactor for *in situ* synthesis of nanoparticles and plasma sputtering process [3, 27, 48-54]. The possibility of loading nanoparticles using exhaustion method started recently [55]. The exhaustion method is the best process for uniform distribution of nanoparticles and is especially appropriate to be used when the simultaneous application of nanoparicles and dye onto fabric is performed, and dyed and antimicrobial effective fabric is achieved at the same time [3, 24, 25, 27]. Depending on the desirable functionality of a functionalized fabric, a treating bath may contain only dye, dye and silver nanoparticles or silver nanoparticles alone. An exhaustion method for loading of silver nanoparticles onto textile substrate was also used on silk fibers [54]. In that research, different concentrations of colloidal silver were used (10, 25, 50 and 100 ppm) and the effect of medium pH on the silver nanoparticles uptake on the fibers was studied. The antimicrobial effectiveness of functionalized fibers was better for samples with higher silver concentration and for samples treated in a medium with a lower pH. Also the use of salt (NaCl) improved a uniform distribution of silver particles on the fibers' surface which consequently improved antimicrobial effectiveness of fabrics. A difference between pad-dry-cure and exhaustion method on adhesion and antimicrobial activity of fabrics was performed with commercial silver nanoparticles and a reactive organic-inorganic binder [45]. Results revealed that using the same initial concentration of silver, the pad-dry-cure method resulted in a much lower quantity of adsorbed silver nanoparticles in comparison to the exhaust method. Another possible method for applying silver onto textiles is plasma polymerization method where surface of textile is functionalized with nanostructured silver film by magnetron sputtering [56].

When dealing with modification of textiles by silver it is important to know how the functionalization will affect the color change of fabric. A reflectance (UV/VIS) spectrophotometry is one of the methods to be used when detecting or controlling the presence of silver nanoparticles in a solution or on a textile substrate. It is a direct measure that the abundance of silver is in the topmost layer of the textile fabric [30]. The instrument analyzes the light being reflected from the sample and produces an absorption spectrum. Some of the electrons in the nanoparticles are not bound to the selected atom of silver, but are forming an electronic cloud. Light falling on these electrons excite the collective oscillations, called surface plasmons. The resonance condition is established when the frequency of light photons matches the natural frequency of surface electrons oscillating against the restoring force of positive nuclei. Surface plasmon resonance is the basis of many standard tools for measuring adsorption of material onto planar metal (typically gold and silver) surfaces or onto the surface of metal nanoparticles. It is the fundamental principle behind many color-based biosensor applications. As a result of the particles growth, an intense absorption band at 400 nm to 415 nm caused by collective excitation of all free electrons in the particles was observed [57]. Increase in diameters of the nanoparticles from 1 to 100 nm induces a shift of the surface plasmon absorption band to higher wavelength [50, 57-59]. That means that the size of nanoparticles is defined by their optical response, therefore by that the color that can be seen [50, 60]. Loading of colloidal silver nanoparticles onto bleached cotton fabric caused yellowish coloring of fabric with absorption maximum at 370 nm [24, 30]. The evaluation of color changes of textiles modified by silver nanoparticles was also determined in CIELAB color space [60]. Loading of colloidal silver nanoparticles in concentration of 10 ppm induced very small, eye insensitive, color change ($\Delta E^* < 1$) to the fabric. When colloidal silver nanoparticles in concentration of 50 ppm were loaded, the color change of fabric was obvious ($\Delta E^* = 15.09$). Loading of silver onto textiles before or after dyeing also causes color changes; however the changes are not as extensive as with the white fabric. When colloidal silver was loaded before dyeing, the color change was $\Delta E^* = 1.44$ and when loaded after dyeing, the color change was $\Delta E^* = 2.73$. Color changes of textiles can be also induced by plasma treatment. When plasma is used for a modification of cotton to improve its hydrophilicity, then cotton has a better dyeability and consequently deeper coloration [61]. Also, a raw cotton fabric can be bleached by using ozone plasma [62]. The whiteness index (CIE WI) of fabric was even higher after plasma treatment than after peroxide bleaching (CIE WI O_3 = 95.3; CIE WI H_2O_2 = 94.5).

Since the modification of substrates by plasma improves the adhesion of metal nanoparticles onto substrates than it is no surprise that plasma modification of textiles has a special significance when applying silver nanoparticles. The chapter presents the influence of plasma treatment on loading capacity of cotton toward different forms of nanosilver. Exhaust dyeing process was used for loading of nanosilver onto plasma-treated cotton, which represents a new approach to textile finishing in achieving multifunctional textile properties.

2. Experimental setup and methodology

Low-pressure plasma of different working gases and air atmospheric corona plasma were used for a modification of textiles. Untreated and plasma-treated textile substrates were additionally modified by loading of silver nanoparticles during dyeing process. Morphological, chemical and physical properties of plasma-treated textile substrates were studied using microscopy (SEM), spectroscopy (XPS) and measuring of the breaking strength and elongation of textiles. The quantity of adsorbed silver was determined using mass spectrometry (ICP-MS), while the antibacterial efficiency of functionalized textiles was determined using microbiological tests.

2.1. Plasma modification of textiles

Cotton substrates were modified using different plasma systems, i.e. atmospheric air corona plasma [63-65] and low-pressure inductively coupled radiofrequency (ICRF) discharge plasma of different working gases, water vapor [3, 26, 66, 67] and tetrafluoromethane [27], respectively. ICRF plasma is particularly suitable for treatment of delicate materials with a large surface for the following reasons: the neutral gas kinetic temperature remains close to the room temperature; the plasma-to-floating potential difference is small; the density of neutral reactive particles is large; and extremely high treatment uniformity is achieved. The use of low-pressure plasma is a contemporary technological process, not yet fully applied in textile industry due to its discontinuous process. For a continuous, on-line processing interfaced to a conventional production line the use of corona atmospheric pressure plasma is recommended. Furthermore, the corona plasma treatment also introduces new functional groups onto fibers surfaces, produces surface cleaning and etching effect of treated textiles. The comparative research of corona plasma treatment of polyester was presented as well [24].

2.1.1. Low-pressure plasma treatment

A RF generator with a nominal power of 5 kW and a frequency of 27.12 MHz was applied. The power absorbed by plasma, however, was much smaller due to poor matching and was estimated to about 500 W. The discharge chamber was a cylindrical Pyrex tube with a diameter of 27 cm and a length of 30 cm. Cotton fabric was put onto a glass holder mounted in the center of the discharge chamber. After closing the chamber the desired pressure of 0.4 mbar was achieved by a two-stage rotary pump with a nominal pumping speed of 65 m³/h. The pressure was fairly stable during the experiment. Although the ultimate pressure of the rotary pump was below 1 Pa, the pressure remained much higher at 40 Pa. The source of water vapor was the cotton fabric itself. When tetrafluoromethane was used as a working gas, it was leaked into the chamber in order to obtain a pressure of about 100 Pa, the pressure where plasma is most reactive. By switching on the RF generator, the gas in the discharge chamber was partially ionized and dissociated, starting the plasma treatment of the fabric. The plasma treatment time was 10 s in both cases. The schematic diagram of low-pressure RF plasma reactor is presented in Fig. 1.

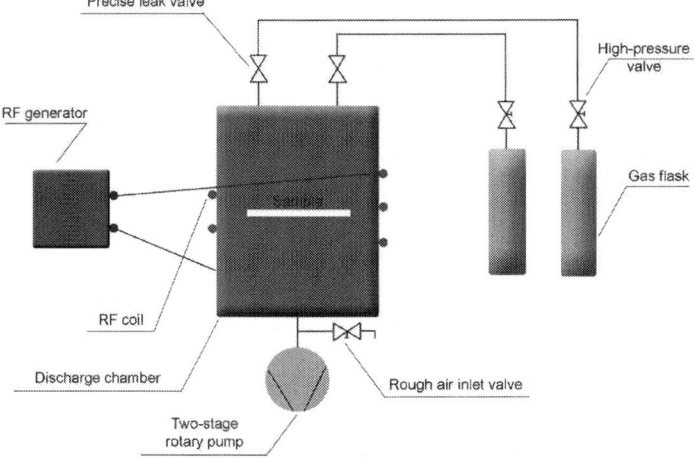

Figure 1. Schematic diagram of low-pressure RF plasma reactor

2.1.2. Atmospheric plasma treatment

Textile samples were treated in a commercial device, Corona-Plus CP-Lab MKII (Vetaphone, Denmark) (Fig. 2). Samples (270 × 500 mm²) were placed on a backing roller (the electrode roll covered with silicon coating), rotating at a working speed of 4 m/min. The distance between electrodes was adjusted with air gap adjusters at both sides of the electrode to 2 mm. Corona discharge was generated within the air gap between the electrode and backing roller. The power was 900 W and the number of passages was set to 30.

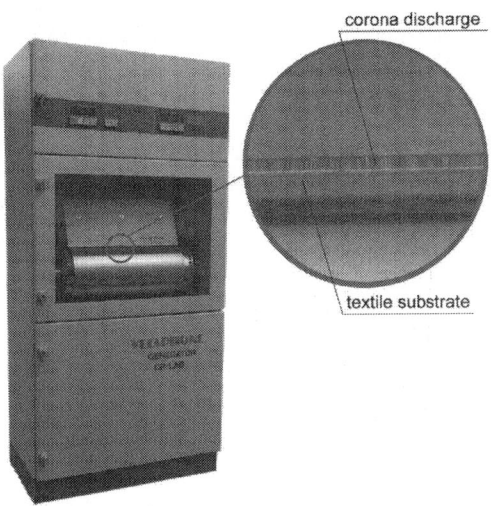

Figure 2. Picture of atmospheric corona plasma reactor and a detail of treating area shown in a red circle

2.2. Morphological, chemical and physical properties of untreated and plasma-treated textiles

2.2.1. X-ray photoelectron (XPS) analysis

Information on the chemical composition and chemical bonds of surface atoms of untreated and plasma-treated textile samples was obtained with XPS analysis. During the XPS analysis, a sample is illuminated with monochromatic X-ray light in an XPS spectrometer and the energy of emitted photoelectrons from the sample surface is analyzed. In the photoelectron spectrum, which represents the distribution of emitted photoelectrons as a function of their binding energy, peaks can be observed that may correspond to the elements present on the sample surface up to about 6 nm in depth. From the shape and binding energy of the peaks within XPS spectra, the chemical bonding of surface elements was inferred with the help of data from the literature.

2.2.2. Scanning Electron Microscopy (SEM)

The morphological surface properties of fibers and their changes after plasma treatment were studied using scanning electron microscopy (JEOL SEM type JSM-6060LV). All samples were coated with carbon and a 90% Au/10% Pd alloy layer.

2.2.3. Dynamometer tensile testing

Breaking strength and elongation of untreated and plasma-treated fabrics were analyzed according to the ISO 2062:1997 standard. Instron 6022 was used for this purpose. 100 mm samples of cotton yarn were analyzed using a pre-loading of 0.5 cN/tex and a speed of 250 mm/min. Samples were conditioned according to the ISO 139 standard. Tensile stress (cN/dtex) and elongation ε (%) are the mean values of the measured tensile strength and the elongation of 10 specimens, respectively, in the warp and in the weft directions.

2.3. Modification by dyeing and nanoparticles

2.3.1. Nanoparticles

For a research three different types of silver nanoparticles were used: silver nanoparticles of known dimensions (Ames Goldsmith Inc.), commercial form RucoBac AGP (Rudolph Chemie) and laboratory synthesized colloidal silver nanoparticles. Silver nanoparticles of known dimensions was medium density mono-dispersed 30 nm silver powder (Silver Nano Powder NP-30) and high density mono-dispersed 80 nm silver powder (Silver Nano Powder NP-80) [68]. The powdered silver nanoparticles are namely intended for a fine line printing (Ink-jet printing) and for electronics use. Because of their purity and known size they are very suitable and important for studies of adhesion to textile materials [69]. The initial concentration of silver powders in the dyeing baths was 20 mg/l. Commercial form of silver nanoparticles RucoBac AGP is a hygienic finish for all fiber types. It is a highly concentrated hygiene and freshness system complying with the Oeko-Tex® standard 100. RucoBac AGP is a nano-dispersion of titanium dioxide (TiO_2) as the carrier of the active component silver chloride (AgCl) [70]. The recommended concentration of RucoBac AGP is 0.2–0.5%. Laboratory synthesized colloidal silver nanoparticles were made by a reduction of silver salt in an aqueous solution at room temperature. An amount of 25.5 mg of $AgNO_3$ was dissolved in 750 ml of bi-distilled water, and argon gas was introduced into the solution for 30 min. The reduction was performed with 375 mg of $NaBH_4$ under constant stirring. The solution was kept under an argon atmosphere for another hour [71].

2.3.2. Dyeing of cotton samples

2.3.2.1. Dyeing with reactive dyes.

Exhaust dyeing with or without the addition of silver nanoparticles [3, 27, 72] was performed on untreated and plasma-treated cotton samples. The dye solutions were separately prepared using two different dyeing concentrations: blank (dyeing without dyestuff) and 0.05% on weight of fabric (owf) of Cibacron Deep Red S-B in liquor ratio 20 : 1. Dyeing baths contained 30 g/l of salt (Na_2SO_4 anhydrous) and 8 g/l of sodium carbonate (Na_2CO_3 anhydrous). After the dyeing process, rinsing with distilled water was performed and neutralization with 1 ml/l of 30% acetic acid (CH_3COOH). As a soaping agent 1 g/l of CIBAPON R was used.

2.3.2.2. Vat dyeing procedure

Vat dyeing was performed on untreated and plasma-treated cotton samples. The dyed cotton fabrics were post-treated in a colloidal silver solution [71]. A dyeing bath was prepared with 4% owf Bezathren Blau BCE, 15 ml/l NaOH 38°Bé and 3.5 g/l $Na_2S_2O_4$. The liquor ratio was 83.3 : 1. The dyeing of cotton fabrics was performed at 60°C for 60 min. The dyed samples were rinsed twice in deionised water for 5 min, post-treated in 2 ml/l HCOOH 85% for 5 min and rinsed in deionised water for 5 min.

2.4. Color measurements

Color measurements of differently modified samples were performed after conditioning them according to the ISO 139 standard. CIE standard illuminant D65/10 was used on a Datacolor Spectraflash SF 600-CT reflection spectrophotometer. The aperture diameter of the measuring port of the spectrophotometer was 9 mm.

2.5. Elemental and antimicrobial analysis

The quantity of adsorbed silver was determined using ICP-MS. This technique combines a high-temperature inductively coupled plasma (ICP) source with a mass spectrometer (MS). The antibacterial efficiency of functionalized textiles was determined using microbiological tests according to the ASTM Designation: E 2149–01 method. Antibacterial activity of the cotton fabrics was tested by a certified laboratory against *Staphylococcus aureus* (ATCC 25923), *Escherichia coli* (ATCC 25922), *Streptococcus faecalis* (ATCC 27853) and *Pseudomonas aeruginosa* (ATCC 27853).

3. Results and discussion

The influence of different plasma systems on the adhesion of nanoparticles is discussed. The emphasis of the study is to use minimal concentrations, initially, of nanoparticles for loading onto textiles and to achieve maximum quantity on the material. Exhaust dyeing process was used for loading of silver nanoparticles onto textiles. Before applying nanoparticles to any material, its surface needs to be adequately prepared and chemically and morphologically well analyzed. Only good conditions on the substrate surface can provide a qualitative deposition of particles [73]. In literature one can find quotations of XPS analysis of plasma modified cotton substrates which were pre-prepared with various procedures prior to plasma modification (i.e. alkaline boiling, scouring, laundering). But to study plasma modification of cotton it is important to know about the surface changes of substrates that were not cleaned or otherwise pre-prepared. Therefore, the chemical surface changes were evaluated for raw, bleached and bleached/mercerized cotton before and after plasma treatment.

The surface of raw untreated cotton fabric contains a high concentration of carbon and a low concentration of oxygen. This is not characteristic for native cellulose [74]. XPS spectrum C 1s of cellulose also does not include C-C/C-H bonds (Figure 3). The surface of raw cotton is rich in C atoms (C-C/C-H bonds), what could indicate the presence of C-C/ C-H bond rich substances, such as waxes, pectin and proteins. The surfaces of bleached and bleached/mercerized cotton fabrics are alike but very different from raw cotton. Bleached and bleached/mercerized cotton samples are pre-oxidized due to the scouring, bleaching and mercerizing process, and therefore more similar to native cellulose. After modifications with atmospheric air corona plasma and water vapor low-pressure plasma cotton samples contain a higher O/C ratio (the samples contain more of the oxygen and less of the carbon atoms) (Figure 4).

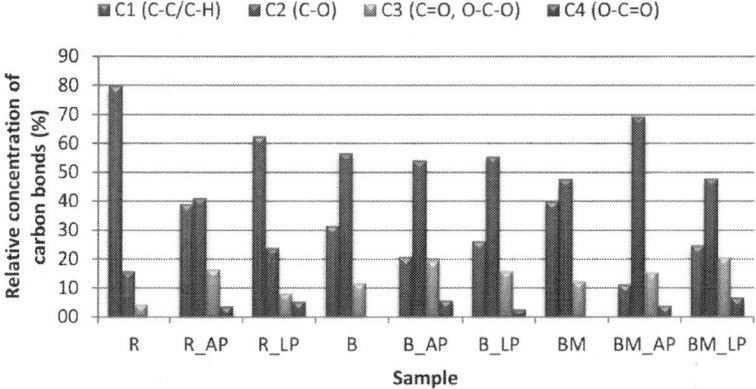

Figure 3. Relative concentration of carbon bonds on the surface of cotton samples: R – raw cotton, B – bleached cotton, BM – bleached/mercerized cotton, AP – atmospheric plasma treatment, LP – low-pressure plasma treatment

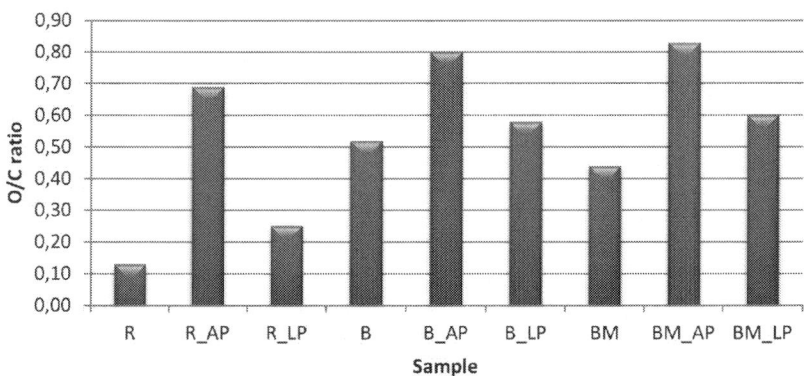

Figure 4. The concentration ratio between oxygen and carbon on the surface of cotton samples: : R – raw cotton, B – bleached cotton, BM – bleached/mercerized cotton, AP – atmospheric plasma treatment, LP – low-pressure plasma treatment

The increase of O/C ratio is expected since plasma interaction with cotton causes the surface oxidation. Changes of C-atom bonds on raw cotton samples after corona and low-pressure plasma treatment are visible in Figure 3 as the ratio of C-O and C=O bonds increased. The results show that increase of relative concentration of C-O bonds is distinctive after corona plasma treatment (41 %) than after low-pressure plasma treatment (23.9 %). It is similar for C=O bonds. The relative concentration of C=O bonds on the surface of raw cotton increases to 16.6 % after corona plasma treatment and to 8.2 % after low-pressure plasma treatment. Both plasma treatments result in a formation of O-C=O bonds. C 1s spectrum of plasma-treated samples resembles to the spectra of native cellulose, which could indicate that plasma selectively removed the noncellulosic parts present on a surface of a raw untreated cotton fabric. After plasma treatment the oxygen content on surface of bleached and bleached/mercerized cotton samples increases. After a treatment with corona plasma the content of oxygen on bleached cotton sample increases from 34.3 at% to 44.5 at%, while the treatment with low-pressure plasma increases oxygen to 36.8 at%. After both plasma treatments the content of C-O remains the same (~55 %), the content of C=O bonds increases and appearance of O-C=O bonds is noticeable. Increase of bonds is distinctive after corona plasma treatment than after low-pressure plasma treatment. Content of C-C/C-H bonds decreases after both plasma treatments. Increase of O/C ratio on plasma-treated bleached/mercerized cotton samples is distinctive after corona plasma treatment, where the oxygen concentration increases to 45.3 at%, while increases to 37.5 at% after low-pressure plasma treatment. Due to oxidation process in plasma, the content of C-C/C-H bonds deceases and the change is more noticeable for bleached/mercerized cotton than for a bleached cotton fabric. For bleached/mercerized cotton sample concentration of C-O bonds is increased in the case of corona plasma treatment, from 47.6 % to 69.3 %, while it remains almost equal after low-pressure plasma treatment (47.8 %). After both plasma treatments the content of C=O bonds increases and appearance of O-C=O bonds is noticeable. These changes are more distinct after low-pressure plasma treatment (20.7 % and 6.8 %) than after corona plasma treatment (15.5 % and 3.9 %). The results summarized in Figure 4 show higher concentration of oxygen on atmospheric air corona and water vapor low-pressure plasma-treated raw cotton than on untreated bleached or bleached/mercerized cotton. From these results it can be concluded that by using plasma technology some of the technological processes of pretreatment of cotton fabrics before dyeing can be avoided. Plasma-treated raw cotton fabric absorbs the same or more of dyestuff as untreated bleached or bleached/ mercerized cotton fabric [3, 65, 65, 76]. Reactive dyes are typical anionic dyes containing reactive systems which in alkaline media make bonds with –OH groups of cotton fibers. Increase of dyeability of cotton treated with atmospheric air corona plasma or water vapor low-pressure plasma is correlated to an increase of the number of hydrophilic groups on the surface of the fibers. In Figure 5 the correlation between color differences (expressed as ΔE^*) of reactive dyed cotton and differences in O/C ratio (expressed as $\Delta O/C$) of untreated and plasma-treated cotton is presented. The most perceptible changes in color differences (ΔE^*) are noticeable when differences in O/C ratio ($\Delta O/C$) are higher. Visible color differences between untreated and water vapor plasma-treated cotton were noticeable for raw ($\Delta E^* = 2.20$) and bleached/ mercerized cotton ($\Delta E^* = 1.07$).

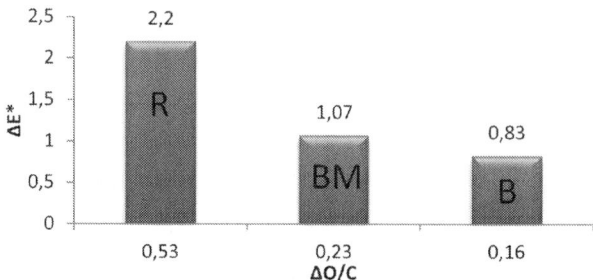

Figure 5. The correlation between color differences (expressed as ΔE*) of reactive dyed cotton and differences in O/C ratio (expressed as ΔO/C) of untreated and plasma-treated cotton (R – raw cotton, B – bleached cotton, BM – bleached/mercerized cotton)

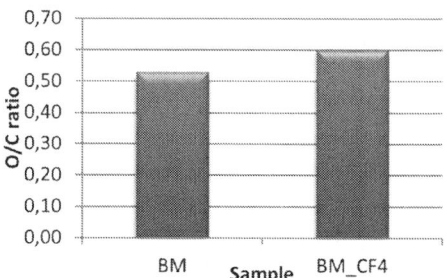

Figure 6. The concentration ratio between oxygen and carbon on the surface of untreated (BM) and CF$_4$ plasma-treated (BM_CF4) cotton samples

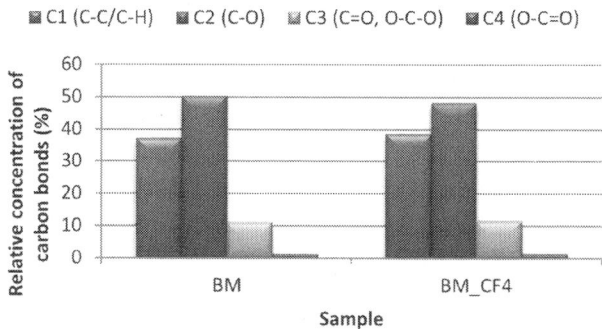

Figure 7. Relative concentration of carbon bonds on the surface of untreated (BM) and CF$_4$ plasma-treated (BM_CF4) cotton samples

The samples modified in tetrafluoromethane low-pressure plasma (CF$_4$ plasma) gave different results. There was no visible color difference between untreated and CF$_4$ plasma-treated bleached/mercerized cotton samples (ΔE* = 0.62). Both samples were practically equally colored. Treating cotton fabrics for 10 sec with CF$_4$ plasma does not influence their dyeing properties. The results of XPS study show a small increase in the concentration of oxygen after CF$_4$ plasma treatment, but the result is hardly remarkable (Figure 6 and 7).

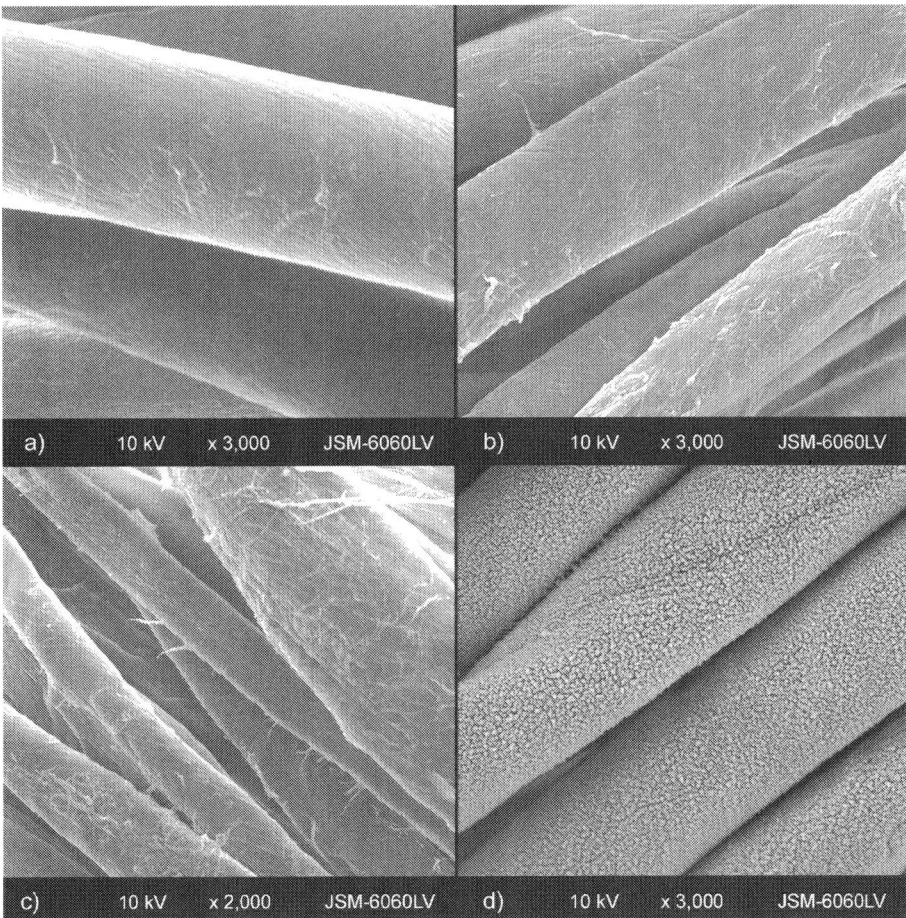

Figure 8. SEM images of untreated and plasma-treated cotton samples

It is interesting, however, that no fluorine was observed in the XPS spectra, but that the additional ISE analysis indicated the presence of <5 ppm of total fluoride on CF$_4$ plasma-treated samples. Here, it is worth mentioning that we are not the first group to observe little or no fluorine on organic materials treated with CF_4 plasma [76, 77]. Although XPS analysis did not show specific chemical surface differences between untreated and CF_4 plasma-treated cotton samples, the SEM analysis showed strongly modified surface morphology of plasma-treated cotton (Figure 8), while smaller changes of the surface morphology of cotton samples treated with atmospheric air corona plasma and water vapor low-pressure plasma are noticed.

SEM image of untreated cotton (Figure 8 a) shows a typical grooved surface morphology with macrofibrils oriented predominantly in the direction of the fiber axis. The outlines of the macrofibrils are still visible, and they are smooth and distinct due to the presence of an amorphous layer covering the fiber. The surface of corona plasma-treated cotton has striped, cleaned and more distinct macrofibrilar structure (Figure 8 b). The same effect can be noticeable on the surface of cotton fibers treated with water vapor low-pressure plasma (Figure 8 c). The plasma-treated fiber surface remains grooved and the macrofibrile structure has gained a much sharper outline. The individual macrofibrils (0.2–1 nm) and their transversal connections are visible in the primary cell wall. Between them, narrow voids thinner than 10 nm are noticeable. The surface morphology of CF_4 plasma-treated cotton (Figure 8 d) is very different comparing to those treated with water vapor low-pressure or corona plasma. CF_4 plasma-treated cotton has an extremely rough and nanostructured surface with dimensions of the grains roughly between 150 and 500 nm. Dissociation energies of plasma molecules are the reason for such rich etched surface. Both CF_4 and H_2O molecules get dissociated in our plasma [78]. The dissociation energy of CF_4 is 12.6 eV [79] while the ionization energy of CF_4 is about 16 eV, which is much higher than the dissociation energy of water vapor or OH molecules, which is at about 5 and 4 eV, respectively. The electrons with energy of few eV are therefore likely to dissociate water and OH molecules rather than dissociate the CF_4 molecules. The result is an extremely high dissociation fraction of water molecules and a moderate dissociation fraction of CF_4 molecules. Since the partial pressures of water and CF_4 are comparable, it can be concluded that the density of O and OH radicals in our plasma is at least as high as the density of CF_x radicals if not more so. The textile sample exposed to plasma is therefore subjected to interact with CF_x, O, OH, F and H radicals. The density of H is probably as high as that of F so extensive recombination to HF is expected. CF_x radicals tend to graft onto the textile, but the grafting probability on cellulose is low compared to the interaction probability with O and OH radicals. O and OH radicals are extremely reactive and cause etching of cellulose. Fluorine atoms are efficient at abstracting hydrogen in the first step of the oxidation reaction. In addition, the presence of fluorine atoms in the plasma enhances the dissociation of the oxygen, further increasing the ashing rates [80]. CF_4 is the gas most commonly added to oxygen to enhance the generation of atomic oxygen in plasma and to increase polymer etch rates [81].

While the surface of plasma-treated cotton was changed, the mechanical properties of cotton fabrics did not alter after plasma modification. The breaking strength and elongation of textiles practically do not change (Figure 9 and 10) [3, 27]. These results are in accordance with the results obtained by other authors [82-85]. Plasma modification of textiles does not impair their mechanical properties.

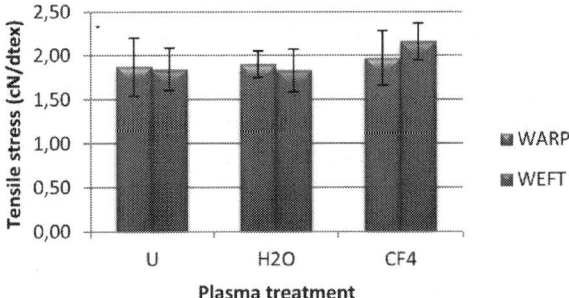

Figure 9. Tensile stress (cN/dtex) of cotton after plasma treatment: U – untreated, H2O - ICRF water vapor plasma-treated, CF4 - ICRF tetrafluoromethane plasma-treated

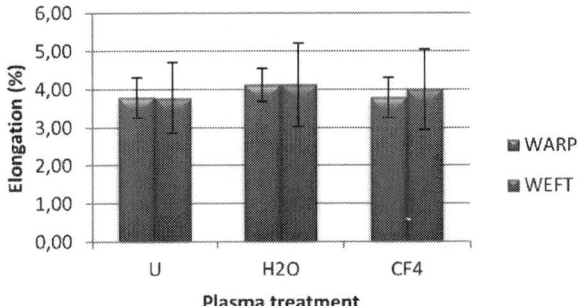

Figure 10. Elongation (%) of cotton after plasma treatment: U – untreated, H2O - ICRF water vapor plasma-treated, CF4 - ICRF tetrafluoromethane plasma-treated

A textile surface with such extremely rich surface morphology (as seen from SEM) is likely to influence higher adsorption of silver nanoparticles, which is in accordance with our ICP-MS results in Table 1, where the quantity of adsorbed silver onto cotton samples is presented.

From the results summarized in Table 1, it is clear that 30 nm silver nanoparticles are more adhesive to the cotton fabric than 80 nm silver nanoparticles. Regarding their volume, nano-particles have a high specific surface area. By decreasing the size of a particle, its surface dis-tribution rate increases [87]. The quantity of adsorbed silver on untreated bleached/ mercerized cotton was 32 ppm when 30 nm silver nanoparticles were used and 13 ppm when 80 nm silver nanoparticles were used. Similar trend can be observed when cotton was modified with plasma. When cotton was treated with water vapor low-pressure plasma the quantity of adsorbed 30 nm silver nanoparticles was 50 ppm and the quantity of adsorbed 80 nm silver nanoparticles was 17 ppm, which means that adsorption of 80 nm silver nano-particles was three times lower than the adsorption of 30 nm silver nanoparticles. The re-sults show that plasma heavily modifies the morphology and surface chemical properties of cotton and by that has a great impact on the adsorption of silver nanoparticles onto fabrics. Atmospheric air corona treatment of cotton fabrics enhanced the quantity of silver onto raw cotton up to 4 times and onto bleached/mercerized cotton up to 2 times [64]. The adsorption of 80 nm silver nanoparticles onto CF_4 plasma-treated cotton was 2 times higher and onto water vapor plasma-treated cotton was 1.3 times higher than on untreated cotton. These fabrics had a sufficient antimicrobial effectiveness against *Escherichia coli* and *Pseudomonas aeruginosa* (Table 1) [3]. The adsorption of 30 nm silver nanoparticles onto CF_4 plasma-treated cotton is 1.7 times higher and onto water vapor plasma-treated cotton was 1.6 times higher than on untreated cotton. This gave a good antimicrobial effectiveness against *Enterococus faecalis* and *Pseudomonas aeruginosa* (Table 1) [27].

Plasma modification of cotton	Functionalization with silver nanoparticles	Ag quantity (ppm) on cotton	Bacterial reduction (%)
Untreated	30 nm	32	–a
	80 nm	13	–a
Corona air	80 nm	39	–a
ICRF water vapor	30 nm	50	52 (*E. coli*) 64 (*P. aeruginosa*)
	80 nm	17	–a
ICRF CF_4	30 nm	54	68 (*E. faecalis*) 77 (*P. aeruginosa*)
	80 nm	26	

–a no bacterial reduction

Table 1. The quantity of silver (ppm) and antimicrobial efficiency expressed as a bacterial reduction (%) of untreated and plasma-treated cotton samples functionalized with powdered silver nanoparticles

Although the purpose of our research was to study the efficiency and appropriateness of plasma modification of textiles in order to achieve a higher adsorption of nanoparticles onto their surfaces, the wash fastness of functionalized textiles was examined as well. Wash fastness test was carried out in a laboratory apparatus Launder-O-Meter [3, 27]. The samples were washed repetitively ten times at 95°C in a solution of 5 g/l of SDC standard detergent and 2 g/l of Na_2CO_3 (where 10 globules were added). The duration of the washing cycles was 30 min. After every wash cycle, the samples were rinsed twice in distilled water and then for 10 minutes under a tap water, which was followed by squeezing and air drying. In Table 2 the results of silver quantity on washed cotton samples are presented. From the results it can be seen that after ten times washing at 95°C the quantity of silver on the cotton samples decreased (Table 2).

Plasma modification of cotton	Functionalization with silver nanoparticles	Ag quantity (ppm) on cotton before washing	Ag quantity (ppm) on cotton after washing
ICRF water vapor	30 nm	50	26
	80 nm	17	10
ICRF CF$_4$	30 nm	54	36
	80 nm	26	22

Table 2. The quantity of silver (ppm) before and after washing

When cotton was modified by water vapor plasma the quantity of 30 nm silver nanoparticles decreased by 48%, and when modified by CF$_4$ plasma the quantity of silver nanoparticles decreased by 34%. The drop in the concentration of silver on the cotton after ten wash cycles is also observed with the 80 nm silver nanoparticles. Cotton that is treated with water vapor plasma loses 41%; however, cotton treated with CF$_4$ plasma loses 15% of the silver nanoparticles. After a 10-second treatment with CF$_4$ plasma, the surface of the bleached/mercerized cotton is more liable to the adsorption of silver nanoparticles than the surface that was modified by the water vapor plasma. The particle adhesion is a complex phenomenon depending on the interaction mechanism between particles and the surface of a material. Adsorption of silver nanoparticles onto cotton from a dyeing bath is much higher for smaller particles than for larger ones. When particles are already on the surface of cotton fibers the attractive interactions and forces are stronger for 80 nm silver nanoparticles than for 30 nm silver nanoparticles. That is the reason for a better wash durability of 80 nm silver particles. From the results presented in Table 1 and 2, it is evident that the quantity of silver on untreated and unwashed cotton is almost the same as on the samples that were modified by plasma and washed 10 times at 95°C.

The further experiments in functionalization of cotton with other forms of silver nanoparticles (i.e. commercial form RucoBac AGP and laboratory synthesized colloidal silver nanoparticles) were based on the obtained adhesion results of silver nanoparticles of known dimensions. The application of RucoBac AGP and synthesized colloidal silver nanoparticles was carried on a bleached/mercerized cotton fabric modified by atmospheric air corona plasma and water vapor low-pressure plasma.

The exhaustion method was used for the deposition of RucoBac AGP onto blank dyed, dyed, plasma-treated blank dyed and plasma-treated dyed cotton samples. 0.1 % of RucoBac AGP was used, which represents a half of the lowest concentration recommended by the agent producer. The reason for such a decision was to verify the possibility in achieving good antibacterial efficiency of a cotton fabric with the use of a very low concentration of silver composite. The liquor ratio was 10 : 1, and the treatment time 30 min at 50 °C. Afterwards, the samples were dried at 130 °C for 4 min. The results of the ICP-MS and antimicrobial analysis of samples are presented in Table 3.

Sample Treatment	Plasma treatment of cotton	Ag quantity (ppm) on cotton	Bacterial reduction (%)
Functionalization with 0.1 % RucoBac AGP	Untreated	1,5	—[a]
	Corona air	1,8	—[a]
	ICRF water vapor	11	58 (E.coli) 98 (S.faecalis)
Dyeing with reactive dye and functionalization with 0.1 % RucoBac AGP	Untreated	8,9	100 (E.coli, S.aureus, S.faecalis and P.aeruginosa)
	Corona air	8,4	100 (E.coli, S.aureus, S.faecalis and P.aeruginosa)
	ICRF water vapor	13	80 (S.aureus and E.coli) 100 (S.faecalis)

—[a] no bacterial reduction

Table 3. The quantity of silver (ppm) and antimicrobial efficiency, expressed as a bacterial reduction (%), of untreated and plasma-treated cotton samples functionalized with RucoBac AGP

The results summarized in Table 3 show low adsorption of silver onto untreated blank dyed cotton. The adsorption of silver onto corona modified cotton did not significantly increase. Both samples do not exhibit the antimicrobial effectiveness. However, treating cotton with water vapor low-pressure plasma increased the adsorption of silver onto cotton up to 7 times, which resulted in an excellent antimicrobial effectiveness against *Streptococcus faecalis* and a sufficient antimicrobial effectiveness against *Escherichia coli*. The adsorption of silver significantly increased when RucoBac AGP was applied onto dyed cotton fabric, regardless of plasma treatment. RucoBac AGP is a nano-dispersion of TiO_2 as the carrier of the active component AgCl. In the presence of moisture, silver cations react with hydroxyl functional cellulosic groups and are attached to each other electrostatically. The presence of a reactive dye on cotton and the introduction of additional covalently bound sulfonic acid groups will facilitate the uptake of a cationic antimicrobial agent [87]. Therefore, it is possible to conclude that RucoBac AGP is bound to the cotton surface through sulfonic groups of a covalently bound dye and through partially ionized hydroxyl and carboxyl groups present on the fiber [72]. The significant increased adsorption of silver can be also noticed with samples modified by water vapor low-pressure plasma and dyeing, while there was no noticeable change when samples were modified by corona plasma. Results obtained from XPS analysis showed that plasma modification of bleached/mercerized cotton increased the content of oxygen on the surface. In addition, after both plasma treatments the appearance of new bonds was noticeable (O-C=O) and the content of C=O bonds increased. That means that oxygen rich functional groups incurred on the surface of plasma modified cotton. These changes were more distinct after low-pressure plasma treatment than after corona plasma treatment. The difference in both plasma treatments is noticeable from the ICP-MS results also (Table 3), since the adsorption of RucoBac AGP onto cotton was greater after modification by water vapor low-pressure plasma. Nevertheless, all modified cotton fabrics had an excellent antimicrobial effectiveness against *Staphylococcus aureus, Escherichia coli, Streptococcus faecalis* and *Pseudomonas aeruginosa*.

The objectives for textile industry have been continuous on-line treatments of fabrics. Low-pressure plasma reactors, such as radio frequency powered plasma, provide greater stability and uniformity but generally require more handling of textile materials through the vacuum system than corona discharges at atmospheric pressures. In this respect, the use of corona plasma is more appropriate for the industry. In our following research when laboratory colloidal silver was used, we focused on treating cotton fabric with atmospheric corona plasma. The synthesis of colloidal silver was performed by reducing silver salt in an aqueous solution at room temperature under argon atmosphere. The procedure was described previously, in section 3.2.1. A synthesis of colloidal silver and loading of silver onto cotton fabric was performed as the second phase after dyeing cotton fabrics with a blue vat dye. To verify whether vat-dyeing influences the adsorption of colloidal silver onto cotton fabric, a blank vat-dyeing procedure (dyeing with all chemicals and no dye) was also performed [71]. Apart from indigo, the vat dyes used in dyeing applications are mainly derivatives of anthraquinone and of higher condensed aromatic ring systems with a closed system of conjugated double bonds. They generally contain two, four or six reducible carbonyl groups [88].

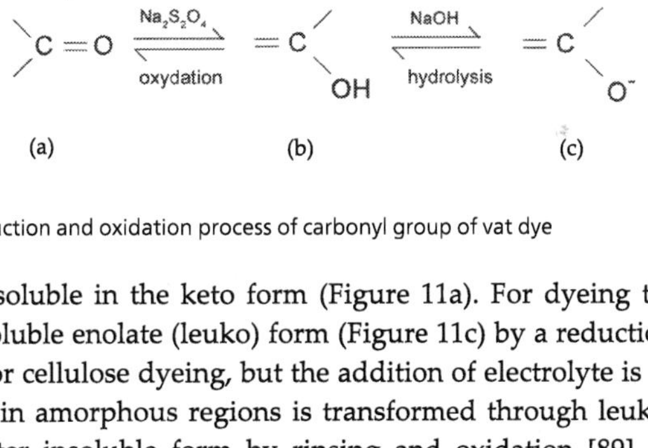

Figure 11. Scheme of reduction and oxidation process of carbonyl group of vat dye

The vat dyes are insoluble in the keto form (Figure 11a). For dyeing they must be transformed to a water soluble enolate (leuko) form (Figure 11c) by a reduction. This form of the dye is appropriate for cellulose dyeing, but the addition of electrolyte is also required. After the dyeing, the dye in amorphous regions is transformed through leuko acid (Figure 11b) into its original water insoluble form by rinsing and oxidation [89]. The vat dyeing of bleached/mercerized cotton fabric before colloidal silver treatment significantly influences the adsorption of silver onto cotton fabrics [71]. Also the adsorption of silver is influenced by an immersion time of cotton into colloidal silver solution. Although it was proven that modification of cotton by corona plasma strongly influences the adsorption of powdered silver nanoparticles by increasing their quantity on the modified cotton, the experiment using synthetized colloidal silver on corona treated cotton had to be conducted. The Table 4 presents the results of ICP-MS and antimicrobial analysis of untreated, corona and vat dyed colloidal silver loaded cotton.

Sample description	Ag quantity (ppm) on cotton	Bacterial reduction (%)
Dyed and functionalized with colloidal silver	24	100 (*E.coli, S.aureus, S.faecalis* and *P.aeruginosa*)
Corona plasma-treated, dyed and functionalized with colloidal silver	43	100 (*E.coli, S.aureus, S.faecalis* and *P.aeruginosa*)
Corona plasma-treated and functionalized with a half of concentration of colloidal silver	1.6	—[a]
Corona plasma-treated, dyed and functionalized with a half of concentration of colloidal silver	4.6	100 (*E.coli, S.aureus, S.faecalis* and *P.aeruginosa*)

—[a] no bacterial reduction

Table 4. The quantity of silver (ppm) and antimicrobial efficiency, expressed as a bacterial reduction (%), of untreated and plasma-treated cotton samples functionalized with synthetized colloidal silver

The results summarized in Table 4 show that atmospheric air corona treatment enhanced the quantity of silver onto dyed cotton up to 1.8 times comparing to the untreated dyed cotton. In addition to the morphological changes induced by plasma (seen by SEM), XPS analysis showed the increase of C-O and C=O bonds and formation of O-C=O bonds on the surface of treated cotton. The increased concentration of oxygen and newly formed bonds contributed to a better adhesion of Ag^+ ions from the colloidal solution onto cellulosic fibers. In addition to the increased number of functional groups containing oxygen, the dyed cotton fabric contains additional anionic sites due to the partial ionization of the molecules of insoluble vat dye. Colloidal silver is produced in a water solution using $AgNO_3$ reduced by $NaBH_4$. $NaBH_4$ also slightly reduces the water insoluble vat dye on the dyed fabric into a slightly soluble form. Although $NaBH_4$ is not a sufficiently strong reducing agent for dyeing cotton with a vat dye [90], it is nevertheless strong enough to enhance the negative charge of the dye on the fabric [91] such that silver ions (Ag^+) from a silver colloidal solution can be electrostatically attracted to the dyed cotton surface. It was reported that while Ag^+ ions from a colloidal silver solution are exhausted to the anionic cotton fibers to a high degree because of the attractive electrostatic interactions, the high increase of the adsorption ability of silver nanoparticles caused by the van der Walls forces resulted from the high surface area to volume ratio of these particles [43].

The important goal of our research was to use minimal concentrations, initially, of silver nanoparticles for loading onto textiles and to achieve a maximum quantity on the material, and thus to achieve functionalized cotton textile with an excellent antimicrobial efficiency. Since the cotton fabrics already had an excellent antimicrobial efficiency when functionalized with rather low concentration of silver, the decision was made to verify the possibility in achieving good antibacterial efficiency of a cotton fabric with the use of half of the initial concentration of $AgNO_3$ and $NaBH_4$. In this case, the corona plasma modified and undyed cotton contained 1.6 ppm of silver. The quantity of silver was so low that the fabric did not have an antimicrobial efficiency. However, when cotton fabric was modified by corona plasma and then vat dyed the quantity of silver was 4.6 ppm, which gave the fabric an excellent antimicrobial efficiency against *Staphylococcus aureus*, *Escherichia coli*, *Streptococcus faecalis* and *Pseudomonas aeruginosa*. The antimicrobial analysis showed the antimicrobial ineffectiveness of dyed cotton sample, therefore the dye itself did not contribute to the antibacterial efficiency of the functionalized cotton sample.

4. Conclusion

Our research shows that plasma treatment is an effective method to be used in achieving surface changes on the textile material by changing the functional groups on the textile surface and by changing the morphology of the fibers. The results of adsorption of different forms of silver nanoparticles on untreated and plasma-treated surfaces of fabrics confirm the fact that, for nanotechnological processes, the surface of the material has to be properly prepared. The adsorption of metal nanoparticles on textile materials depends on specific chemical and morphological properties of fibers. Plasma modification of cotton had a positive influence on the increased adsorption of silver nanoparticles loaded during exhaust dyeing process. From the bath, which contained a low concentration of silver nanoparticles, we have successfully applied a greater quantity of silver onto plasma modified cotton (up to four times). In some cases using plasma the dyeability of cotton was also improved. We succeeded to create a cotton fabric containing minimal quantity of silver with an excellent antimicrobial efficiency. This is very interesting from the technological point of view since by this method the quantity of silver in wastewater can be dramatically reduced. Another important result of the research was that plasma modification did not impair the mechanical properties of textiles. By using plasma technology new and improved properties of materials can be created that cannot be achieved by standard procedures, where nanostructuring of natural and synthetic fibers is emphasized. The use of plasma for modification of textiles brings to the textile industry many novelties, since plasma technology can be used as a substitute or as a support to the existing technologies, and by that positively influences the economy and ecology of the industrial processes. The knowledge of using plasma technology enables an introduction of contemporary (state-of-the-art) and ecological process of a textile modification into the textile industry and a development of highly technological products with improved or new properties.

Acknowledgements

This work was financially supported by Slovenian Research Agency (Programme P2-0213 Textile and Ecology), and Eureka Nanovision project.

Author details

Marija Gorjanc[1], Marija Gorenšek[1], Petar Jovančić[2] and Miran Mozetič[3]

*Address all correspondence to: marija.gorjanc@ntf.uni-lj.si

1 University of Ljubljana, Faculty of Natural Sciences and Engineering, Department of Textiles, Ljubljana, Slovenia

2 University of Belgrade, Faculty of Technology and Metallurgy, Belgrade, Serbia

3 "Jozef Stefan" Institute, Ljubljana, Slovenia

References

[1] Kang J.Y., Sarmadi M. Textile Plasma Treatment Review – Natural Polymer-Based Textiles. AATCC Review 2004; 4(10) 28-32.

[2] Gorjanc M., Recelj P., Gorenšek M. Plasma Technology for Textile Purposes. Tekstilec 2007; 50(10-12) 262-266.

[3] Gorjanc M., Bukosek V., Gorensek M., Vesel A. The Influence of Water Vapor Plasma Treatment on Specific Properties of Bleached and Mercerized Cotton Fabric. Textile Research Journal 2010; 80(6) 557-567.

[4] Mozetic M., Cvelbar U., Sunkara M.K., Vaddiraju S. A Method for the Rapid Synthesis of Large Quantities of Metal Oxide Nanowires at Low Temperatures. Advanced Materials 2005; 17(17) 2138-2142.

[5] Levchenko I., Rider A. E., Ostrikov K. Control of Core-shell Structure and Elemental Composition of Binary Quantum Dots. Applied Physics Letters 2007; 90(19) 193110.

[6] Drenik A., Vesel A., Mozetič M. Controlled Carbon Deposit Removal by Oxygen Radicals. Journal of Nuclear Materials 2009; 386-88, 893-895.

[7] Huang S.Y., Arulsamy A.D., Xu M., Xu S., Cvelbar U., Mozetič M. Customizing Electron Confinement in Plasma-Assembled Si/AlN Nanodots for Solar Cell Applications. Physics of Plasmas 2009; 16(12) 123504-1-123504-5.

[8] Mozetic M. Reactive Plasma Technologies in Electronic Industry, Informacije MID-EM 2003; 33(4) 222-227.

[9] Ferreira J.A., Tabares F.L., Tafalla D. Removal of Carbon Deposits in Narrow Gaps by Oxygen Plasmas at Low Pressure. Journal of Vacuum Science and Technology A 2007; 25(4) 746-750.

[10] Shishoo R. Plasma Conquering the Textile Industry. Magazine for European Research 1999; 24, 22-23.

[11] Fridman, A. Plasma chemistry. New York : Cambridge University Press. 2008.

[12] Morent R., De Geyter N., Verschuren J., De Clerck K., Kiekens P., Leys C. Non-thermal Plasma Treatment of Textiles. Surface and Coatings Technology 2008; 202, 3427-3449.

[13] Franceschini D.F., Achete C.A., Freire F.L., Beyer W., Mariotto G. Structural Modifications in a-c-h Films Doped and Implanted With Nitrogen. Diamond and Related Materials 1994; 3(1-2) 88-93.

[14] Chen P., Wang J., Wang B., Li W., Zhang C., Li H., Sun B. Improvement of Interfacial Adhesion for Plasma-Treated Aramid Fiber-Reinforced Poly(Phthalazinone Ether Sulfone Ketone) Composite and Fiber Surface Aging Effects. Surface and Interface Analysis 2009; 41(1) 38-43.

[15] Guo F., Zhang Z.-Z., Liu W.-M., Su F.-H., Zhang H.-J. Effect of Plasma Treatment of Kevlar Fabric on the Tribological Behavior of Kevlar Fabric/Phenolic Composites. Tribology International 2009; 42(2) 243-249.

[16] Jia C., Chen P., Li B., Wang Q., Lu C., Yu Q. Effects of Twaron Fiber Surface Treatment by Air Dielectric Barrier Discharge Plasma on the Interfacial Adhesion in Fiber Reinforced Composites. Surface and Coatings Technology 2010; 204(21-22) 3668-3675.

[17] Liu L., Jiang Q., Zhu T., Guo X., Sun Y., Guan Y., Qiu Y. Influence of Moisture Regain of Aramid Fibers on Effects of Atmospheric Pressure Plasma Treatment on Improving Adhesion with Epoxy. Journal of Applied Polymer Science 2006; 102(1) 242-247.

[18] Bhoj A.N., Kushner M.J. Repetitively Pulsed Atmospheric Pressure Discharge Treatment of Rough Polymer Surfaces: I. Humid Air Discharges. Plasma Sources Science and Technology 2008; 17(3), art. no. 035024.

[19] KinlocH A.J. The Science of Adhesion: Part 1: Surface and Interfacial Aspects. Journal of Materials Science 1980; 15, 2141-2166.

[20] Wu S. Polymer Interface and Adhesion. New York : Marcel Dekker. 1982.

[21] Marsh D.H., Riley D.J., York D., Graydon A. Sorption of Inorganic Nanoparticles in Woven Cellulose Fabrics. Particuology 2009; 7(2) 121-128.

[22] Bozzi A., Yuranova T., Guasaquillo I., Laub D., Kiwi J. X Self-cleaning of Modified Cotton Textiles by TiO₂ at Low Temperatures under Daylight Irradiation. Journal of Photochemistry and Photobiology A: Chemistry 2009; 174, 156-164.

[23] Cunko R., Varga K. Application of Ceramics for the Production of High-Performance Textiles. Tekstil 2006; 55(6) 267-278.

[24] Gorensek M., Gorjanc M., Bukosek V., Kovac J., Jovancic P., Mihailovic D. Functionalization of PET Fabrics by Corona and Nano Silver. Textile Research Journal 2010; 80(3) 253-262.

[25] Gorenšek M., Gorjanc M., Bukošek V., Kovač J., Petrović Z., Puač N. Functionalization of Polyester Fabric by Ar/N$_2$ Plasma and Silver. Textile Research Journal 2010; 80(16) 1633-1642.

[26] Gorjanc M., Mozetič M., Gorenšek M. Low-temperature plasma for pretreatment of cotton fabric for better adhesion on nano silver. In: Simončič, B., Gorjanc, M. (eds.). Conference proceedings 19. June 2009. Ljubljana; 2009.

[27] Gorjanc M., Bukošek V., Gorenšek M., Mozetič M. CF$_4$ Plasma and Silver Functionalized Cotton. Textile Research Journal 2010; 80(20) 2204-2213.

[28] Mihailovic D, Saponjic Z., Molina R., Puac N., Jovancic P., Nedeljkovic J., Radetic M. Improved Properties of Oxygen and Argon RF Plasma-Activated Polyester Fabrics Loaded with TiO$_2$ Nanoparticles. ACS Applied Materials and Interfaces 2010; 2(6) 1700-1706.

[29] Tao D., Feng Q., Gao D., Pan L., Wei Q., Wang D. Surface Functionalization of Polypropylene Nonwovens by Metallic Deposition. Journal of Applied Polymer Science 2010; 117(3) 1624-1630.

[30] Yuranova T., Rincon A.G., Bozzi A., Parra S., Pulgarin C., Albers P., Kiwi J. Antibacterial Textiles Prepared by RF-Plasma and Vacuum-UV Mediated Deposition of Silver. Journal of Photochemistry and Photobiology A-Chemistry 2003; 161 (1) 27-34.

[31] Pedrosa P., Chappe J.-M., Fonseca C., Machado A. V., Nobrega J.M., Vaz F. Plasma Surface Modification of Polycarbonate and Poly(propylene) Substrates for Biomedical Electrodes. Plasma Processes and Polymers 2010; 7(8) 676-686.

[32] Junkar I., Vesel A., Cvelbar U., Mozetič M., Strnad S. Influence of Oxygen and Nitrogen Plasma Treatment on Polyethylene Terephthalate (PET) Polymers. Vacuum 2009; 84 (1) 83-85.

[33] Borcia C., Dumitrascu N. Adhesion Properties of Polyamide-6 Fibres Treated by Dielectric Barrier Discharge. Surface and Coatings Technology 2006; 201, 1117-1123.

[34] Qi K., Xin J.H., Daoud W.A., Mak C.L. Functionalizing Polyester Fiber with a Self-cleaning Property using Anatase TiO$_2$ and Low-Temperature Plasma Treatment. International Journal of Applied Ceramic Technology 2007; 4(6) 554-563.

[35] Mejia M.I., Marin J.M., Restrepo G., Pulgarin C., Mielczarski E., Mielczarski J., Arroyo Y., Lavanchy J.-C., Kiwi J. Self-cleaning Modified TiO$_2$-cotton Pretreated by UVC-light (185 nm) and RF-plasma in Vacuum and also under Atmospheric Pressure. Applied Catalysis B-environmental 2009; 91(1-2) 481-488.

[36] Szymanowski H., Sobczyk A., Gazicki-Lipman M., Jakubowski W., Klimek L. Plasma enhanced CVD deposition of titanium oxide for biomedical applications. Surface and coatings technology, 2005; 200(1-4) 1036-1040.

[37] Ilic V., Saponjic Z., Vodnik V., Molina R., Dimitrijevic S., Jovancic P., Nedeljkovic J., Radetic M. Antifungal Efficiency of Corona Pretreated Polyester and Polyamide Fab-

rics Loaded with Ag Nanoparticles. Journal of Materials Science 2009; 44(15) 3983-3990.

[38] Gorenšek M., Recelj P. Reactive Dyes and Nano-silver on PA6 Micro Knitted Goods. Textile Research Journal 2009; 79(2) 138-146.

[39] Kvitek L., Panacek A., Soukupova J., Kolar M., Vecerova R., Prucek R., Holecova M., Zboril R. Effect of Surfactants and Polymers on Stability and Antibacterial Activity of Silver Nanoparticles (NPs). Journal of Physical Chemistry C 2008; 112(15) 5825-5834.

[40] Tomšič B., Simončič B., Cvijn D., Orel B., Zorko M., Simončič A. Elementary Nano Sized Silver as Antibacterial Agent. Tekstilec 2008; 51(7-9) 199–215.

[41] Vertelov G., Krutyakov Y., Efremenkova O., Olenin A., Lisichkin G. A Versatile Synthesis of Highly Bactericidal Myramistin Stabilized Silver Nanoparticles. Nanotechnology 2008; 19(35) art.no. 355707.

[42] Choi O., Deng K., Kim N., Ross L., Surampalli R., Hu Z. The Inhibitory Effects of Silver Nanoparticles, Silver Ions and Silver Chloride Colloids on Microbial Growth. Water Research 2008; 42(12) 3066-3074.

[43] Klemenčič D., Simončič B., Tomšič B., Orel B. Biodegradation of Silver Functionalised Cellulose Fibers. Carbohydrate Polymers 2010; 80, 426-435.

[44] Tomšič B., Simončič B., Žerjav M., Simončič A. A Low Nutrition Medium Improves the Determination of Fungicidal Activity of AgCl on Cellulose Fibres. Tekstilec 2008; 51(7-9) 231-241.

[45] Tomšič B., Simončič B., Orel B., Žerjav M., Schroers H.-J., Simončič A., Samardžija Z. Antimicrobial Activity of AgCl Embedded in a Silica Matrix on Cotton Fabric. Carbohydrate Polymers 2009; 75(4) 618-626.

[46] Gavriliu S., Lungu M., Enescu E., Gavriliu L. Composite Nanopowder for Antibacterial Textiles. Industria Textila 2010; 61(2) 86-90.

[47] Yuranova T., Rincon A. G., Pulgarin C., Laub D., Xantopoulos N., Mathieu H.-J., Kiwi J. Performance and Characterization of Ag-cotton and Ag/TiO$_2$ Loaded Textiles during the Abatement of E. coli. Journal of Photochemistry and Photobiology A-chemistry 2006; 181(2-3) 363-369.

[48] Rimai D.S., Quesnel D.J., Busnaina A.A. The Adhesion of Dry Particles in the Nanometer to Micrometer-size Range. Colloids and Surfaces A: Physicochemical and Engineering Aspects 2000; 165(1-3), 3-10.

[49] Dastjerdi R., Mojtahedi M.R.M., Shoshtari A.M. Investigating the Effect of Various Blend Ratios of Prepared Masterbatch Containing Ag/TiO$_2$ Nanocomposite on the Properties of Bioactive Continuous Filament Yarns. Fibers and Polymers 2008; 9(6) 727-734.

[50] Dubas, S.T., Kumlangdudsana, P., Potiyaraj, P. Layer-by-layer Deposition of Antimicrobial Silver Nanoparticles on Textile Fibers. Colloids and Surfaces A: Physicochemical and Engineering Aspects 2006; 289(1-3) 105-109.

[51] Hadad L., Perkas N., Gofer Y., Calderon-Moreno J., Ghule A., Gedanken A. Sonochemical Deposition of Silver Nanoparticles on Wool Fibers. Journal of Applied Polymer Science 2007; 104(3) 1732-1737.

[52] He J.H., Kunitake T., Nakao A. Facile in situ Synthesis of Noble Metal Nanoparticles in Porous Cellulose Fibers. Chemistry of Materials 2003; 15(23) 4401-4406.

[53] Hegemann D., Hossain M.M., Balazs D.J. Nanostructured Plasma Coatings to Obtain Multifunctional Textile Surfaces. Progress in Organic Coatings 2007; 58(2-3) 237-240.

[54] Moazami A., Montazer M., Rashidi A., Rahimi M.K. Antibacterial Properties of Raw and Degummed Silk with Nanosilver in Various Conditions. Journal of Applied Polymer Science 2010; 118(1) 253-258.

[55] Gorensek M., Recelj P. Nanosilver Functionalized Cotton Fabric. Textile Research Journal 2007; 77(3) 138-141.

[56] Jiang S.X., Qin W.F., Zhang L. Surface Functionalization of Nanostructured Silver-Coated Polyester Fabric by Magnetron Sputtering. Surface and Coatings Technology 2010; 204(21-22) 3662-3667.

[57] Malinsky M.D., Kelly K.L., Schatz G.C., Van Duyne R.P. Chain length dependence and sensing capabilities of the localized surface plasmon resonance of silver nanoparticles chemically modified with alkanethiol self-assembled monolayers, J. Am. Chem. Soc., 123 (2001), pp. 1471–1482.

[58] Bosbach J., Hendrich C., Stietz F., Vartanyan T., Trager F. Ultrafast Dephasing of Surface Plasmon Excitation in Silver Nanoparticles: Influence of Particle Size, Shape, and Chemical Surrounding. Physical Review Letters 2002; 89(25) art.no. 257404.

[59] Mock J.J., Barbic M., Smith D.R., Schultz D.A., Schultz S. Shape Effects in Plasmon Resonance of Individual Colloidal Silver Nanoparticles. Journal of Chemical Physics 2002; 116(15) 6755-6759.

[60] Ilic V., Saponjic Z., Vodnik V., Potkonjak B., Jovancic P., Nedeljkovic J., Radetic M. The Influence of Silver Content on Antimicrobial Activity and Color of Cotton Fabrics Functionalized with Ag Nanoparticles. Carbohydrate Polymers 2009; 78 (3) 564-569.

[61] Carneiro N., Souto A.P., Silva E., Marimba A., Tena A., Ferreira H., Magalhaes V. Dyeability of Corona-Treated Fabrics. Coloration Technology 2001; 117, 298-302.

[62] Navarro, A., Bautista, L. Surface modification and characterization in cotton fabric bleaching. Proceedings of 5th world textile conference AUTEX 2005. Maribor; 2005.

[63] Gorjanc M., Jovančić P., Bukošek V., Gorenšek M. Study of adsorption of nano silver on cotton pretreated with plasma. Proceedings of the 9th Autex, Izmir, Turkey; 2009.

[64] Gorjanc M., Jovančić P., Gorenšek M. Nano silver on cotton. In: Mihailović D., Kobe S., Remškar M., Jamnik J., Čopič M., Drobne D. (eds.). Hot nano topics 2008 : incorporating SLONANO 2008, 3 overlapping workshops on current hot subjects in nanoscience, 23-30 May, Portorož, Slovenia : abstract book. Ljubljana; 2008.

[65] Gorjanc M., Jovančić P., Kovač J., Bukošek V., Gorenšek M. Plasma treatment of cotton: correlation between dyeability of cotton with reactive dye and chemical composition of surfaces. In: XXII IFATCC International congress (International Federation of the Association of Textile Chemists and colorists), Lago Magiore (Italy), May 5-7th, 2010. From textile chemistry to fashion : multifunctionality, sustainability, competitivity : proceedings of the XXII IFATCC International congress, Stresa, Lago Magiore. Milano: Associazione Italiana di Chimica Tessile e Coloristica: = AICTC Italia; 2010.

[66] Gorjanc M., Mozetič M., Bukošek V., Gorenšek M. Impact of low-pressure plasma on dyeability of cotton. In: Dragčević Z. (ed.) 4th International Textile, Clothing & Design Conference [also] ITC&DC, October 5th to October 8th, 2008, Dubrovnik, Croatia. Magic world of textiles : book of proceedings. Zagreb: Faculty of Textile Technology, University of Zagreb; 2008.

[67] Gorjanc M., Mozetič M., Gorenšek M. Low-pressure plasma for pretreatment of cotton fabric for better adhesion of nanosilver. Tekstilec 2009; 52(10-12) 263-269.

[68] Ames Goldsmith - Silver nano powders. Data sheet, 2005. http://www.amesgoldsmith.com/products/silver-nano-powders.asp/ (accessed 25 February 2011).

[69] Gorjanc M. Plasma Modification of Cotton Surfaces for a Deposition of Nanosilver. PhD thesis. University of Ljubljana, Faculty of Natural Sciences and Engineering; 2011.

[70] The ultimate hygenic finish for textiles. Rudolph Chemie, Rudolph Info 5/07; 2007.

[71] Gorjanc M., Kovač F., Gorenšek M. The Influence of Vat Dyeing on the Adsorption of Synthesized Colloidal Silver onto Cotton Fabrics. Textile Research Journal 2012; 82(1) 62-69.

[72] Gorjanc M., Gorenšek M. Influence of Dyeing Cotton with Reactive Dye on Adsorption of Silver. Tekstilec 2011; 54(10-12) 228-237.

[73] Doering R., Nishi Y. Handbook of Semiconductor Manufacturing Technology. Boca Raton : CRC Press; 2000.

[74] Johansson L.S., Campbell J.M., Stenius P., Pere J., Buchert T. Surface Characterization of Cellulosic Materials with XPS. Abstracts of Papers of the American Chemical Society 2001; 221 (Part I) U180-U180.

[75] Gorjanc M., Mozetič M., Jazbec K., Gorenšek M. Influence of ICRF water vapour plasma on dyeability and UPF of cotton fabrics. In: Simončič B., Hladnik A., Pavko-Čuden A., Ahtik J., Luštek Preskar B., Demšar A., Urbas R. (eds.). 41st International Symposium on Novelties in Textiles and 5th International Symposium on Novelties

in Graphics and 45th International Congress IFKT, Ljubljana, Slovenia, 27-29 May 2010. Symposium proceedings. Ljubljana: Faculty of Natural Sciences and Engineering, Department of Textiles; 2010.

[76] Tsafack M.J., Levalois-Grutzmacher J. Towards Multifunctional Surfaces using the Plasma-Induced Graftpolymerization (PIGP) Process: Flame and Waterproof Cotton Textiles. Surface and Coatings Technology 2007; 201(12): 5789-5795.

[77] McCord M.G., Hwang Y.J., Qui Y., Hughes L.K. Bourham M.A. Surface Analysis of Cotton Fabrics Fluorinated in Radio-Frequency Plasma. Journal of Applied Polymer Science 2003; 88(8), 2038-2047.

[78] Cvelbar U., Krstulovic N., Milosevic S., Mozetic M. Inductively Coupled RF Oxygen Plasma Characterization by Optical Emission Spectroscopy. Vacuum 2007; 82(2) 224-227.

[79] Kurihara M., Petrovic Z., Makabe T. Transport Coefficients and Scattering Cross-Sections for Plasma Modelling in CF_4-Ar Mixtures: a Swarm Analysis. Journal of Physics D-Applied Physics 2000; 33(17); 2146-2153.

[80] Flamm D.L. Mechanisms of Radical Production in CF_3Cl, CF_3Br, and Related Plasma Etching Gases: The role of Added Oxidants. Plasma Chemistry and Plasma Processing 1981; 1(1) 37-52.

[81] d'Agosiono R. Plasma Deposition, Treatment, and Etching of Polymers. San Diego: Academic Press Inc; 1990.

[82] Poll, H.U., Schladitz, U., Schreiter, S. Penetration of Plasma Effects into Textile Structures. Surface and Coatings Technology 2001; 142, 489-493.

[83] Hwang Y.J., McCord M.G., An J.S., Kang B.C., Park S.W. Effects of Helium Atmospheric Pressure Plasma Treatment on Low-Stress Mechanical Properties of Polypropylene Nonwoven Fabrics. Textile Research Journal 2005; 78(11) 771-778.

[84] Yip J., Chan K., Sin K.M., Lau K.S. Study of Physico-Chemical Surface Treatments on Dyeing Properties of Polyamides. Part 1: Effect of Tetrafluoromethane Low Temperature Plasma. Coloration Technology 2002; 118(1) 26-30.

[85] Sun D., Stylios G.K. Investigating the Plasma Modification of Natural Fiber Fabrics - the Effect on Fabric Surface and Mechanical Properties. Textile Research Journal 2005; 75(9) 639-644.

[86] Owens F.J., Poole Jr., C.P., editors. The physics and chemistry of nanosolids. New Jersey: Wiley & Sons, Inc.; 2008.

[87] Kawabata A., Taylor J.A. Effect of Reactive Dyes upon the Uptake and Antibacterial Action of Poly(Hexamethylene Biguanide) on Cotton. Part 1: Effect of Bis(Monochlorotriazinyl) Dyes. Coloration Technology 2004; 120(5) 213-219.

[88] Preston C., editor. The Dyeing of Cellulosic Fibers. London: Dyers' Company Publications Trust; 1986.

[89] Gorenšek M. Dyes and textile colour. In: Jeler S. (ed.) Textile and colour. Firenze: Edizioni Tassinari; 2004. P5-19.

[90] Shore J. Cellulosics Dyeing. Bradford : Society of Dyers and Colourists; 1995.

[91] Nair, G.P., Shah, R.C. Sodium Borohydride in Vat Dyeing. Textile Research Journal 1970; 40(4) 303-312.

[92] Klemenčič D., Simončič B., Tomšič B., Orel B. Biodegradation of Silver Functionalized Cellulose Fibres. Carbohydrate Polymers 2010; 80, 426-435.

Chapter 2

Surface Modification Methods for Improving the Dyeability of Textile Fabrics

Sheila Shahidi, Jakub Wiener and
Mahmood Ghoranneviss

Additional information is available at the end of the chapter

http://dx.doi.org/10.5772/53911

1. Introduction

Polymer and textiles have a vast number of advantages and attractiveness as a material. However, despite these advantageous, polymers have limitations. In general, special surface properties with regard to chemical composition, hydrophilicity, roughness, crystallinity, conductivity, lubricity, and cross-linking density are required for successful application of polymers in such wide fields as adhesion, membrane filtration, coatings, friction and wear, composites, microelectronic devices, thin-film technology and biomaterials, and so on. Unfortunately, polymers very often do not possess the surface properties needed for these applications. In fact, polymeric fibers that are mechanically strong, chemically stable, and easy to process usually will have inert surfaces both chemically and biologically. Vice versa, those polymers having active surfaces usually do not possess excellent mechanical properties which are critical for their successful application.

Due to this dilemma, surface modification of the polymeric fibers without changing the bulk properties has been a classical research topic for many years, and is still extensive studies as new applications of polymeric materials emerge, especially in the fields of biotechnology, bioengineering, and most recently in nanotechnology.

Modification is used to designate a deliberate change in composition or structure leading to an improvement in different type of fiber properties.

The challenge is, however, that there does not exist an ideal modification that eliminates all the negative properties and preserves all the positive properties of the fibers. This is why there are a great number of different single-purpose modifications. [1]

In spite of the great number of existing modification methods no consistent classification is available as yet. Some authors divide the methods into two groups depending on whether they involve changes in fiber composition (chemical modification) or changes in fiber structure (physical modification).

Surface modification of polymers has become an important research area in the plastic industry. Because polymers are inert materials and usually have a low surface energy, they often do not possess the surface properties needed to meet the demands of various applications. Advances in surface treatment have been made to rather chemical and physical properties of polymer surfaces without affecting bulk properties. Technologies such as surface modifications, which convert inexpensive materials into valuable finished goods, will become even more important in the future as material cost becomes a significant factor in determining the success of an industry. [2]

There are a few factors to consider when modifying a surface:

1. Thickness of the surface is crucial. Thin surface modifications are desirable, otherwise mechanical and functional properties of the material will be altered. This is more so when dealing with nanofibers as there is less bulk material present.

2. Sufficient atomic or molecular mobility must exist for surface changes to occur in reasonable periods of time. The driving force for the surface changes is the minimization of the interfacial energy.

3. Stability of the altered surface is essential, achieved by preventing any reversible reaction. This can be done by cross-linking and/or incorporating bulky groups to prevent surface structures from moving.

4. In some cases a transparent scaffold is desired, especially in optical sensors or ophthalmology; after surface treatment they should remain transparent. Any cloudiness introduced is of real concern.

5. Uniformity, reproducibility, stability, process control, speed, and reasonable cost should be considered in the overall process of surface modification. The ability to achieve uniform surface treatment of complex shapes and geometries can be essential for sensor and biomedical applications.

6. Precise control over functional groups. This is a challenging yet difficult scope. Many functional groups might bond to the surface such as hydroxyl, ether, carbonyl, carboxyl, and carbonate groups, instead of one desired functional group.

This review is focused on the application of recent methods for the modification of textiles using physical methods (corona discharge, plasma, laser, electron beam and neutron irradiations, Ion beam), chemical methods (ozone-gas treatment, surface grafting, enzymatic modification, sol-gel technique, micro-encapsulation method and treatment with different reagents). Nowadays, surface functionalization of synthetic fibers for various applications is considered as one of the best methods for modern textile finishing processes especially for improving the dyeability of fabrics. [3] Combination of physical technologies and nano-science enhances the durability of textile materials against washing, ultraviolet radiation, friction, abrasion, tension and fading. Textile fibers typically undergo a variety of pre-treatments before dyeing and printing is feasible. Nonetheless, these treatments still create undesirable process conditions which can result in increased waste production, unpleasant working conditions and higher energy consumption. Therefore reducing pollution in textile production is becoming of utmost importance for manufacturers worldwide. In coming years, the textile industry must implement sustainable technologies and develop environmentally safer methods for textiles processing to remain competitive. [4, 5]

1.1. Physical methods for modification of textile fabrics

1.1.1. Plasma types and applications

Plasma is by far the most common form of matter. Plasma in the stars and in the tenuous space between them makes up over 99% of the visible universe and perhaps most of that which is not visible.

The coupling of electromagnetic power into a process gas volume generates the plasma medium comprising a dynamic mix of ions, electrons, neutrons, photons, free radicals, metastable excited species and molecular and polymeric fragments, the system overall being at room temperature. This allows the surface functionalisation of fibres and textiles without affecting their bulk properties. These species move under electromagnetic fields, diffusion gradients, etc. on the textile substrates placed in or passed through the plasma. This enables a variety of generic surface processes including surface activation by bond breaking to create reactive sites, grafting of chemical moieties and functional groups, material volatilisation and removal (etching), dissociation of surface contaminants/layers (cleaning/ scouring) and deposition of conformal coatings. In all these processes a highly surface specific region of the material is given new, desirable properties without negatively affecting the bulk properties of the constituent fibres.

Plasmas are acknowledged to be uniquely effective surface engineering tools due to:

- Their unparalleled physical, chemical and thermal range, allowing the tailoring of surface properties to extraordinary precision.
- Their low temperature, thus avoiding sample destruction.
- Their non-equilibrium nature, offering new material and new research areas.
- Their dry, environmentally friendly nature. [5, 6]

Plasma reactors

Different types of power supply to generate the plasma are:

Low-frequency (LF, 50–450 kHz)

Radio-frequency (RF, 13.56 or 27.12 MHz)

Microwave (MW, 915 MHz or 2.45 GHz)

The power required ranges from 10 to 5000 watts, depending on the size of the reactor and the desired treatment.

Low-pressure plasmas

Low-pressure plasmas are a highly mature technology developed for the microelectronics industry. However, the requirements of microelectronics fabrication are not, in detail, compatible with textile processing, and many companies have developed technology of low pressure reactors to achieve an effective and economically viable batch functionalisation of fibrous products and flexible web materials.

A vacuum vessel is pumped down to a pressure in the range of 10^{-2} to 10^{-3} mbar with the use of high vacuum pumps. The gas which is then introduced in the vessel is ionised with the help of a high frequency generator.

The advantage of the low-pressure plasma method is that it is a well controlled and reproducible technique.

Atmospheric pressure plasmas

The most common forms of atmospheric pressure plasmas are described below.

Corona treatment

Corona discharge is characterised by bright filaments extending from a sharp, high-voltage electrode towards the substrate. Corona treatment is the longest established and most widely used plasma process; it has the advantage of operating at atmospheric pressure, the reagent gas usually being the ambient air. Corona systems do have, in principle, the manufacturing requirements of the textile industry (width, speed), but the type of plasma produced cannot achieve the desired spectrum of surface functionalisations in textiles and nonwovens. In particular, corona systems have an effect only in loose fibres and cannot penetrate deeply into yarn or woven fabric so that their effects on textiles are limited and short-lived. Essentially, the corona plasma type is too weak. Corona systems also rely upon very small interelectrode spacing (1 mm) and accurate web positioning, which are incompatible with 'thick' materials and rapid, uniform treatment.

Dielectric barrier discharge (silent discharge)

The dielectric barrier discharge is a broad class of plasma source that has an insulating (dielectric) cover over one or both of the electrodes and operates with high voltage power ranging from low frequency AC to 100 kHz. This results in non-thermal plasma and a multitude of random, numerous arcs form between the electrodes. However, these microdischarges are nonuniform and have potential to cause uneven treatment.

Glow discharge

Glow discharge is characterised as a uniform, homogeneous and stable discharge usually generated in helium or argon (and some in nitrogen). This is done, for example, by applying radio frequency voltage across two parallel-plate electrodes. Atmospheric Pressure Glow Discharge (APGD) offers an alternative homogeneous cold-plasma source, which has many of the benefits of the vacuum, cold-plasma method, while operating at atmospheric pressure. [5]

Cold plasmas can be used for various treatments such as: plasma polymerisation (gaseous monomers); grafting; deposition of polymers, chemicals and metal particles by suitable selection of gas and process parameters; plasma liquid deposition in vaporised form.

In general, reactions of gas plasmas with polymers can be classified as follows:

1. Surface reactions: Reactions between gas-phase species and surface species and reactions between surface species produce functional groups and cross-links, respectively, at the surface. Examples of these reactions include plasma treatment by argon, ammonia, carbon monoxide, carbon dioxide, fluorine, hydrogen, nitrogen dioxide, oxygen, and water.

2. Plasma polymerization: The formation of a thin film on the surface of polymer via polymerization of an organic monomer such as CH_4, C_2H_6, C_2F_4, or C_3F_6 in plasma. It involves reactions between gas-species, reactions between gas-phase species and surface species, and reactions between surface species.

3. Etching: Materials are removed from a polymer surface by physical etching and chemical reactions at the surface to form volatile products. Oxygen plasma and oxygen and fluorine-containing plasmas are frequently used for the etching of polymers. [6, 7]

Plasma treatment play very important role for improving the dyeing properties of textile fabrics. Some of these improvements are discussed as follow.

1.1.1.1. Surface modification of polypropylene non-woven fabrics by atmospheric-pressure plasma activation followed by acrylic acid grafting

Polypropylene (PP) non-woven fabrics have been activated by an atmospheric-pressure plasma treatment using surface dielectric barrier discharge in N_2 and ambient air. Subsequently, the plasma activated samples were grafted using catalyst-free water solution of acrylic acid. Surface properties of the activated and polyacrylic acid post-plasma grafted non-woven were characterized by scanning electron microscopy, Fourier transform infrared spectroscopy, electron spin resonance spectroscopy, surface energy and dyeability measurements.

The grafted non-woven exhibit improved water transport and dyeing properties.

The plasma activation in nitrogen plasma gas was more efficient than in air. Post-plasma surface grafting lead to a stable and homogeneous grafting of pAA onto PP non-woven fabrics, which made PP fabrics easily coloured by conventional water-soluble acid dye. Supposedly, peroxy radicals formed at a short ambient air exposure of the plasma activated fabrics were responsible for initiating the grafting process. Regarding the surface peroxy radicals generation, the nitrogen plasma gas was superior to ambient air and provided better grafting. [7-9]

1.1.1.2. One-bath one-dye class dyeing of pes/cotton blends after corona and chitosan treatment

The feasibility of one-bath one-dye class dyeing of PES/cotton blend with direct (Direct Red 80) or reactive (Reactive Red 3) dye after pre-treatment with corona discharge (CD) and chitosan has been investigated by Ristic and his coworkers. It has been confirmed that corona discharge treatment enhances hydrophilicity of cotton and PES fibers due to surface modification of the material and formation of C-O, C=O, and COOH groups. In-creased hydrophilicity of PES and cotton induced only slightly increased color intensity (K/S) with both dyes investigated. Nevertheless, subsequent treatment with biopolymer chitosan noticeably enhanced the color intensity obtained, especially with PES fiber. The dyeability improved proportionally with the concentration of chitosan treatment solutions. However, the highest values of color intensity were obtained for the PES/cotton fabric subjected to the combined CD and chitosan treatment, suggesting that CD pre-treatment enhances efficiency of chitosan application.

Satisfactory values of dye fastness and fixation degree of reactive dye were obtained. [10]

1.1.1.3. Surface and bulk cotton fibre modifications: plasma and cationization. influence on dyeing with reactive dye

In similar research work, the single or combined effects of corona air plasma and cationising with an epihalohydrin have been evaluated on the surface and dyeing properties of open-work twill cotton fabrics. Dyeing was performed with a hetero-bis functional reactive dye. Wetting properties of cotton fabrics were improved with a very short corona plasma treat-ment and a double side-effect was observed on the dyed fabric by contact angle analysis, be-cause of the low penetration of the plasma on the fabric. Exhaustion of the dye and colour intensity of the cotton fabrics were increased due to the plasma treatment. This is well ex-plained by the functionalisation of the surface with oxygenated moieties, without any signif-icant alteration in surface topography of the fibres. Cationising of the cotton fabrics using an epihalohydrin as cationising agent increases the exhaustion of the dyestuff as high as 90%, and produces a dramatic improvement (80% increase) in the colour intensity (K/Scorr) on both sides of the fabrics. The improvement in colour intensity of the cationised cotton fabrics can be explained taking into account that the hydrolised reactive dye has high anionic char-acter which can be bound to the cationic amine of the cationic agent on the cotton fabrics. It has been observed that plasma treatment previous to cationising increases the impregnation of the fabrics. [11]All theses possible effects are schematically represented in Figure 1.

1.1.1.4. The impact of corona modified fibres' chemical changes on wool dyeing

Corona/plasma treatment is an environmentally friendly process applied to wool fabrics. The main contribution of the present work was to study the impact of Corona on dyeability of wool fibers. First, the different chemical aspects of a woven wool fabric's surface were de-termined using two different analytical skills (XPS and polyelectrolyte titration).The results show that, low-temperature plasma treatment has ability to change wool fibre morphology which could have an impact on sorption properties. fabrics were dyed with blue acid and blue metal-complex dyes, and dyeing behaviour were studied by means of on-line VIS spectrophotometry. Finally, dyed samples were colourimetricaly evaluated and colour differences were calculated. The results provided evidence that the overall carbon content was decreased while oxygen and nitrogen atoms were increased when using ionized air for fabric modification. It has also been noted that the amount of positive-charged functional groups in various pH ranges are higher for Corona-treated wool fabric in comparison with the untreated sample.

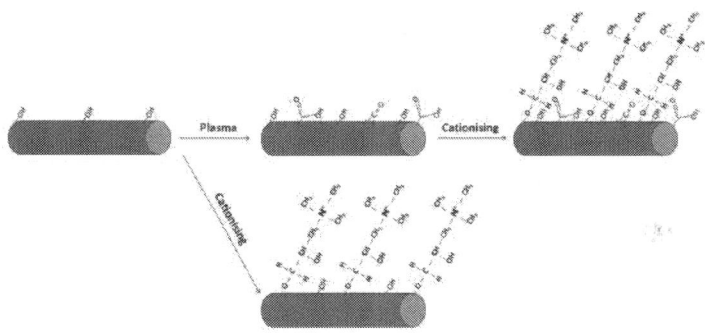

Figure 1. Effects of plasma and cationising processes in the surface functionalization of the cellulose constituting the cotton fibre

The surface performances of untreated and Corona-treated wool fabrics were studied both morphologically and chemically. Corona treatment is confirmed as inducing chemical and physical changes on the surface such as oxidizing/removing external fatty-acid monolayer, enlarging positively charged functional groups, creating new dyesites, and therefore, improving the exhaustion rate during dyeing.

Corona treatment applied in the pre-treatment stages of wool production can lead to an optimization of different dyeing procedures, implying lower dyeing temperatures and shorter dyeing time, achieving the same or even better colour exhaustion in comparison to conventional pre-treated wool fabric. For these reasons, the energy consumption can be reduced, thus also enhancing environmental protection. [12]

1.1.2. Surfaces modification with ionization radiation

Ionization radiation such as high-energy electrons, X-rays, and gamma-rays can displace electrons from atoms and molecules, producing ions. It differs from other types of radiation such as infrared, visible, and UV in that it is highly energetic and delivers to the irradiated material a large amount of energy, much greater than that associated with chemical bonds. Common industrial ionization radiation sources are high-energy electrons (0.1-10 MeV) and cobalt-60 sources, and gamma radiation.

Electron beams from 0.1 to several mega electron volts are used for high doses and high speeds in various industrial processes, with penetration up to several millimeters for polymeric materials.

Surface modification by UV and IR lasers is useful in some specific applications. One key advantage of laser treatment is that the area to be treated can be very small and localized. Depending on the level of power chosen, ablation or chemical and physical changes can occur. Various chemical changes occur on photon-irradiated polymer surfaces. When PTFE was irradiated with ArF laser at high fluencies, defluorination and surface oxidation occurred. For polypropylene, formation of oxygen functional groups such as C-O and C=O groups was detected after UV laser irradiation in air and water, and in ozon. The treated surfaces were shown to have improved bond ability with an epoxy adhesive. The surface of poly(vinyl chloride) becomes electrically conductive after successive UV irradiation in chlorine and nitrogen and argon laser irradiation in air. These types of surface modification are very useful for increasing the dyeability of polymeric textile fabrics. [2] Some of irradiation dyeability modification is discussed below.

1.1.2.1. Effect of UV and gamma radiation on the fabric dyed with natural dyes

The irradiation of fabric is also another factor which affects the colour strength of the fabric. Previous studies show that UV irradiation adds value to colouration and also increases the dye uptake ability of the cotton fabrics through oxidation of surface fibers of cellulose. Gamma rays are ionizing radiations that interact with the material by colliding with the electrons in the shells of atoms. They lose their energy slowly in material being able to travel through significant distances before stopping. The free radicals formed are extremely reactive, and they will combine with the material in their vicinity. Upon irradiation the cross linking changes the crystal structure of the cellulose, which can add value in colouration process and causes photo modification of surface fibers. The irradiated modified fabrics can allow: more dye or pigment to become fixed, producing deeper shades, more rapid fixation of dyes at low temperature and increases wet ability of hydrophobic fibers to improve depth of shade in printing and dyeing. [13]

1.1.2.2. Surface modification of meta-aramid films by UV/ozone irradiation

Meta-aramid films surface was modified by UV/O_3 irradiation, and surface properties have been investigated by reflectance, ATR, ESCA, and surface energy. Upon UV/O_3 treatment, the surface roughness and the O1s/C1s atomic ratio obviously improved, resulting from the implantation of carbonyl and hydroxyl groups. The surface energy of the meta-aramid films increased substantially due to the significantly enhanced Lewis acid parameter, which promoted the acid-base interaction of the surfaces with increase in UV energy. Also meta-aramid films became hydrophilic as indicated by substantially decreased water contact angle.

The dyeability of aramid films to cationic dyes was significantly increased due to higher hydrophilic surface and strong electrostatic attraction between the cationic dyes and anionic dyeing sites of the meta aramid. [14]

*1.1.2.3. Modification of polypropylene fibers by electron beam irradiation. i. evaluation of dyeing
properties using cationic dyes*

The dyeing properties of hydrophobic polypropylene fibers using cationic dyes were investigated to improve dyeability by electron beam irradiation and sulfonic acid incorporation.
The best dyeing result was obtained when polypropylene fibers incorporated by sulfonic
acid group after electron beam irradiation were dyed with cationic dyes at alkaline conditions and 30~75 kGy irradiation ranges. In order to improve the dyeing properties of electronic beam irradiated polypropylene, sulfonic acid group which has good reactivity was
introduced on the fiber. To incorporate sulfonic acid with the electronic beam irradiated
(70.5 kGy) polypropylene fiber, the fibers were added and reacted to the solution of 1,4-di-
oxane and $ClSO_3H$ at 70 °C (Figure 2).

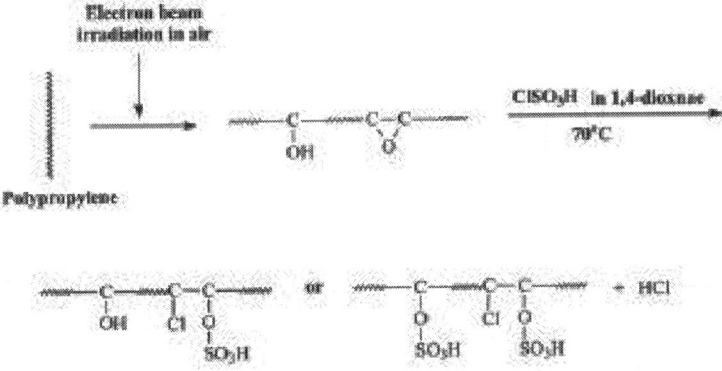

Figure 2. Introduction of sulfone groups on electron beam irradiated PP.

In order to make hydrophobic polypropylene fibers dyeable, it was shown that functional
group such as carboxylate was formed on fiber substrates by electronic beam irradiation.

Concerning the pH and amount of absorbed electronic beam irradiation, the color strength
increased as pH increased in alkaline conditions, and also increased as the absorbed dose
increased to 30~75 kGy. As a result, it was confirmed that the pH of the dyebath and the
amount and the range of the absorbed irradiation could be important variables for color
strength but it seemed difficult to get deeper colors.

In the case of polypropylene fibers incorporated by sulfonic acid group to improve dyeability, the introduction of sulfonic acid group was confirmed by ESCA analysis and it was
judged that such introduction has some advantages in color strength over only electronic
beam irradiated fibers. Finally, the wash fastness of dyed fabrics using cationic dyes showed
satisfactory ratings of 4~5 on both electronic beam irradiated fibers and sulfonic acid incorporated fibers. [15]

1.1.2.4. UV Excimer laser modification on polyamide materials: effect on the dyeing properties

Polyamide materials irradiated with 193 nm ArF Excimer laser developed ripple-like structures of micron size on the surface. (Figure 3)

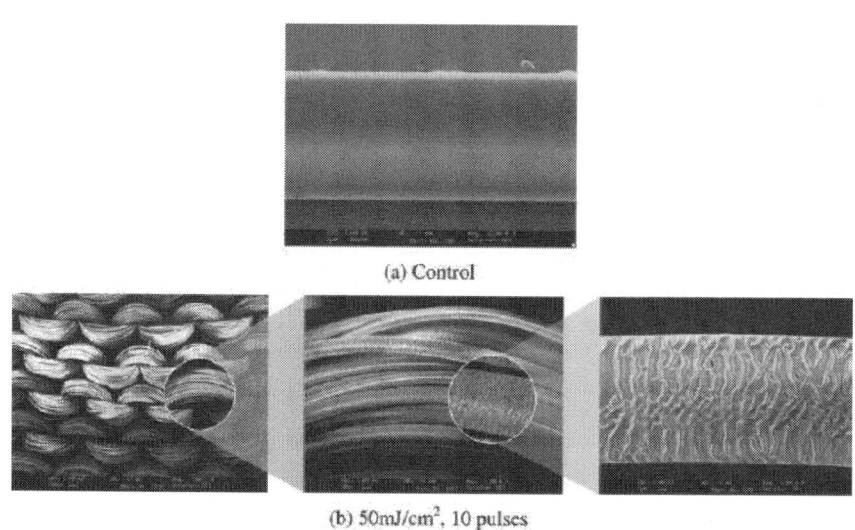

(a) Control

(b) 50mJ/cm², 10 pulses

Figure 3. Morphological features of untreated and laser treated nylon 6 fabric

These structures are strictly perpendicular to the stress direction of the fiber. Dyeing results revealed that the dyeing properties of all dyes on polyamide fabrics changed remarkably after the treatment.

Results suggest that the change in coloration closely correspond with the ripple-like structures and the changes in chemical properties induced by laser treatment. It should be noted that the increased rate of exhaustion of acid dyes by laser treatment is not beneficial for dyeing polyamides since this will enhance the non-uniformity of the dye. While deeper dyeing is a greater advantage for disperse and reactive dyes since darker shades are obtainable using only the usual amounts of dyestuffs. To conclude, the excimer laser modification process has a high industrial potential, as it is an environmentally friendly dry process not involving any of the solvents required for a wet chemical process. After laser treatment, the dyeing properties of disperse and reactive dyes are improved and this provides an alternative choice for dyeing polyamide materials. (Figure 4) [16]

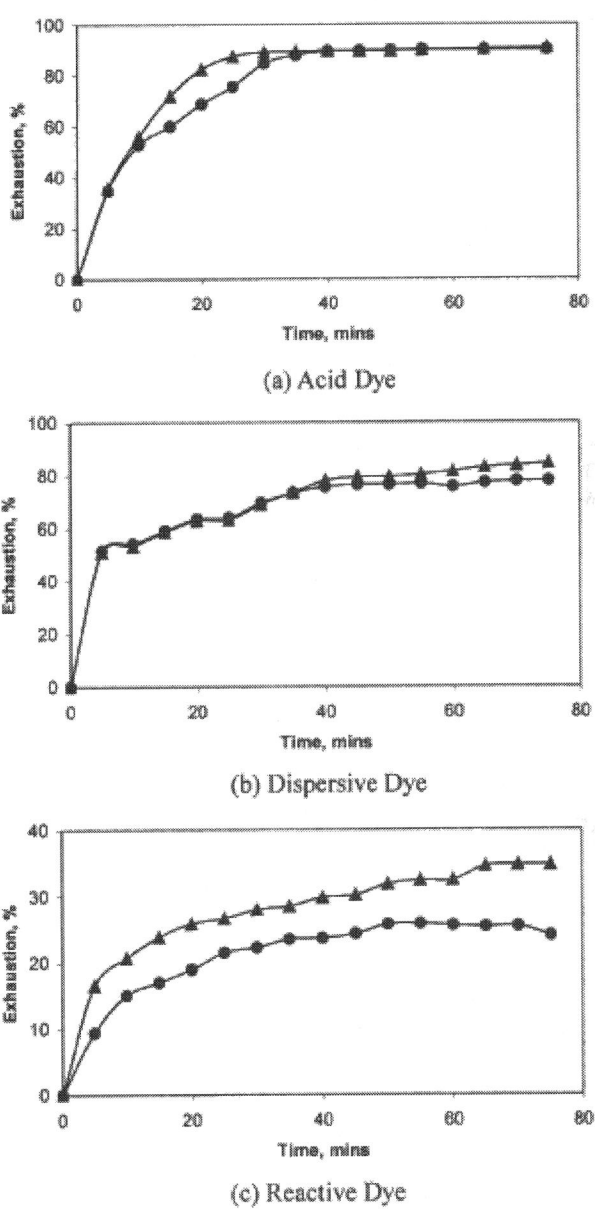

(a) Acid Dye

(b) Dispersive Dye

(c) Reactive Dye

Figure 4. Dye bath exhaustion study of laser treated nylon 6 fabrics; (●) Control (▲) Laser treated

1.2. Chemical modification of fabrics

Chemical treatment has been used in industry to treat large objects that would be difficult to treat by other commonly used industrial technique such as flame and corona-discharge treatments. Chemical etchants are used to convert smooth hydrophobic polymer surfaces to rough hydrophilic surfaces by dissolution of amorphous regions and surface oxidation. Chromic acid is the most widely used etchant for polyolefins and other polymers. [2]

Alkaline, acidic and solvents hydrolysis is another method to improve various physical and chemical properties of synthetic fibers. The alkaline hydrolysis of PET fibers is usually carried out with an aqueous alkaline solution, such as sodium hydroxide. In the alkaline hydrolysis process, PET undergoes a nucleophilic substitution. Chain scission of PET occurs, resulting in a considerable weight loss and the formation of hydroxyl and carboxylate end groups, which improves the handling, moisture absorption and dyeability of the fabric with enhanced softness.

There are different kinds of auxiliaries that are used to modify the surface properties of textile fabrics. Some of new chemical and their applications in textile industry are described bellow:

1.2.1. Aminolysis

Several studies have assessed the effects of amine interaction with polyester. Early studies assessed the aminolysis of polyester as a means of examining fiber structure without regard to maintaining the integrity of the polymer.

The degradation effects on polyester of a monofunctional amine versus alkaline hydrolysis have been studied. These studies, which again involved high levels of fiber degradation, demonstrated that alkaline hydrolysis has a more substantial effect on fiber weight without extensive strength loss. In contrast, aminolysis had less effect on fiber weight but decreased fiber strength, indicative of a reaction within the polymer structure rather than simply at the surface. It was later demonstrated that bifunctional amine compounds could be reacted with the polymer with minimal loss in strength while generating amine groups at the fiber surface. The early stages of the reaction were largely confined to the fiber surface and the resulting fiber had modified wetting properties and improved adhesion with the matrix when used in composites. A recent paper has re-examined the interaction of untreated and alkali hydrolyzed polyester with a range of aliphatic diamines. 1,6-Hexanediamine, 2-methylpentamethylene diamine, 1,2- diaminocyclohexane, tetraethylenepentamine, and ethylene diamine were applied to untreated polyester. Ethylene diamine was also applied for a range of solution concentrations in toluene. The treatment generated amine groups on the fiber surface and was revealed by staining with anionic dyes under conditions in which the amine group was protonated. Unexpectedly, the reaction resulted in the simultaneous formation of carboxylic acid groups in a manner similar to alkaline hydrolysis, revealed by staining with Methylene Blue

(Figure 5). The reaction thus resulted in a bifunctional polyester surface. The ratio of amine and carboxylic acid groups differed with unhydrolyzed and hydrolyzed starting materials (Figure 5). Strength loss was somewhat greater than with alkaline hydrolysis. [17]

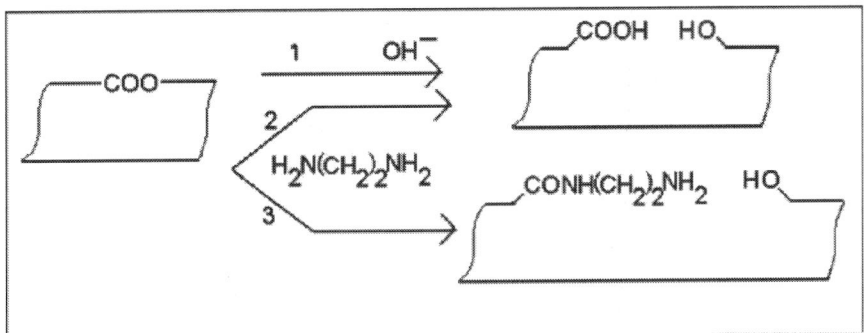

Figure 5. Hydrolysis and aminolysis of polyester

1.2.2. Sol gel pre modification of fabrics

The sol-gel process is an excellent tool to obtain ordered hybrid organic-inorganic nanocomposites. The method involves the mixing of precursors in an aqueous or alcoholic medium. Precursors are molecules, which contain a central metal or semimetal atom, to which reactive alkoxy groups and/or organic groups are bonded. These reactive groups are subjected to an acidic or alkaline catalyzed hydrolysis and condensation reaction, thus forming a sol and subsequently a gel. Aging or drying step enables the production of powders, xerogels, aerogels, fibers, or coatings [18]. The latter procedure renders possible surface modification of textiles, thus imparting novel properties to the material.

The sol-gel technology was also applied to influence the dyeing properties. Luo et al. [19] succeeded in improving the wash fastness of cotton materials that had been dyed with direct dyes using GPTMS and TEOS. Mahltig incorporated dyes into a silica matrix [20-22]. Du investigated the fixation properties and mechanism of direct dyes on silk, nylon 6 and cotton applying various organotrialkoxysilanes. They found that the most suitable precursors for silk and nylon 6 are amide- and vinyl-containing sols [14]. Yin also managed to enhance the fastness properties of cotton material that had been dyed with direct dyes [22].

1.2.2.1. Dyeing Treatment of Sol-gel Pre-treated Cotton Fabrics

Schramm and his coworker investigated the impact of alkoxysilanes (TEOS, GPTMS, APTES, and TESP-SA) on the dyeing process of cotton substrates, which were dyed with 4 % owf C.I. Reactive Red 141 and 4 % owf C.I. Reactive Black 5. (Figure 6) For this purpose cotton samples were pre-treated with the alkoxysilanes and subsequently dyed. The results show that TESP-SA are lowering the L* values significantly, whereas TEOS, GPTMS, and

APTES give rise to a moderate change of L*. The after-treatment of dyed cotton fabrics with alkoxysilane causes almost no effect with respect to the colorimetric data. The direct incorporation of the alkoxysilanes into the dyeing bath resulted in a reduction of the color properties, when APTES or TESP-SA was employed. The crease-proof finishing treatment caused an increase in the a* value of red-dyed samples. [23]

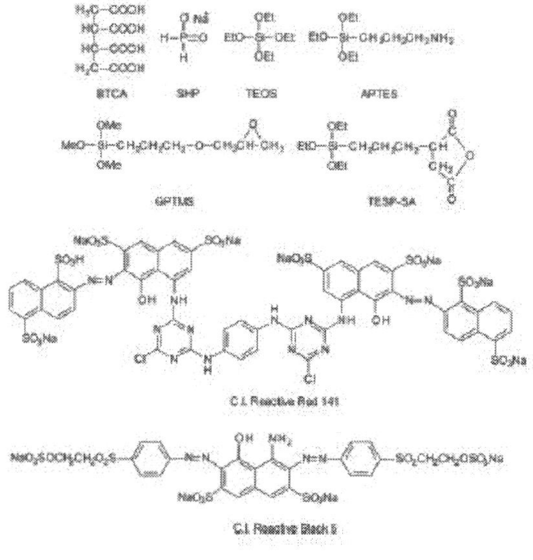

Figure 6. Chemical formula of the substances of interest.

1.2.2.2. An evaluation of the dyeing behavior of sol–gel silica doped with direct dyes

Direct dyes are important dyes used on cellulose fiber directly. They can be applied in the same dyeing bath with other dyes. Moreover, the price is much lower than that of other dyes. However, several problems often occur in dyed fabric with direct dyes, such as lower washing and rubbing fastnesses. Some direct dyes perform rather poorly with respect to washing and rubbing fastnesses. They are mainly caused by the water-soluble groups, sulphonic and/ or carboxyl groups. Without an appropriate treatment, direct dyes bleed a little with each washing, lose their color on fabric and endanger other clothes washed in the same bath. At present, copper salt fixing reagent and cationic fixing reagent are widely used in textile industry to improve the fastness by cross-linking reactions between metal ion or formaldehyde and other molecules. But metal ions in fixing reagent such as cupric cation ($Cu2+$) will aggravate the difficulty of waste processing. Also some reagents and intermediates used in diazotization or crosslinking reactions such as formaldehyde, nitroaniline seriously affect the human health and the environment.

Moreover, the vividness after copper salt or diazotization finishing usually fades away, so that the application scope of direct dyes have been seriously limited in textile industry. A novel dyeing solution containing silica and direct dyes has been prepared by the sol–gel process. During this process, EtOH, tetraethoxysilane (TEOS), H2O and 3-glycidoxypropyl-trimethoxysilane (GPTMS) were added in turn. The molar ratio of TEOS:H2O:EtOH was 1:5:8 and the concentration of GPTMS was 0.05 mol/L. Fabric was dyed at 90°C for 40 min. The concentration of NaCl added into the dyeing solution was 10 g/L. The dyed fabrics were baked at 150°C for 5 min. With this process, the results indicate that the K/S value is enhanced by more than 10%, the rubbing fastness and the washing change fastness are improved by one grade and the washing staining fastness is improved by half a grade. Not only is the K/S value enhanced from 9.3 to 11.5, but also the wet rubbing fastness and the washing change fastness are increased half a grade. Using a video microscope, a smoother fiber surface is observed. The calculated sol–gel weight gain is 4.6%. As a nonpolluting process, the sol–gel technology shortens the dyeing process and brings a better fixation property, meeting the needs of energy-saving and pollution-free processes.

1.2.3. Application of nanoparticles for surface modification of fibers

The dyeability of synthetic fibers depends on their physical and chemical structure. Dyeing process consist of three steps including the diffusion of dye through the aqueous dye bath on to the fiber, the adsorption of dye into the outer layer of the fiber and the diffusion of dye from the adsorbed surface into the fiber interior. It was shown by researchers that functional groups of PET and water molecules play a great role in this process. The terminal carboxylic and hydroxyl groups in PET chains interact with water molecules. This makes a swelled fiber resulting to increase the attraction of disperse dye by these functional groups of fiber.

The proportion of crystalline and amorphous regions of polymer is another factor influencing the dyeability. Researchers are concerned with the development and implementation of new techniques in order to fulfill improvement in dyeability of various polymers. Blending of polymeric fibers with nanoclays as inexpensive materials is still claimed as cost effective method to enhance dyeability. Up to now, only two research articles are focused on dyeing properties of polypropylene- and polyamide 6- layered clay incorporated nanocomposites prepared by melt compounding. Toshniwal et al. suggested that polypropylene fibers could be made dyeable with disperse dyes by addition of nanoclay particles in polymer matrix [25].

The previous study on dyeability of PET/clay nanocomposites stated the following type of interactions between the disperse dye and clay surfaces:

- Hydrogen bonding between OH groups of modified clays and the NH2 and CO groups of disperse dye molecules.

- Electrostatic bonding between the negatively charged oxygen atom of carbonyl groups in disperse dye molecule and positively charged nitrogen atom of quaternary ammonium salt in modified clays.

- Direct π interactions and van der Waals forces between methyl and ethyl groups of modified clays on one hand and methoxy group and benzene rings of disperse dye molecule on the other hand.

1.2.4. Application of cyclodextrins in textile dyeing

Cyclodextrins can be considered as a new class of auxiliary substances for the textile industry. Cyclodextrins can be used for textile application because of their natural origin and their biodegradability. Cyclodextrins play on important role in textile scientific research area and should play a significant role in the textile industry as well to remove or substitute various auxiliaries or to prepare textile materials containing molecular capsules which can immobilize perfumes, trap unpleasant smells, antimicrobial reagents and flame retardants. As cyclodextrins can incorporate different dyes into their cavity, they should be able to act as retarders in a dyeing process. Various auxiliary products are used in wet finishing processes, especially in dyeing and washing. One of the dyeing auxiliary products are leveling agent. Leveling agents help to achieve uniform dyeing by slowing down the dye exhaustion or by dispersing the dye taken by the fibre in a uniform way. They can be classified into two groups: agents having affinity to the dye and agents having affinity to the fibre. Agents having affinity to the dyes slow down the dyeing process by forming complexs with the dyes. The complex compound moves slower compared to the dye itself; at higher temperature the dye is released and it can be fixed to the fibre. Application of cyclodextrins as leveling agents having affinity to dyestuffs has been investigated in research work about dyeing of cellulose fibres with direct dyes by exhaust method where β-cyclodextrin was tested as a dye complexing agent. cyclodextrins as a dye retardant in the dyeing of PAN fibres with cationic dyes was studied; further it was reported that some azo disperse dyes formed inclusion complexes with cyclodextrins. [26]

1.2.5. Chitosan applications in textile dyeing

Nowadays, the surface modification of textile fibres is considered as the best route to obtain modern textile treatments.

Among various available biopolymers, the polysaccharide chitosan (CHT) is highly recommendable, since it shows unique chemical and biological properties and its solubility in acidic solutions makes it easily available for industrial purposes. The polysaccharide-based cationic biopolymer chitosan is poly(1,4)-2-amino-2-deoxy-b-D-glucan, usually obtained by deacetylation of chitin that is widely present in the nature as a component of some fungi, exoskeleton of insects and marine invertebrates (crabs and shrimp). The chemistry of chitosan is similar to that of cellulose, but it reflects also the fact that the 2-hydroxyl group of the cellulose has been replaced with a primary aliphatic amino group. Among many other uses, it has been recently shown that chitosan improves the dye coverage of immature fibres in cotton dyeing and that it could be successfully used as a thickener and binder in pigment printing of cotton.

Also Gupta et al showed that, chitosan treated cotton has better dyeability with direct and reactive dyes and treatment with modified chitosan makes it possible to dye cotton in bright shades with cationic dyes having high wash fastness. Treated samples showed good antimicrobial activity against Escherichia coli and Staphylococcus aureus at 0.1% concentration as well as improved wrinkle recovery. [28]

In similar research work, Jocic and his coworkers assessed the interactions that could occur during dyeing of the chitosan treated wool fibres, by measuring the absorbance values of the solutions containing dye and chitosan. It has been shown that there is a 1:1 stoichiometry between protonated amino groups and sulfonate acid groups on the dye ions in low concentrated chitosan solutions. This interaction forms an insoluble chitosan/dye product. With the excess of chitosan in the solution, the dye can be distributed between the different chitosan molecules and the soluble chitosan/dye products remain in the solution. It is suggested that the mechanism of the interaction involves the possibility of adsorbed dye molecules to be desorbed and redistributed between other components present in the system, depending on system parameters (pH, temperature and electrolyte presence). This fact is important in explanation of dyeing behaviour of chitosan treated wool and enables the assessment of the mechanism of dyeing of accordingly modified textile fibres [29].

Also Kitkulnumchai et al, showed that, the KIO4 oxidation of cellulose fabrics created more aldehydic groups on the fabric surface and the following reductive ammination with chitosan afforded stable C–N bonds between cellulose and chitosan chains. The attachment of chitosan to the fabric considerably improved dye uptake of mono chlorotriazine and vinyl sulfone reactive dyes resulting in greater dye exhaustion and the color yield (K/S).

The enhanced dyeability of the modified fabric is likely resulted from the reduction of the coulombic repulsion between the fabric surface and the anionic dye molecules in the presence of the positively charged chitosan on the surface. The oxidation step can cause some drop in the fabric strength. This oxidation reductive amination with chitosan is thus a convenient method for modifying the surface activity of cellulose fabrics whereas the fabric strength is not of the great priority.[30]

In another research, Cotton fabrics have been successfully dyed by green tea extract upon chitosan mordanting, and UV protection property of the chitosan mordanted green tea dyed cotton was increased.

The following conclusions have been made from this study;

1. Chitosan mordanting can effectively increase the ΔE and the K/S, that is, the dyeing efficiency of green tea dyeing onto cotton fabrics. As chitosan concentration increased, the ΔE and the K/S of cotton fabrics by green tea extract increased gradually.

2. Chitosan mordanting can effectively increase the UV protection property of both UV-A and UV-B of green tea dyed cotton fabrics. Chitosan mordanted undyed cotton and chitosan unmordanted dyed cotton did not show an increase in UV protection property. Therefore, it can be assumed that chitosan increased the uptake of active moiety, catechin, in green tea, which would be responsible for the UV protection and subsequently increased UV protection property of the chitosan mordanted green tea dyed fabric.

3. As chitosan mordanting concentration increased, UV protection property increased in both UV-A and UV-B.

Around 7% UV protection increase from control was observed upon chitosan mordanting, which is similar value of the green tea dyed cellulose fabric using a specific metal mordant.

Therefore, it can be concluded that green tea dyeing can be used in developing UV protective cotton textiles, and the chitosan mordanting process would be necessary in green tea dyeing of cotton to increase not only the dyeing efficiency but also the UV protection property of cotton fabrics. [31]

Author details

Sheila Shahidi[1*], Jakub Wiener[2] and Mahmood Ghoranneviss[3]

*Address all correspondence to: sh-shahidi@iau-arak.ac.ir

1 Textile Department, faculty of Engineering, Arak Branch, Islamic Azad University, Arak, Iran

2 Department of Textile Chemistry, Faculty of Textile, Technical University of Liberec, Liberec, Czech Republic

3 Plasma Physics Research Center, Science and Research Branch, Islamic Azad University, Tehran, Iran

References

[1] J. Militky, J. Vanicek, J.Krystufek, V. Hartych, Modified Polyester Fibers, Elsevier, 1991

[2] Chan, Polymer Surface Modification and Characterization, Hanser Publisher, Munich Vienna New York, 1994

[3] Renuga Gopal, Ma Zuwei, Satinderpal Kaur, and Seeram Ramakrishna Surface Modification and Application of Functionalized Polymer Nanofibers, Functionalized Polymer Nanofibers, 75

[4] Mazeyar Parvinzadeh Gashti, Julie Willoughby and Pramod Agrawal, Surface and Bulk Modification of Synthetic Textiles to Improve Dyeability, textile dyeing, Intech, 2011 with F. solani pisi cutinase and pectate lyase. Enzyme & Microbial Technology, Vol. 42, No. 6, (May 2006), pp. 473-482, ISSN 0141-0229

[5] R.Shishoo, Plasma Technologies for Textiles, Woodhead Publishing for textiles, Cambridge England, 2006

[6] T.Oktem, Modification of polyester and polyamide fabrics by different in situ plasma polymerization methods, Turk J Chem, 24 (2000)-275-285.

[7] Peter. P, Surface modification of fabrics using one atmosphere glow discharge plasma to improve fabric wettabilitty, Textile Research Journal, 67(5), 1997, 359-396.

[8]   L'. ˇCern´akov´a, D. Kov´aˇcik, A. Zahoranov´a M. ˇCern´ak, and M. Maz´ur, Surface Modification of Polypropylene Non-Woven Fabrics by Atmospheric-Pressure Plasma Activation Followed by Acrylic Acid Grafting Plasma Chemistry and Plasma Processing, Vol. 25, No. 4, August 2005

[9]   Nebojša Risti , Petar Jovan i , Cristina Canal, and Dragan Joci, One-bath One-dye Class Dyeing of PES/Cotton Blends after Corona and Chitosan Treatment , Fibers and Polymers 2009, Vol.10, No.4, 466-475

[10] Alejandro Patin˜o Cristina Canal Cristina Rodrı´guez Gabriel Caballero Antonio Navarro Jose´ Ma Canal, Surface and bulk cotton fibre modifications: plasma and cationization. Influence on dyeing with reactive dye Cellulose (2011) 18:1073–1083

[11] Darinka Fakin, Alenka Ojstrsˇek, Sonja Cˇ elan Benkovicˇ The impact of corona modified fibres' chemical changes on wool dyeing, journal of materials processing technology 2 0 9 (2 0 0 9) 584–589

[12] Javed, I., Bhatti, I.A., Adeel, S., 2008.Effect of UV radiation on dyeing of cotton fabric with extracts of henna leaves. Indian Journal of fiber and textile research, 33,157-162.

[13] Eun-Min Kim and Jinho Jang , Surface Modification of Meta-aramid Films by UV/ ozone Irradiation Fibers and Polymers 2010, Vol.11, No.5, 677-682

[14] Hongje Kim and Jin-Seok Bae, Modification of Polypropylene Fibers by Electron Beam Irradiation. I.Evaluation of Dyeing Properties Using Cationic Dyes Fibers and Polymers 2009, Vol.10, No.3, 320-324

[15] Joanne Yip Kwong Chan Kwan Moon Sin Kai Shui Lau, UV Excimer laser modification on polyamide materials: effect on the dyeing properties, Mat Res Innovat (2002) 6:73–78

[16] Martin Bide, Matthew Phaneuf, Philip Brown, Geraldine McGonigle, and Frank Lo-Gerfo, MODIFICATION OF POLYESTER FOR MEDICAL USES, Modified Fibers with Medical and Specialty Applications, 91–124, 2006

[17] C. J. Brinker and G. W. Scherer, "Sol-gel Science, The Physics and Chemistry of Sol-gel Processing", Academic Press, San Diego, 1990.

[18] M. Luo, X. Zhang, and S. Chen, Color. Technol., 119, 297 (2003).

[19] B. Mahltig, H. Böttcher, D. Knittel, and E. Schollmeyer, Text. Res. J., 74, 521 (2004).

[20] J. Du, Z. Li, and S. Chen, Color. Technol., 121, 29 (2005).

[21] Y. Yin, C. Wang, and C. Wang, J. Sol. Gel. Sci. Technol., 48, 308 (2008).

[22] Christian Schramm and Beate Rinderer, Dyeing and DP Treatment of Sol-gel Pre-treated Cotton FabricsFibers and Polymers 2011, Vol.12, No.2, 226-232

[23] Yunjie Yin , Chaoxia Wang , Chunying Wang, An evaluation of the dyeing behavior of sol–gel silica doped with direct dyes , J Sol-Gel Sci Technol (2008) 48:308–314

[24] Toshniwal, L., Fan, Q., Ugbolue, S.C., (2007) Dyeable polypropylene fibers via nano-technology, Journal of Applied Polymer Science, 1, pp. 706–711.

[25] Bojana Voncina, Application of Cyclodextrins in Textile Dyeing, Textile dyeing, In-tech

[26] Dragan Jocica,*, Susana Vı'lchezb, Tatjana Topalovicc, Antonio Navarroa, Petar Jo-vancicc, Maria Rosa Julia`b, Pilar Errab, Chitosan/acid dye interactions in wool dye-ing system Carbohydrate Polymers 60 (2005) 51–59

[27] Deepti Gupta a,*, Adane Haile, Multifunctional properties of cotton fabric treated with chitosan and carboxymethyl chitosan, Carbohydrate Polymers 69 (2007) 164–171

[28] Dragan Jocic, Susana Vı'lchez, Tatjana Topalovic, Antonio Navarro, Petar Jovancic, Maria Rosa Julia`, Pilar Erra, Chitosan/acid dye interactions in wool dyeing system, Carbohydrate Polymers 60 (2005) 51–59

[29] Yupaporn Kitkulnumchai, Anawat Ajavakom, Mongkol Sukwattanasinitt, Treatment of oxidized cellulose fabric with chitosan and its surface activity towards anionic re-active dyes, Cellulose (2008) 15:599–608

[30] Sin-hee Kim*, Dyeing Characteristics and UV Protection Property of Green TeaDyed Cotton Fabrics–Focusing on the Effect of Chitosan Mordanting Condition–Fibers and Polymers 2006, Vol.7, No.3, 255-261

Chapter 3

Cyclodextrins in Textile Finishing

Bojana Voncina and Vera Vivod

Additional information is available at the end of the chapter

http://dx.doi.org/10.5772/53777

1. Introduction

Chemical finishing is crucial for giving textiles new functionalities and making them appropriate for special applications, such as antimicrobial resistance, flame retardancy and others. Textile finishing is also an important process as it improves appearance, performance or hand.

Cyclodextrins can act as hosts and form inclusion compounds with various small molecules. Such complexes can be formed in solutions, in a solid state as well as when cyclodextrins are linked to various surfaces where they can act as permanent or temporary hosts for small molecules that provide certain desirable attributes. This characteristic makes cyclodextrin a promising reagent in textile finishing.

2. Cyclodextrins

Cyclodextrins(CDs) are relevant molecules used in different applications and industries from pharmacology, cosmetics, textiles, filtration to pesticide formulations. They comprise a family of three well-known industrially produced substances. The practically important, industrially produced CDs are the α-, β-, and γ-CDs. There are some seldom used cyclic oligosaccharides as well but because of their cost are not applicable to industrial applications [1]. CDs are cyclic oligomers of a-D-glucopyranose that can be produced with the transformation of starch by certain bacterias such as Bacillus macerans [2, 3].

The preparation process of CDs consists of four principal phases: (i) culturing of the microorganism that produces the cyclodextrin glucosyl transferase enzyme (CGT-ase); (ii) separation, concentration and purification of the enzyme from the fermentation medium; (iii) enzymatical conversion of prehydrolyzed starch in mixture of cyclic and acyclic dextrins;

and (iv) separation of CDs from the mixture, their purification and crystallization. CGT-ase enzymes degrade the starch and starts intramolecular reactions without the water participation. In the process, cyclic (CDs) and acyclic dextrins are originated, which are oligosaccharides of intermediate size. CDs are formed by the link between units of glucopyranose. The union is made through glycosidic oxygen bridges by α-(1,4) bonds. The purification of α- and γ-CDs increases the cost of production considerably, so that 97% of the CDs used in the market are β-CDs [2].

CDs ring structures act as hosts and form inclusion compounds with various small molecules. Such complexes can be formed in solutions, in a solid state as well as when cyclodextrins are linked to various surfaces. In all forms they can act as permanent or temporary hosts to small molecules that provide certain desirable attributes.

In the textile field CDs may have many applications such as: absorption of unpleasant odours; they can complex and release fragrances, "skin-care-active" and bioactive substances. Further, various textile materials treated with cyclodextrins could be used as selective filters for adsorption of small pollutants from waste water [4].

After the discovery of CDs scientists considered them poisonous substances and their ability for complexes formation was only considered a scientific curiosity. Later on, research on CDs proved that they are not only non-toxic but they can be helpful for protecting flavours, vitamins and natural colours [2]. CDs already play a significant role in the textile industry and can be used in dyeing, surface modification, encapsulation, washing, and preparation of polymers and in fibre spinning.

Since year 2000, β-CD has been introduced as a food additive in Germany. With respect to OECD experiments, this compound has shown no allergic impact. In the USA α-, β- and γ-CDs have obtained the GRAS list (FDA list of food additives that are 'generally recognized as safe') status and can be commercialized as such. In Japan α-, β- and γ-CDs are recognized as natural products and their commercialization in the food sector is restricted only by considerations of purity. In Australia and New Zealand α- and γ-CDs are classified as Novel Foods from 2003 and 2004, respectively. The recommendation of Joint FAO/WHO Expert Committee on Food Additives (JECFA) for a maximum level of β-CDs in foods is 5 mg/kg per day. For α- and γ-CDs no Acceptable Daily Intake (ADI) was defined because of their favourable toxicological profiles [2, 5].

Natural cyclodextrins and their hydrophilic derivatives are only able to permeate lipophilic biological membranes, such as the eye cornea, with considerable difficulty. All toxicity studies have demonstrated that orally administered cyclodextrins are practically non-toxic, due to lack of absorption from the gastrointestinal tract. The main properties of β-CD, the most important cyclodextrin in textile application are: less irritating than α-CD after i.m. injection, binds cholesterol, small amount (1-2%) is adsorbed in the upper intestinal tract, no metabolism in the upper intestinal tract, metabolised by bacteria in caecum and colon, LD50 oral rat > 5000 mg/kg, LD50 i.v., rat: between 450-790 mg/kg, however, application of high doses may be harmful and is not recommended [6, 7].

2.1. Structure of cyclodextrins

The three mayor CDs are crystalline, homogeneous, and non-hygroscopic substances, which are torus-shaped oligosaccharides [1, 8-10], composed in the more common forms of six to eight (α-1,4)-linked α-D-glucopyranose units (α-, β- and γ-cyclodextrin), Figure 1 schematically present the main three cyclodextrins [11]. They are of circular and conical conformation, where the height is about 800 pm. The inner diameter of the cavity varies from 500 to 800 pm.

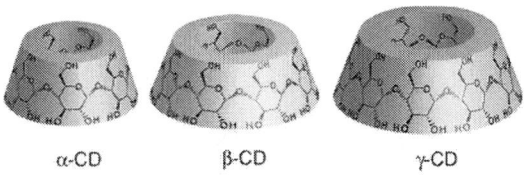

α-CD β-CD γ-CD

Figure 1. Schematic presentation of the main three CDs [12].

All glucopyranose units in the torus-like ring possess the thermodynamically favoured chair conformation because all substituents are in the equatorial position. As a consequence of the 4C1 conformation of the glucopyranose units, all secondary hydroxyl groups are situated on the lager side of the ring, whereas all the primary ones are placed on the narrower side of the ring. Hydroxyl groups on the outside of the CDs ensure good water solubility. The cavity is lined with hydrogen atoms of C3, by the glycosidic oxygen bridges and hydrogen atoms of C5. The nonbonding electron pairs of the glycosidic oxygen bridges are directed toward the inside of the cavity producing a high electron density and because of this the inner side of the cavity has some Lewis base characteristics. The C-2-OH group of one glucopyranoside unit can form a hydrogen bond with the C-3-OH group of the adjacent glucopyranose unit. In the CD molecule, a complete secondary belt is formed by these hydrogen bonds, therefore the β-CD has a rather rigid structure. Because of this arrangement, the interior of the toroids is not hydrophobic but considerably less hydrophilic than the aqueous environment and thus able to host other hydrophobic molecules. CDs behave more or less like rigid compounds with two degrees of freedom, rotation at the glucosidic links C4-O4 and C1-O4 and rotations at the O6 primary hydroxyl groups at the C5-C6 band. The intramolecular hydrogen bond formation is probably the explanation for the observation that β-CD has the lowest water solubility of all CDs. The hydrogen-bond belt is incomplete in the α-CD molecule, because one glucopyranose unit is in a distorted position. Consequently, instead of the six possible H-bonds, only four can be established fully. γ-CD is a noncoplanar with more flexible structure; therefore, it is the most soluble of the three CDs. Figure 2 shows a sketch of the characteristic structural features of CDs. On the side where the secondary hydroxyl groups are situated, the diameter of the cavity is larger than on the side with the primary hydroxyls, since free rotation of the primary hydroxyls reduces the effective diameter of the cavity [13, 14].

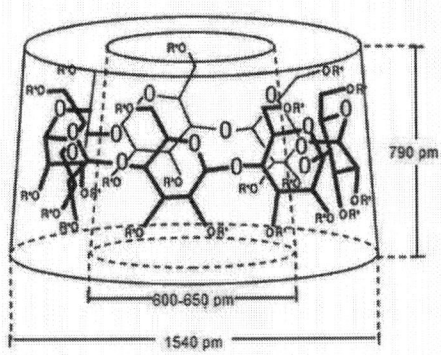

Figure 2. Schematic presentations of characteristic structural features of CDs.

By far, β-cyclodextrin (β-CD), with 7 sugar units, has been commercially the most attractive (more than 95% consumed) due to its simple synthesis, availability and price.

A β-CD molecule has a molecular weight of 1135 and a height of 750–800 pm. The internal diameter of the molecule's hole is between 600 and 680 pm, and the external diameter is 1530 pm[1, 15]. The volume of the hole is 260–265 Å^3, and the dissolution is 1.85 g/100 mL of water. The cavity is hydrophobic; the external section is hydrophilic in nature. β-CD is stable in alkali solutions and it is sensitive to acid hydrolysis [16].

2.2. Properties of cyclodextrins

The most notable feature of CDs is their ability to form solid inclusion complexes ("host–guest" complexes) with a very wide range of solid, liquid and gaseous compounds by a molecular complexation. The phenomenon of CD inclusion compound formation is a complicated process involving many factors playing an important role.

Complex formation is a dimensional fit between host cavity and guest molecule. The lipophilic cavity of CD molecules provides a microenvironment into which appropriately sized non-polar moieties can enter to form inclusion complexes. No covalent bonds are broken or formed during formation of the inclusion complex.

According to some authors [11, 17] hydrophobic interactions are the main driving forces for CD-based host-guest compounds. Other requirements such as steric hindrance and relation between sizes of host and guest cavities are also important. This is illustrated by the fact that not only hydrophobic interaction will lead to an association between a guest molecule and a CD but ionic solutes, such as non-associated inorganic salts, can also be involved in these complexes.

Some researchers [7] claim that the main driving force of complex formation is the release of enthalpy-rich water molecules from the cavity. The water molecules located inside the cavity cannot satisfy their hydrogen bonding potentials and therefore are of higher enthalpy.

The energy of the system is lowered when these enthalpy–rich water molecules are replaced with suitable guest molecules which are less polar than water. In an aqueous solution, the slightly apolar CD cavity is occupied by water molecules which are energetically unfavoured, and therefore can be readily substituted by appropriate "guest molecules" which are less polar than water. This apolar–apolar association decreases the CD ring strain resulting in a more stable lower energy state. The dissolved CD is the "host" molecule, and the "driving force" of the complex formation is the substitution of the high-enthalpy water molecules by an appropriate "guest" molecule.

The binding of guest molecules within the host CD is not fixed or permanent but rather is a dynamic equilibrium. Binding strength depends on how well the 'host–guest' complex fits together and on specific local interactions between surface atoms. Complexes can be formed either in solution or in the crystalline state and water is typically the solvent of choice. Inclusion complexation can be accomplished in a co-solvent system and in the presence of any non-aqueous solvent [7]. Generally, one guest molecule is included in one CD molecule, although in the case of some low molecular weight molecules, more than one guest molecule may fit into the cavity, and in the case of some high molecular weight molecules, more than one CD molecules may bind to the guest. In principle, only a portion of the molecule must fit into the cavity to form a complex. CD inclusion is a stoicmetric molecular phenomenon in which usually only one guest molecule interacts with the cavity of the CD molecules to become entrapped. 1:1 complex is the simplest and most frequent case. However, 2:1, 1:2, 2:2, or even more complicated associations, and higher order equilibrium exist almost always simultaneously.

Inclusion in CDs has a profound effect on the physicochemical properties of guest molecules as they are temporarily included within the host cavity.

These properties are:

- solubility enhancement of highly insoluble guests,

- stabilisation of labile guests against the degradative effects of oxidation, visible or UV light and heat,

- control of volatility and sublimation,

- physical isolation of incompatible compounds,

- chromatographic separations,

- taste modification by masking of flavours, unpleasant odours,

- controlled release of drugs and flavours,

- removal of dyes and auxiliaries from dyeing effluents,

- retarding effect in dyeing and finishing,

- protection of dyes from undesired aggregation and adsorption.

2.3. Application of cyclodextrins

Complexes can be formed in solutions, in the solid state, as well as when CDs are linked to a solid surface where they can act as permanent or temporary hosts to those small molecules that provide certain desirable attributes such as adsorption of dyestuff molecules, fragrances or antimicrobial agents. This "molecular encapsulation" is already widely utilized in many industrial products, technologies, and analytical methods [7, 18].

Due to the relatively non-polar character of the cavity in comparison to the polar exterior, CD can form inclusion complexes with a wide variety of guest molecules, such as drugs [19, 20, 21], ionic and non-ionic surfactants [23, 24, 25], dyes [26, 27] and polymers [28], etc. The use of CDs has increased annually around 20–30%, of which 80-90% was in food products [29]. In the pharmaceutical industry, CDs and their derivatives have been used either for complexation of drugs or as auxiliary additives such as solubilizers, diluents, or ingredients for improving of drugs physical and chemical properties, or to enhance the bioavailability of poorly soluble moieties [30]. In the chemical industry, CDs and their derivatives are used as catalysts to improve the selectivity of reactions, as well as for the separation and purification of industrial-scale products [31]. In the food, cosmetics, toiletry, and tobacco industries, CDs have been widely used either for stabilization of flavours and fragrances or for the elimination of undesired tastes, microbiological contaminations, and other undesired compounds [7]. For the last 30 years, the use of CDs and their derivatives in the textile domains has captivated a lot of attention. Many of the papers and patents report the use of CDs for finishing and dyeing processes. For instance, they discuss the capture of unpleasant smells due to perspiration, or how to do the controlled release of perfumes, insecticides and antibacterial agents [4, 32-45].

3. Cyclodextrins in textiles

In the textile field CDs may have many applications such as: they can absorb unpleasant odours; they can complex and release fragrances or "skin-care-active" substances like vitamins, caffeine and menthol as well as bioactive substances such as biocides and insecticides. Further, various textile materials treated with CDs could be used as selective filters for adsorption of small pollutants from waste waters - "preparation of textile nanosponges".

3.1. Cyclodextrins in textile finishing

One of the new concepts for modification of textile substrates is based on the permanent fixation of supramolecular compounds on the material's surface and, thus, imparts new functionality to the fabric. [46]. One of the most promising supramolecular moieties applied to textiles are CDs. Covalent bonding of CDs onto textile fibres was firstly patented in 1980 by Szejtli [47]. He and co-workers reported to bond CD via crosslinking reagent epichlorhydrin onto alkali-swollen cellulose fibres. They found out that CD covalently bonded to cellulose retained the ability to form complexes; when cellulose was treated with a drug it complexed with CD. The complexed drug was upon the contact with the skin released; cellulose textile substrate containing covalently bound β-CD was treated with solution of iodine, potassium iodide and methanol as a solvent to prepare a medical bandage [48].

Szejtli published [4] a very extensive review about CDs in the textile industry. In his review he divided the application of CDs in textile sectors in the following areas: binding of CDs to fibre surfaces, CDs in textile dyeing, in textile finishing, CDs and detergents and miscellaneous applications of CDs in textile industry and textile care. Due to the fact that application of CDs in textile dyeing processes was extensively reported in a chapter of the book edited by Hauser [18], we will emphasis in current publication about the application of CDs in textile finishing.

3.1.1. Binding of cyclodextrins to fibre surfaces

The attachment of CD molecules on textile substrate provides hosting cavities that can include a large variety of guest molecules for specific functionality. There are two possible approaches to bond CDs onto textile fibres such as chemical bonding of modified CDs on the fibre surfaces or to use bifunctional reagents to link CDs covalently on fibre surfaces.

The most promising approach to bond modified CDs onto textile fibres is the modification of CDs with trichlorotriazines to prepare monochlorotriazinyl-cyclodextrin (CD-MCT) [49, 50]. Analogues to reactive dyes the CD-MCT can be fixed to the fabric by well-known methods and with common equipment. CD-MCT can be applied to fibre surfaces by an exhaustion method or by thermofixation. Moldenhauer with co-workers found out [51] that the fixation was the best when textile substrate was cotton. Mixed fibre materials like cotton/polyurethane or cotton/polyamide can be finished with β-CD-MCT in good yields. Ibrahim et.al [52] reported the improvement of UV protective properties of cotton/wool and viscose/wool blends via incorporating of reactive β-CD-MCT in the easy care finishing formulations, followed by subsequent treatment with copper-acetate or post-dyeing with different classes of dyestuffs (acid, basic, direct and reactive). They found out that post-dyeing of the prefinished textile blends results in a significant increase in the UPF (UV-protection factor) values as a direct consequence of a remarkable reduction in UV radiation transmission through the plain weave fabric. β-CD modified with monochlorotriazine was applied to the cotton fabrics for entrapping of sandalwood oil as an aroma-finishing agent by Sricharussin [53]. The Fourier transform infrared, tensile stress tests and gas chromatography-mass spectroscopy measurements were used to investigate the effects of the treatment. It was found that β-CD-MCT can be fixed to cotton fabrics with the pad-dry-cure method at high temperature. No loss of tensile strength of the treated fabrics was reported. The fragrance disappeared from untreated cotton after 8 days when stored at ambient temperature (30°C) but on other hand, the fragrance was retained in β-CD-MCT-treated cotton fabrics for 21 days in the same conditions. Agrawal et.al compared the efficiency of enzymatic treatments and existing chemical techniques for bonding β-CD and its derivatives to cotton surface. Novel chemical based crosslinking with homo-bi-functional reactive dye (C.I. reactive black 5) and grafting with reactive β-CD-MCT show maximum attachment to cotton surface. Innovative, enzymatic coupling of especially synthesized 6-monodeoxy-6-mono(N-tyrosinyl)-β-cyclodextrin was performed on cotton textile surface at low temperature. Alteration in surface topography has been observed for all β-CD treated samples [54]. Martel with co-workers coupled β-CD-MCT to chitosan, to obtain a chitosan derivative bearing cyclodextrin. Because the average degree of substitution of the CD derivative was 2.8, the reaction yielded crosslinked insoluble products. The structure of these materials has been investigated by high-resolution magic-angle spinning (HRMAS) with gradients. For the first time, HRMAS spectra of chitosan polymers containing β-CD were obtained. This NMR technique produced one- and two-dimensional well-resolved solid-state spectra. Decontamination of waters containing textile dyes were carried out with the crosslinked derivatives. Report by Martel showed that the new chitosan derivatives are characterized by a rate of sorption and a global efficiency superior to that of the parent chitosan polymer and of the well-known cyclodextrin-epichlorohydrin gels [55]. El-Tahlawy with co-workers carried out a novel technique for preparation of cyclodextrin-grafted chitosan. β-CD citrate was synthetized by esterifying of β-CD with citric acid (CA) in presence or absence of sodium hypophosphite as a catalyst in a semidry process. β-CD/grafted chitosan was prepared by coupling β-CD citrate with chitosan dissolved in different formic acid solutions having different concentrations. The reacting ingredients were subjected to various reaction conditions to attain the optimum condition. β-CD/grafted chitosan were evaluated by measuring the nitrogen content of both chitosan and grafted chitosan. Chitosan and β-CD/grafted chitosan, having different molecular weights, were evaluated as antimicrobial agents for different microorganisms [56].

Very effective bonding of CDs on cellulose fibres can be achieved by a high-performance resin finish [57] or with non-formaldehyde reagents such as polycarboxylic acids [58, 59] which can covalently esterify hydroxyl groups of cellulose and CDs and therefore the grafting of CD on cellulose can occur. The same linking/crosslinking reagents can be used in the treatment of different synthetic fibres. Polyester fibres were modified by β-CD using citric acid [7, 59] in research work of our group [60], 1,2,3,4-butane tetracarboxylic acid was used as a linker. Odour control is a very important topic in the apparel and underwear items. Odour can be controlled by applying an antimicrobial finish, removing the odour molecules as they are formed or covering up the odour with a fragrance. The odour molecules being hydrophobic become trapped in the cavities of the CDs and are removed during laundering.

Within our research work PET fibres were treated in aqueous solution of different concentrations of β–CD and BTCA. We reported that BTCA molecules react via anhydride formation with hydroxyl groups of β-CD and form nano-assembly which can be physically attached to the PET fibres surface at the elevated temperature. Such assembly could be schematically presented as shown in Figure 3.

For reducing the termofixation temperature the catalyst cyanamide was used. We concluded that the treatment of PET with β-CD/BTCA was very successful even at temperature as low as 115°C when CA as a catalyst has been used. After 10 washings the gain on mass remained as high as 7.8% (Figure 4).

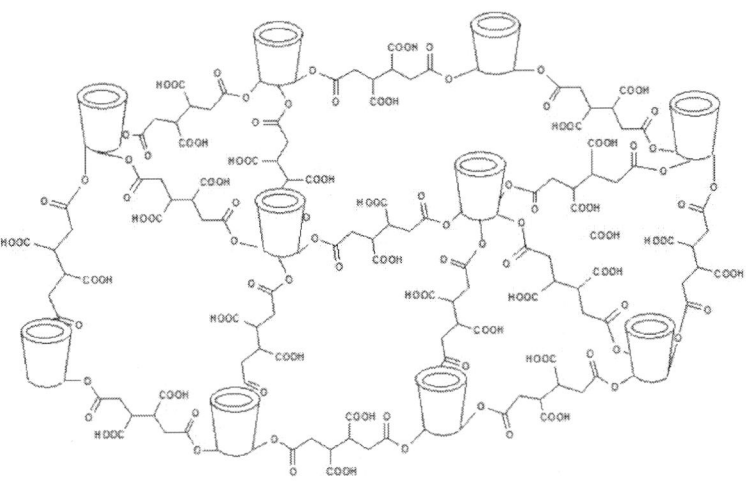

Figure 3. Nano-assembly of β-CD crosslinked with BTCA on textile surface.

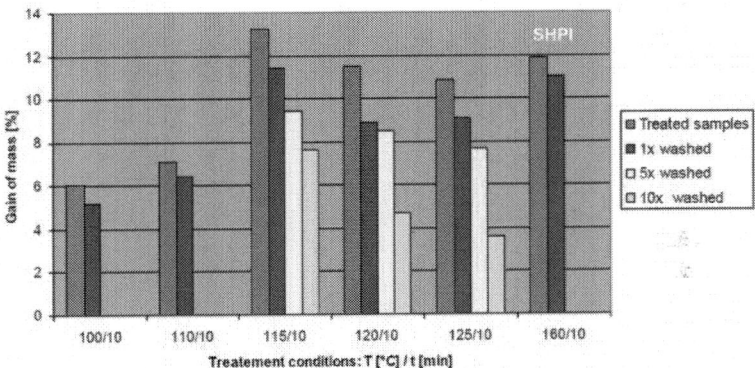

Figure 4. Samples 100/10, 110/10, 115/10, 120/10 and 125/10 are PET samples treated with β-CD, BTCA, CA at 100, 110, 115, 120 and 125°C, respectively; sample 160/10 was treated with β-CD, BTCA and SHPI at 160°C; all samples were termofixed for 10 minutes.

We were able to reduce the curing temperature to 115ºC and prepared nanoencapsulated textile materials with increased adsorption capacity (the adsorption of ammonia gas onto treated and untreated PET textile materials was measured using Japan standard test method - JIS K0804) and with postponed release of volatile compounds. From the ammonia gas adsorption measurements (Table 1), it is possible to conclude that the adsorption of ammonium gas increased when PET fabric was treated with β-CD/BTCA/CA at 115ºC/10min - after one hour of exposure to ammonia gas the concentration of gas in the chamber was zero, compare to the concentration when untreated PET fabric was exposed to the ammonium gas, where the concentration in the chamber was changed from the initial value of 125ppm to 77ppm.

	PET treated with β-CD/BTCA/ CA at 115°C/10 min	Untreated PET
Initial concentration (ammonia)	125ppm	125ppm
One hour concentration (ammonia)	0ppm	77ppm

Table 1. Decrease of ammonium gas concentration due to the adsorption.

In order to quantify the odour-releasing behaviour of β-CD treated PET fabrics, we organized a sensory panel of nine people to whom the odour was presented under controlled conditions. In order to study the postponed release of the volatile compounds from the β-CD treated textile substrate the following was performed: β-CD/BTCA/CA treated PET textile substrate was sprayed with perfume and dried; the intensity of the perfume from the untreated PET fabric spayed with perfume was also monitored for comparison purposes. Both treatments were performed in triplets. The size of the clothes was 10 by 10 cm. All with perfume treated textile samples were stored separately in dark places. Samples were stored in open conditions so that the perfume was able to evaporate constantly. The odour release was measured once per week. The smell intensity was evaluated from 0 to 4, where 0 means no smell and 4 means very intensive smell. From Figure 5 it is possible to see that the odour release intensity of untreated PET fabrics sprayed with perfume (BLIND, spray) starts to decrease after 6 weeks; the odour intensity of perfume sprayed on β-CD treated fabrics remains constant, but there is slight indication that the intensity of the perfume starts to increase after 6 weeks. We can conclude that, in the case of β-CD treated PET fabrics, some postponed release of the fragrance occurs.

Glycidyl methacrylate is widely used in the production of polymer coatings and finishes, adhesives, plastics and elastomers, it can be grafted to various textiles substrates as well. Desmet and co-workers functionalized cotton-cellulose by gamma-irradiation-induced grafting of glycidyl methacrylate (GMA) to obtain a hydrophobic cellulose derivative with epoxy groups suitable for further chemical modification. Two grafting techniques were applied. In pre-irradiation grafting (PIG) cellulose was irradiated in air and then immersed in a GMA monomer solution, whereas in simultaneous grafting (SG) cellulose was irradiated in an inert atmosphere in the presence of the monomer. In the paper authors claimed that the PIG led to a more homogeneous fibre surface, while SG resulted in higher grafting yield but showed clear indications of some GMA-homopolymerization. Effects of the reaction parameters (grafting method, absorbed dose, monomer concentration, solvent composition) were evaluated by SEM, gravimetry (grafting yield) and FT-IR spectroscopy. It is reported that water uptake of the cellulose decreased while adsorption of pesticide molecules increased upon grafting. The adsorption was further enhanced by β-CD immobilization during SG.

This method can be applied to produce adsorbents from cellulose based agricultural wastes [61]. CDs can be incorporated into fibres during the spinning processes [62, 63].

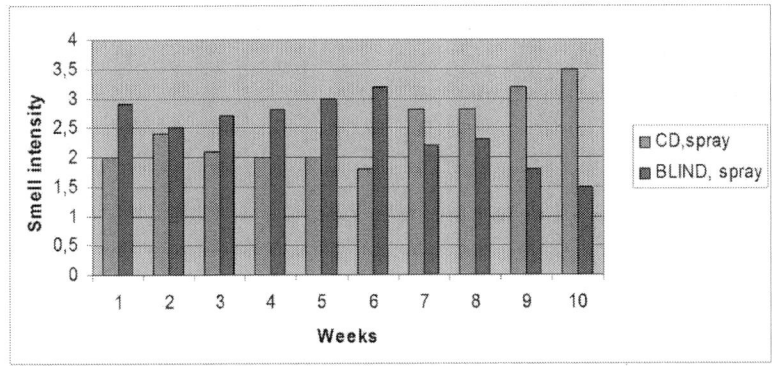

Figure 5. Odour intensity of PET fabrics pre-treated with β-CD (blue) and untreated PET fabrics sprayed with the perfume.

Electrospinning is proven to be an effective method for producing non-woven mats of fibres with high aspect ratios. Manasco and co-workers reported the preparation of submicron hydroxypropyl-beta-cyclodextrin (HP-β-CD) fibres by electrospinning without the addition of a carrier polymer. They focused on exploring solution properties that make fibre formation possible contrary to the widely accepted premise that molecular entanglement of macromolecules is required for electrospinning. The ability to electrospin from these solutions was attributed to hydrogen-bonded aggregation between HP-β-CD molecules at high concentrations [64]. Further it is reported by Uyar and co-workers that poly(methyl methacrylate) (PMMA) nanofibres containing the inclusion complex forming β-CD were successfully produced by means of electrospinning in order to develop functional nanofibrous webs. Electrospinning of uniform PMMA nanofibres containing different loadings of β-CD (10%, 25% and 50% (w/w)) was achieved. The surface sensitive spectroscopic techniques; X-ray photoelectron spectroscopy (XPS) and time-of-flight secondary ion mass spectrometry (ToF-SIMS) showed that some of the β-CD molecules are present on the surface of the PMMA nanofibres, which is essential for the trapping of organic vapours by inclusion complexation. Direct pyrolysis mass spectrometry (DP-MS) studies showed that PMMA nanowebs containing β-CD can entrap organic vapours such as aniline, styrene and toluene from the surroundings due to inclusion complexation with beta-CD that is present on the fibre surface. The study showed that electrospun nanowebs functionalized with CDs may have the potential to be used as molecular filters and/or nanofilters for the treatment of organic vapour waste and air filtration purposes [65]. Polyvinyl alcohol (PVA) nanowebs incorporating vanillin/cyclodextrin inclusion complex (vanillin/CD-IC) were produced via electrospinning technique by Kayaci and Uyar [66]. The vanillin/CD-IC was prepared with three types of CDs; α-CD, β-CD and γ-CD to find out the most favourable CD type for the stabilization of vanillin. PVA/vanillin/CD-IC nanofibres, having fibre diameters similar to 200 nm, were electrospun from aqueous mixture of PVA and vanillin/CD-IC. The results indicated that vanillin with enhanced durability and high temperature stability was achieved for PVA/vanillin/CD-IC nanowebs due to complexation of vanillin with CD. Additionally, they reported that PVA/vanillin/γ-CD-IC nanoweb was more effective for the stabilization and slow release of vanillin suggesting that the strength of interaction between vanillin and the γ-CD cavity is stronger when compared to α-CD and β-CD.

3.1.2. Cosmetotextiles

Although still in its infancy, the market for cosmetotextiles - often referred to as "wearable skincare" - is set to grow rapidly, and the textile industry is optimistic that further technical developments will open up new markets and create growing business opportunities [67]. Cosmetotextile became a fast growing new branch of specialized micro- or nano-encapsulation textile products, with most patents issued after the 1990 [68]. The physical and chemical properties of the guest molecules encapsulated in CDs can change due to complex formation. Thus, for example, the stability of the complexed molecule against light and oxygen increases and the vapour pressure is reduced. The solubility of slightly soluble molecules increases in a CD complex. All these and further advantages of CDs and their complexes can be used for the formulation of cosmetic products [69]. Cosmetotextile allows the administration of active molecules simply and controllably. It can also be used to change the surface properties of a fabric in order to make it self-cleaning, hydrophobic or lipophobic. The article prepared by Ripoll and co-workers reviews the current state of the art concerning functionalization techniques and the methods used to characterize various functionalized fabric. This review also reveals the surprising lack of publications on the functionalization of textile supports [70].

Moist oils (essential oils, herbal oils, oils from flower seeds) have skin care benefits in that they provide an occlusive layer that lubricates the epidermis, together with a moisturizing effect that helps to prevent excess water loss. Essential oils attributed with a range of properties that help to achieve physical and emotional balance. Besides, one additional advantage of molecular encapsulation is the possibility to reload them 4, 58, 71. Cosmetotextile applications can be used in the treatment of chronic venous insufficiency in legs by means of elastic bandages loaded with natural products which possess flebotonic properties. Cravotto and co-workers 72 have developed an efficient synthetic procedure for the preparation of β-CD-grafted viscose by means of the 2-step ultrasound-assisted reaction. The highly grafted fabric bearing bis-urethane bridged β-CD has been characterized by ATR FT-IR and CP-MAS spectra and by an empiric colorimetric method which used phenolphthalein as the CD guest. They have also developed a suitable cosmetic preparation containing natural substances and extracts (aescin, menthol, Centella asiatica and Ginkgo biloba) to recharge the CD-grafted textile. The efficacy of the new cosmetotextile has been corroborated by in vitro studies of diffusion through membranes, cutaneous permeation and accumulation in porcine skin. Aescin was taken as a reference compound and its concentration in the different compartments was monitored by HPLC analysis. They reported that this cost effective cosmetotextile showed excellent application compliance and was easily recharged.

Electrospun functional polystyrene (PS) fibres containing CD-menthol inclusion complexes are new and advance class of cosmetotextiles. Uyar and co-workers developed functional electrospun fibres containing fragrances/flavours with enhanced durability and stability assisted by CD inclusion complexation. As a model fragrance/flavour molecule they used menthol. CD-menthol inclusion complexes were incorporated in electrospun PS fibres by using three types of CDs: α-CD, β-CD and γ-CD. It is reported that due to complexation of menthol with CDs, the stabilization of menthol was achieved up to 350°C whereas the PS fibres without the CD complex could not preserve volatile menthol molecules. This study suggested that the electrospun fibres functionalized with CD are very effective for enhancing the temperature stability of volatile fragrances/flavours and therefore show potentials for the development of functional fibrous materials [73].

Today the subject of well-being is an area which is receiving much interest, with scent being one of the most important aspects of personal care. The definition of the word aromatherapy is the following: therapeutic uses of fragrances which at least mere volatilize to cure and to mitigate or cure diseases, infection and indisposition by means of inhalation alone [74]. The term aromachology was coined in 1982 to denote the science that is dedicated to the study of the relationship between psychology and fragrance technology to elicit a variety of specific feelings and emotions – such as relaxation, exhilaration, sensuality, happiness and well-being-through odours.

CDs linked on cellulose do not affect the cellulose's properties, and CDs keep their ability to form inclusion complexes with other suitable molecules thus, CDs are the first choice in preparing aromatherapy textiles. Lavender is the most used and most versatile of all the essential oils. It is very useful oil, especially when symptoms are due to a nervous problem. The effects of lemon, camomile, rose, cardamom, clove, and jasmine fragrance oils on human have been confirmed by many research works. The sedative effects for the pharmaceutical and emotional effects of essential oils are listed in Tables 2 and 3 respectively [75, 76].

3.1.3. Miscellaneous applications of cyclodextrins in textile industry

An insect repellent is a substance applied to skin, clothing, or other surfaces which discourages insects from landing or climbing on that surface. Synthetic repellents (such as paradichlorobenzene) tend to be more effective than 'natural' repellents, but on the other hand they are usually toxic. Cedar oil is often used as a natural insect repellent or it is used for its aromatic properties, especially in aromatherapy. Essential oil repellents tend to be short-lived in their effectiveness due to their volatile nature. To prevent the essential oils from evaporating form the textile materials we can encapsulate them in β-CD. In our research group [80, 81] with β-CD, nanoencapsulated wool and PET/wool blend fibres were further treated with cedar oil, which is known for being a natural insect repellent. The complex formation of cedar oil with β-CD was determined by ATR FT-IR spectroscopy. Textile material containing β-CD showed, after being treated with cedar oil, a prolonged moths oppression compared to those textile materials treated only with cedar oil. Table 4 presents the damages to wool and larval conditions according to the time of exposure. No visible damage was observed to the naked eye when β-CD/cedar oil treated wool was exposed to a moth's colony for 2 months. In the control (wool samples treated only with cedar oil) no damage was observed for the first few days, but when the cedar oil had evaporated, the wool cloth was not protected anymore. In contrast, when the cedar oil was encapsulated in the β-CD cavity, evaporation was hindered and resistance to insect pests' activities regarding cedar oil remained.

Effects	Essential Oil
Sedation	Mint, Onion, Lemon, Metasequoia
Coalescence	Pine, Clove, Lavender, Onion, Thyme
Diuresis	Pine, Lavender Onion. Thyme, Fennel, Lemon, Metasequoia
Facilitating Menses	Pine, Lavender. Mint, Rosemary, Thyme, Basil, Chamomile, Cinnamon, Lemon
Dismissing sputum	Onion, Citrus, Thyme, Chamomile
Allaying a fever	Ginger, Fennel, Chamomile, Lemon
Hypnogenesis	Lavender, Oregano, Basil, Chamomile
Curing Hypertension	Lavender, Fennel, Lemon, Ylangylang
Be good for stomach	Pine, Ginger, Clove, Mint, Onion, Citrus, Rosemary, Thyme, Fennel, Basil, Cinnamon
Diaphoresis	Pine, Lavender, Rosemary, Thyme, Chamomile, Metasequoia
Expelling wind	Ginger, Clove, Onion, Citrus, Rosemary, Fennel, Lemon
Losing weigh	Onion, Cinnamon, Lemon
Relieving pain	Vanilla, Lavender. Mint, Onion, Citrus, Rosemary, Chamomile, Cinnamon, Lemon
Detoxification	Lavender
Curing diabetes	Vanilla, Onion, Chamomile, Lemon
Stopping diarrhea	Vanilla, Ginger, Clove, Lavender, Mint, Onion, Oregano, Rosemary, Thyme, Chamomile, Cinnamon, Lemon
Curing flu	Pine, Lavender, Mint, Onion, Citrus, Rosemary, Thyme, Chamomile, Cinnamon, Metasequoia
Curing rheumatism	Lavender, Onion, Citrus, Rosemary, Thyme, Metasequoia
Urging sexual passion	Pine, Ginger, Clove, Mint, Onion, Rosemary, Thyme, Fennel
Relieving spasm	Cinnamon
Promoting appetite	Clove, Lavender, Mint, Onion, Citrus, Rosemary, Fennel, Basil, Chamomile, Cinnamon, Lemon, Metasequoia
Relieving cough	Rosemary

Table 2. Pharmaceutical effect of essential oils [75, 77, 78].

Emotion	Essential Oils with the Sedative Effects
Anxiety	Benzoin, Lemon, Chamomile, Rose, Cardamom, Clove, Jasmine
Lament	Rose
Stimulation	Camphor, Balm oil
Anger	Chamomile, Balm oil, Rose, Ylangylang
Wretchedness	Basil, Cypress, Mint, Patchouli
Allergy	Chamomile, Jasmine, Balm oil
Distrustfulness	Lavender
Tension	Camphor, Cypress, Vanilla, Jasmine, Balm oil, Lavender, Sandalwood
Melancholy	Basil, Lemon, Chamomile, Vanilla, Jasmine, Lavender, Mint, Rose
Hysteria	Chamomile, Balm oil, Lavender, Jasmine
Mania	Basil, Jasmine, Pine
Irritability	Chamomile, Camphor, Cypress, Lavender
Desolation	Jasmine, Pine, Patchouli, Rosemary

Table 3. The sedative or emotion effects of essential oils [75, 79].

Time	β-CD/cedar oil treated wool	Cedar oil treated wool	Untreated wool
48h	No detectable damage	No detectable damage	Very slight visible damage
	Larval conditions: live	Larval conditions: live	Larval conditions: live
72h	No detectable damage	No detectable damage	Moderate visible damage
	Larval conditions: dead	Larval conditions: dead	Larval conditions: live
7 days	Addition of new larvae	Addition of new larvae	/
14 days	No detectable damage	Very slight visible damage	Very heavy damages
	Larval conditions: dead	Larval conditions: live	Larval conditions: live
56 days	No detectable damage	Moderate visible damages	Very heavy damages
	Larval conditions: dead	Larval conditions: live	Larval conditions: live, pupating

Table 4. The estimation of wool damage and the condition of larval colony in accordance to elapsed time.

To maintain antimicrobial activity, frequent administration of conventional formulations of many antibiotics with short half-life is necessary. To enhance release properties, many materials have been introduced into the matrix and coating extended-release system in the past few years. The review by Gao [82] highlights the development of materials used in extended-release formulation and nanoparticles for antibiotic delivery. CDs are mentioned as nanoparticles/nanocarriers which allow the antibiotic extended-release.

A textile polyester vascular graft can be modified by methyl-β-CD to obtain a new implant capable of releasing antibiotics directly in situ at the site of operation over a prolonged period and thereby preventing post-operative infections [83]. Wang reported the inclusion complex of miconazole nitrate with β-CD formation by the co-precipitation method. The DSC curve and X-ray diffraction verified the inclusion complex formation between β-CD and miconazole nitrate. The skin-care textiles can be obtained by treating fabrics with inclusion complexes using the sol-gel method [84].

Novel nano-porous polymers or nanosponges can be prepared for removal of organic pollutants from waste water. The polymeric «nanosponge» materials are not durable (usually they are in gel form), they do not have high mechanical strength, so they must be impregnated onto the pore structure of a ceramic or some other porous surfaces [85, 86]. This technology is very specific for the target pollutant, it is very expensive and the removal of the adsorbed pollutant from the nanosponge is not possible. Textile materials are very important as filter materials. The cost of textile materials is acceptable (polyester, viscose), they have sufficient mechanical strength; the pore size, especially the macro-pore size can vary and it depends on the type of textile (the density of non-woven material) and on the diameter of the fibres. Textile materials can be further modified to prepare filtration materials with additional adsorption.

The amount of aromatic organic pollutants (phenols, aniline, formaldehyde and others) can be reduced from dyeing wastewater by using CDs which can be immobilized on a water insoluble organic support. The new concept for modification of textile substrates based on permanent fixation of supramolecular compounds - CDs on the material surface thus imparts new functionality to the fabric [87]. The guest molecules could be various organic molecules and some metal ions as well. The assembly of nanocapsules on textile materials acts as selective filtration/adsorption media for various pollutants. Prabaharan and Mano further reported that CDs have recently been recognized as useful adsorbent matrices. Due to its hydrophobic cavity, CDs can interact with appropriately sized molecules to result in the formation of inclusion complexes. These complexes are of interest for scientific research as they exist in aqueous solution and can be used to study the hydrophobic interactions which are important in the biomedical and environmental fields. The grafting of CD onto chitosan can result in the formation of a molecular carrier that possess the cumulative effects of inclusion, size specificity and transport properties of CDs as well as the controlled release ability of the polymeric matrix. In this review, different methods of CD grafting onto chitosan are discussed [88]. Electrospinning has been used to create polystyrene (PS) nanofibres containing any of the three different types of cyclodextrin (CD); α-CD, β-CD, and γ-CD [89]. These three CDs are chosen because they have different sized cavities that potentially allow for selective inclusion complex (IC) formation with molecules of different sizes or differences in affinity of IC formation with one type of molecule. The comparative efficiency of the PS/CD nanofibres/nanoweb for removing phenolphthalein, a model organic compound, from solution was determined by UV-Vis spectrometry, and the kinetics of phenolphthalein capture was shown to follow the trend PS/α-CD > PS/β-CD > PS/γ-CD. Direct pyrolysis mass spectrometry (DP-MS) was also performed to ascertain the relative binding strengths of the phenolphthalein for the CD cavities, and the results showed the trend in the interaction strength was β-CD > γ-CD > α-CD. Results of their research demonstrated that nanofibres produced by electrospinning that incorporate CDs with different sized cavities can indeed filter organic molecules and can potentially be used for filtration, purification, and/or separation processes.

4. Conclusion

Since 1980, when Szejtli first patented the bonding of CDs onto textile fibres, a lot of research has gone into the application of CDs on to textile substrates. But there is still a gap between original high level basic science and commercial applications of CDs in all industrial sectors.

Nevertheless the use of CDs in the textile industry has increased in the last years. Grafted CDs on textile substrates or spun fibres which contain CDs can be used to obtain special functionality of textiles such as absorption; they can complex and release fragrances or "skin-care-active" substances like vitamins, caffeine and menthol as well as bioactive substances such as biocides and insecticides and drugs. Furthermore, various textile materials treated with CDs could be used for adsorption of small pollutants from waste waters - for filtration, purification, and/or separation treatments of waste waters.

Author details

Bojana Voncina* and Vera Vivod

*Address all correspondence to: bojana.voncina@um.si

Faculty of Mechanical Engineering, University of Maribor, Maribor, Slovenia

References

[1] Vögtle F. Supramolecular Chemistry, an introduction. New York: John Wiley & Sons; 1991.

[2] Astray G, Gonzalez-Barreiro C, Mejuto JC, Rial-Otero R, Simal- Gándara J. A review on the use of cyclodextrins in foods. Food Hydrocolloids 2009;23(7) 1631–1640.

[3] Jeang CL, Lin DG, Hsieh SH. Characterization of cyclodextrin glycosyltransferase of the same gene expressed from Bacillus macerans, Bacillus subtilis, and Escherichia coli. Journal of Agricultural and Food Chemistry 2005;53(16) 6301-6304.

[4] Szejtli J. Cyclodextrins in the Textile Industry. Starch/Stärke 2003;55(5) 191-196.

[5] Cravotto G, Binello A, Baranelli E, Carraro P, Trotta F. Cyclodextrins as food additives and in food processing. Current Nutrition & Food Science 2006;2(4) 343–350.

[6] Dastjerdi R, Montazer M. A review on the application of inorganic nano-structured materials in the modification of textiles: Focus on anti-microbial properties. Colloids and Surfaces B: Biointerfaces 2010;79(1) 5-18.

[7] Martin del Valle EM. Cyclodextrins and their uses: a review. Process Biochemistry 2004;39(9) 1033–1046.

[8] Li S, Purdy WC. Cyclodextrins and their Applications in Analytical-Chemistry. Chemical Reviews 1992;92(6) 1457-1470.

[9] Szejtli J. Introduction and General Overview of Cyclodextrin Chemistry. Chemical Reviews 1998;98(5) 1743–1753.

[10] Uekama K, Hirayama F, Irie T. Cyclodextrin Drug Carrier Systems. Chemical Reviews 1998;98(5) 2045-2076.

[11] Ribeiro ACF, Valente AJM, Lobo VMM. Transport Properties of Cyclodextrins: Intermolecular Diffusion Coefficients. Journal of the Balkan Tribological Association 2008;14(3) 396-404.

[12] Harada Laboratory. Department of Macromolecular Science. Osaka University. http://www.chem.sci.osaka-u.ac.jp/lab/harada/eng/eng/research/01.html (accessed 2 July 2012)

[13] Saenger W. Sterochemistry of circularly closed oligosaccharides: cyclodextrins structure and function. Biochemical Society Transactions 1983;11(2) 136-139 Connors, K. A. (1997). The Stability of Cyclodextrin Complexes in Solution. Chemical Reviews, Vol.97, No.5, 325-1357, ISSN: 0009-2665

[14] Connors KA. The Stability of Cyclodextrin Complexes in Solution. Chemical Reviews 1997;97(5) 325-1357.

[15] Weber E. Molecular Inclusion and Molecular Recognition - Clathrates I. Topics in Current Chemistry 1987;140 1-20.

[16] Montazer M, Jolaei MM. β-Cyclodextrin stabilized on three-dimensional polyester fabric with different crosslinking agents. Journal of Applied Polymer Science 2010;116(1) 210–217.

[17] Lo Meo P, D'Anna F, Riela S, Gruttadauria M, Noto R. Spectrophotometric study on the thermodynamics of binding of α- and β-cyclodextrin towards some p-nitrobenzene derivatives. Organic & Biomolecular Chemistry 2003;1(9) 1584-1590.

[18] Voncina B. Application of cyclodextrins in textile dyeing. In: Hauser PJ (ed.) Textile dyeing. Rijeka: InTech; 2011. p373-392.

[19] Fernandes CM, Carvalho RA, da Costa SP, Veiga FJB. Multimodal molecular encapsulation of nicardipine hydrochloride by β-cyclodextrin, hydroxypropyl-β-cyclodex-

trin and triacetyl-β-cyclodextrin in solution. Structural studies by 1H NMR and ROESY experiments. European Journal of Pharmaceutical Sciences 2003;18(5) 285-296.

[20] Figueiras A, Sarraguca JMG, Carvalho RA, Pais AACC, Veiga FJB. Interaction of Omeprazole with a Methylated Derivative of β-Cyclodextrin: Phase Solubility, NMR Spectroscopy and Molecular Simulation. Pharmaceutical Research 2007;24(2) 377-389.

[21] Monti S, Sortino S. Photoprocesses of photosensitizing drugs within cyclodextrin cavities. Chemical Society Reviews 2002;31(5) 287-300.

[22] Nilsson M, Cabaleiro-Lago C, Valente AJM, Soderman O. Interactions between gemini surfactants, 12-s-12, and beta-cyclodextrin as investigated by NMR diffusometry and electric conductometry. Langmuir 2006;22(21) 8663-8669.

[23] Cabaleiro-Lago C, Nilsson M, Soderman O. Self-diffusion NMR studies of the host-guest interaction between beta-cyclodextrin and alkyltrimethylammonium bromide surfactants. Langmuir 2005;21(25) 11637-11644.

[24] Valente AJM, Nilsson M, Soderman O. Interactions between n-octyl and n-nonyl β-d-glucosides and α- and β-cyclodextrins as seen by self-diffusion NMR. Journal of Colloid and Interface Science 2005;281(1) 218-224.

[25] Garcia-Rio L, Godoy A. Use of spectra resolution methodology to investigate surfactant/β-cyclodextrin mixed systems. Journal of Physical Chemistry B 2007;111(23) 6400-6409.

[26] Voncina B, Vivod V, Jausovec D. [Beta]-cyclodextrin as a retarding reagent in polyacrylonitrile dyeing. Dyes and pigments 2007;74(3) 642-646.

[27] Costa T, Seixas de Melo JS. The effect of γ-cyclodextrin addition in the self-assembly behavior of pyrene labeled poly(acrylic) acid with different chain sizes. Journal of Polymer Science Part a-Polymer Chemistry 2008;46(4) 1402-1415.

[28] Denter U, Schollmeyer E. Surface modification of synthetic and natural fibres by fixation of cyclodextrin derivatives. Journal of Inclusion Phenomena and Molecular Recognition in Chemistry 1996;25(1-3) 197-202.

[29] Szejtli J. Utilization of cyclodextrins in industrial products and processes. Journal of Materials Chemistry 1997;7(4) 575–587.

[30] Frömming KH, Szejtli, J. Cyclodextrins in Pharmacy. Dordrecht: Kluwer Academic Publishers; 1994.

[31] Hedges AR. Industrial applications of cyclodextrins. Chemical Reviews 1998;98(5) 2035-2044.

[32] Buschmann HJ, Knittel D, Schollmeyer E. Textile Materialien Mit Cyclodextrinen, Textile Materials with Cyclodextrins. Melliand Textilberichte 2001;82(5) 368-370.

[33] Buschmann HJ. Cosmetic textiles: Clothes with a skin care action. Cossma 2001;2(7) 38-39.

[34] Buschmann HJ, Knittel D, Beermann K, Schollmeyer E. Cyclodextrins and textiles. Nachr. Chem. 2001;49(5) 620-620.

[35] Murthy CN, Shown I. Grafting of Cotton Fiber by Water-Soluble Cyclodextrin-Based Polymer. Journal of Applied Polymer Science 2009;111(4) 2056-2061.

[36] Savarino P, Viscardi G, Quagliotto P, Montoneri E, Barni E. Reactivity and effects of cyclodextrins in textile dyeing. Dyes and Pigments, 1999;42(2) 143-147,.

[37] Poulakis, K, Buschmann HJ, Schollmeyer E. Ger. Offen. 4035378. 1992; WO 2002046520. 2002.

[38] Buschmann HJ, Knittel D, Schollmeyer E. Resin finishing of cotton in the presence of cyclodextrins for depositing fragrances. Melliand Textilberichte 1991;72(3) 198-199.

[39] Fujimura T. Jpn. Kokai Tokkyo Koho 60259648. 1985.

[40] Ritter W, Delney J, Volz W, Kerr I. DE 10101294. 2002.

[41] Akasaka M, Shibata T, Ochia H. Jpn. Kokai Tokkyo Koho 03059178. 1991.

[42] Akasaka, M, Sawai Y, Iwase K, Moriishi H. EP 488294. 1992.

[43] Yamamoto K, Saeki T. Jpn. Kokai Tokkyo Koho 09228144. 1997.

[44] Yamamoto K, Saeki T. Jpn. Kokai Tokkyo Koho 10007591. 1998.

[45] Voncina B, Majcen N. Application of cyclodextrin for medical and hygienic textiles. Tekstil 2004;53(1) 1-9.

[46] Knittel D, Schollmeyer E. Technologies for a new century. Surface modification of fibres. Journal of the Textile Institute 2000;91(3) 151-165.

[47] Szejtli J, Zsadon B, Fenyvesi E, Otta K, Tudos F. Hungarian Patent: 181733. 1980. US Patent 4,357,468 (1982).

[48] Szejtli J, Zsadon B, Horvath OK, Ujhazy A, Fenyvesi E. Hungarian Patent: 54506. 1991.

[49] Reuscher H, Hirsenkorn R, Haas W. German Patent: DE 19520967. 1996.

[50] Grechin AG, Buschmann HJ, Schollmeyer E. Quantification of Cyclodextrins fixed onto Cellulose Fibres. Textile Research Journal 2007;77(3) 161-164.

[51] Moldenhauer JP, Reuscher H. Textile finishing with MCT-beta-cyclodextrin. In: Torres Labandeira JJ, Vila-Jato JL. (eds.) Proceedings of the 9th International Symposium on Cyclodextrins, 31 May – 3 June 1998, Santiago de Compostela, Spain. Dordrecht: Kluwer Academic Publishers; 1999.

[52] Ibrahim NA, Allam EA, El-Hossamy MB, El-Zairy WM. UV-Protective Finishing of Cellulose/Wool Blended Fabrics. Polymer-Plastics Technology and Engineering 2007;46(9), 905-911.

[53] Sricharussin W, Sopajaree C, Maneerung T, Sangsuriya N. Modification of cotton fabrics with beta-cyclodextrin derivative for aroma finishing. Journal of the Textile Institute 2009;100(8) 682-687.

[54] Agrawal PB, Warmoeskerken MMCG. Permanent fixation of β-cyclodextrin on cotton surface: An assessment between innovative and established approaches. Journal of Applied Polymer Science 2012;124(5) 4090-4097.

[55] Martel B, Devassine M, Crini G, Weltrowski M, Bourdonneau M, Morcellet M. Preparation and sorption properties of a beta-cyclodextrin-linked chitosan derivative. Journal of Polymer Science Part A-Polymer Chemistry 2001;39(1) 169-176.

[56] El-Tahlawy K, Gaffar MA, . Novel method for preparation of beta-El-Rafie Scyclodextrin/grafted chitosan and it's application. Carbohydrate Polymers 2006;63(3) 385-392.

[57] Ostertag H. Anwendung von β-Cyclodextrinen in der CO-Gewebeveredlung. Melliand Textilberichte 2002;83(11-12) 872-878.

[58] Voncina B, Le Marechal AM. Grafting of cotton with beta-cyclodextrin via poly(carboxylic acid). Journal Of Applied Polymer Science 2005;96(4) 1323-1328.

[59] Martel B, Weltrowski M, Ruffin D, Morcellet M. Polycarboxylic acids as crosslinking agents for grafting cyclodextrins onto cotton or wool fabrics: Study of the process parameters. Journal of Applied Polymer Science 2002;83(7) 1449-1456.

[60] Vončina B, Le Marechal AM. [Beta]-cyclodextrin in medical and hygienic textiles. In: Chen X, Ge Y, Yan X. (eds.). 83rd TIWC. Quality textiles for quality life: proceedings of the Textile Institute 83rd World Conference (83rd TIWC), 23-27 May 2004, Shanghai, China. Manchester: Shanghai: Textile Institute, Donghua University; 2004.

[61] Desmet G, Takacs E, Wojnarovits L, Borsa J. Cellulose functionalization via high-energy irradiation-initiated grafting of glycidyl methacrylate and cyclodextrin immobilization. Radiation Physics and Chemistry 2011;80(12) 1358-1362.

[62] He Y, Inoue Y. Effect of alpha-cyclodextrin on the crystallization of poly(3-hydroxybutyrate). Journal of Polymer Science Part B-Polymer Physics 2004;42(18) 3461-3469.

[63] Vogel R, Tandler B, Haussler L, Jehnichen D, Brunig H. Melt spinning of poly(3-hydroxybutyrate) fibers for tissue engineering using alpha-cyclodextrin/polymer inclusion complexes as the nucleation agent. Macromolecular Bioscience 2006;6(9) 730-736.

[64] Manasco JL, Saquing CD,Tang C,Khan SACyclodextrin fibers via polymer-free electrospinning. RSC Advances 2012;2(9) 3778-3784

[65]   Uyar T, Havelund R,Nur Y,Balan A , Hacaloglu JToppare L, Besenbacher F,Kingshott P. functionalized poly(methyl methacrylate) (PMMA) electrospun nanofibers for organic vapors waste treatment. Journal of Membrane Science 2010;365(1-2) 409-417.

[66]   Kayaci F, Uyar T. Encapsulation of vanillin/cyclodextrin inclusion complex in electrospun polyvinyl alcohol (PVA) nanowebs: Prolonged shelf-life and high temperature stability of vanillin Food Chemistry 2012;133(3) 641-649

[67]   Hague L. Cosmetotextiles market has exciting potential. Aroq Ltd http://www.just-style.com/analysis/cosmetotextiles-market-has-exciting-potential_id111926.aspx (accessed 3 July 2012).

[68]   CEN standard CEN/TC 248 Textiles and textile products; WG 25 Cosmeto-textiles. CEN - ISO standardization committees on textiles Cheng S.Y. "Development of Cosmetic Texiles Using Microencapsulation Technology", RJTA, vol.12, no.4, 2008

[69]   Buschmann HJ, Schollmeyer E. Applications of cyclodextrins in cosmetic products. Journal of Cosmetic Science 2002;53(3) 185-191.

[70]   Ripoll L, Bordes C, Etheve S, Elaissari A, Fessi H. Cosmeto-textile from formulation to characterization: an overview. e-Polymers 2010;040 http://www.e-polymers.org/journal/papers/hfessi_010410.pdf (accessed 5 July 2012).

[71]   Buschmann HJ, Knittel D, Schollmeyer E. New Textile Applications of Cyclodextrins. Journal of Inclusion Phenomena and Macrocyclic Chemistry 2001;40(3) 169-172.

[72]   Cravotto G, Beltramo L, Sapino S, Binello A, Carlotti ME. A new cyclodextrin-grafted viscose loaded with aescin formulations for a cosmeto-textile approach to chronic venous insufficiency. Journal of Materials Science-Materials in Medicine 2011;22(10) 2387-2395

[73]   Uyar T, ,Hacaloglu J .Besenbacher F Electrospun polystyrene fibers containing high temperature stable volatile fragrance/flavor facilitated by cyclodextrin inclusion complexes. Reactive & Functional Polymers 2009;69(3) 145-150.

[74]   Buchbauer G. Aromatherapy: Use of fragrance and essential oils as medicaments. Flavour and Fragrance Journal 1994;9 217-222.

[75]   Wang CX, Chen SH. Aromachology and Its Application in the Textile Field. Fibres & Textiles in Eastern Europe 2005;13(6(54)) 41-44.

[76]   Buschmann HJ. Removal of Residual Surfactant Deposits from Textile Materials with the Aid of Cyclodextrins. Melliand Textilberchite. 1995(9).

[77]   Jellinnek JS. Aromachology: A Status Review. Perfume & Flavorist 1994;19 25-49.

[78]   Wu CH. Essential Oil and Aromatherapy. Kexue Nong Ye 1999;47(3) 1-3.

[79]   Mazzaro D. The home fragrance market. Chemical Market Reporter 2000;4 14.

[80]   Vraz Kresevic S, Voncina B, Vukusic Bischof S, Katovic D. Textile Materials Treated With Eco - Friendly Insecticide Agents. In: Dragcevic Zvonko (ed.): 4th International

Textile, Clothing & Design Conference: Magic world of textiles: book of proceedings, ITC&DC, 05-08 October, 2008, Dubrovnik, Croatia. Zagreb: Faculty of Textile Technology, University of Zagreb; 2008.

[81] Vraz Kresevic S, Voncina B, Gersak J. Insect resistant and eco-friendly textile materials. In: Dragcevic Zvonko (ed.): 4th International Textile, Clothing & Design Conference: Magic world of textiles : book of proceedings, ITC&DC, 05-08 October, 2008, Dubrovnik, Croatia. Zagreb: Faculty of Textile Technology, University of Zagreb; 2008.

[82] Gao P, Nie X, Zou MJ, Shi YJ, Cheng G. Recent advances in materials for extended-release antibiotic delivery system. Journal of Antibiotics 2011;64(9) 625-634.

[83] Blanchemain N, Karrout Y, Tabary N, Neut C, Bria M, Siepmann J, Hildebrand HF, Martel B. 2011Methyl-beta-cyclodextrin modified vascular prosthesis: Influence of the modification level on the drug delivery properties in different media. Acta Biomaterialia 2011;7(1) 304-314.

[84] Wang JH, Cai ZS. A study of inclusion complex of miconazole nitrate with beta-cyclodextrin and its application on protein fabrics. In: Bai L, Rui Y. (eds.) Researches and Progresses of Modern Technology on Silk, Textile And Mechanicals I: Conference publication of the 6th China International Silk Conference and the 2nd International Textile Forum, 13-14 September 2007, Suzhou, Jiangsu, P.R. China. Chemical Industry Press; 2007.

[85] Salipira KL, Krause RW, Mamba BB, Malefetse TJ, Cele LM, Durbach SH. Cyclodextrin polyurethanes polymerised with multi-walled carbon nanotubes: Synthesis and characterisation. Materials Chemistry and Physics 2008;111(2-3) 218-224.

[86] Salipira KL, Mamba BB, Krause RW, Malefetse TJ, Durbach SH. Cyclodextrin polyurethanes polymerised with carbon nanotubes for the removal of organic pollutants in water. Water SA 2008;34 113-118. http://www.wrc.org.za/downloads/watersa/2008/2198.pdf (accessed 3 July 2012).

[87] Mamba BB, Krause RW, Malefetse TJ, Mhlanga SD, Sithole SP, Salipira KL, Nxumalo EN. Removal of geosmin and 2-methylisorboneol (2-MIB) in water from Zuikerbosch Water Treatment Plant (Rand Water) using β-cyclodextrin polyurethanes. Water SA 2007;33 223-228. http://www.wrc.org.za/downloads/watersa/2007/Apr%2007/2002.pdf (accessed 3 July 2012).

[88] Prabaharan M, Mano JF. Chitosan derivatives bearing cyclodextrin cavities as novel adsorbent matrices. Carbohydrate Polymers 2006;63(2) 153-166.

[89] Uyar T, Havelund R, Hacaloglu J, Besenbacher F, Kingshott P. Functional Electrospun Polystyrene Nanofibers Incorporating alpha-, beta-, and gamma-Cyclodextrins: Comparison of Molecular Filter Performance. ACS Nano 2010;4(9) 5121-5130.

Chapter 4

Preparation, Characterization and Application of Ultra-Fine Modified Pigment in Textile Dyeing

Chaoxia Wang and Yunjie Yin

Additional information is available at the end of the chapter

http://dx.doi.org/10.5772/53489

1. Introduction

Pigments are kinds of insoluble colorants used for fibers, plastics and other polymeric materials, which can retain stable chemical structure throughout the coloration process in its dispersed solution [1]. The pigments were obtained from mineral materials, vegetable materials, animal waste materials since 1200 BC. The significant development of pigment in textile field from 18th century attributed to a huge expansion in the range of synthetic pigments [2,3]. Currently, pigment perhaps is one of the most commonly and extensively colorations in fiber dyeing, due to its easy application to a variety of fibers and environmental friendly aspects [4].

As a result of the different physical and chemical characteristics between pigments and dyes, dyes are able to dissolve in water and penetrate into the substrate in the soluble form, while pigments are insoluble and difficult to be dispersed in water without the aid of dispersants [1]. The major drawback of pigments is the poor dispersing stability, which is affected by the particle size and surface charge on account of the imperfection of the dispersants [5]. Pigment dispersing system includes two main types according to the medium: solvent-based pigment and water-based pigment.

Solvent-based pigment systems, in which the organic solvent is used as dispersion medium, are more suitable for certain applications [6]. However, it is difficult to obtain pigment particles that are smaller than 100 nm in organic solvents. When the particle size of pigment is large, the interactions between pigments and organic solvents (e.g., hydrogen bonding) can disrupt the interactions with stabilizing dispersants. As a result, distinct particle aggregation that influences optical properties and viscosity occurs. To obtain a stable solvent-based pigment system, appropriate pigment amount, dispersants and solvents are the key factors [7].

Water-based pigment systems are environment-friendly pathways. Pigments are dispersed into water with the aid of auxiliaries, such as dispersants, emulsifiers, anti-setting agents, etc. Water-based pigment has been widely applied in coloration for textiles, paints, architecture, wood and so on [8,9]. But the unmodified pigment dispersion that contains pigments with large particle size is unstable and also suffers from the problems as precipitation and floating color. And then the color, fastness, handle and uniformity of the fabrics dyed with the unmodified pigment dispersion are influenced.

Ultra-fine modified pigment (UMP) is referred to nanometer or micrometer pigment particle by physical and/or chemical modifications in a composite disperse system with attaching functional groups for pigment dispersion. These functional groups often have contribution to appear the vivid color for the chromophoric groups in the pigment particle by selectively reflecting and absorbing certain wavelengths of visible light [6]. The UMP dispersion usually shows relative stability and higher color strength (K/S value) which closely approaches to that of dyes [10]. When UMP is used to color, the finely divided, insoluble particle remains throughout the coloration process [1]. Without the binder or less the amount of binder added in printing paste, the good handle will be achieved. And the better fastness is attained with the stronger attraction between the UMP and the fiber [10,11]. The UMP systems are divided into nonionic pigment dispersion, anionic pigment dispersion and cationic pigment dispersion according to the charged particles. Compared to the nonionic pigment dispersion, the pigment dispersion with ionic dispersant may produce the stronger combining power to the fabric [12,13].

The cationic disperser has been widely used in many fields. It had been reported in Japanese patents that cationic disperser was applied in electrodeposits coat. Being prepared with cationic disperser, the cationic pigment dispersion is able to decrease the migration of the pigment, and favorable dyeing deepness and color fastness are also obtained when such pigment is applied to pretreated cotton fabric and terylene-cotton fabric in pad dyeing [14]. The cationic surfactant is grafted onto the acrylic acid copolymer to attain disperser which is then applied to the anthraquinone pigment to prepare ink for ink-jet printing on paper [15]. In addition, there are many applications of cationic disperser in other fields, such as papermaking, image thermomigration, ink-jet printing paper, inorganic nano-scale powder and preparing hydrophilic ethylene copolymer dispersion. Moreover, pigment dyeing by exhaust process is possible by imparting substantivity with cationic reagents to induce the necessary affinity between pigments and fibers. The chemical modification of fiber with cationic reagents has been carried out to improve the dyeing properties [1,16].

In this chapter, the preparation of UMP, such as dispersing process, grinding process, ultrasonic wave process, microencapsulated process and microfluidics process are all summarized. And the particle size, disperse stability, Zeta potential and color properties, which are used to characterize the UMP are also analyzed. Moreover, the dyeing properties of UMP on cotton, silk, wool and acrylic yarns are mainly reviewed.

2. Preparation of UMP

Different from dyes, UMPs are totally insoluble in organic mediums due to its strong intermolecular aggregation, and they are required to be finely ground and dispersed in an organic medium to warrant their gloss appearance, lighting efficiency and maximizing material utilization. In order to prepare stable UMP, the particle size of UMP needs diminishing sharply to smaller than 1 μm and the process can be achieved via dispersant, microencapsulation, grinding, ultrasonic wave, microfluidizer processes and their combined process [17-19].

The stability of UMP is determined by forces among the UMP particles, when the repulsive force is higher than attractive force, the UMP particles will stably disperse into the media, otherwise, the UMP particles will aggregate together. In UMP with polymer dispersant, the repulsion forces among particles including static-electronic and steric repulsion are produced by the adsorbed polymer. The former can be measured by the Zeta potentials. The latter is closely connected with the thickness of adsorbed polymer layer [14]. The dispersant and microencapsulation processes are usually utilized together with other methods in the preparation process.

2.1. Dispersing process

2.1.1. Low molecular weight dispersant

The UMP dispersion is a surface modification technique, typical of which is a chemical bonding of low molecular weight hydrophilic moieties on UMP surfaces. Hydrophilic moieties work by electrostatic interaction between UMP particles and vehicle solvents. Usually there are no polymeric substances, surfactants or dispersion to aid in surface-modified UMP dispersions, it is necessary to add binder resins before it is applied to textiles [20].

Low molecular weight dispersants are commonly used for wetting and dispersing the UMP particles. The effect of Triton X-100 on the colloidal dispersion stability of CuPc pigment nanoparticles was investigated by Dong et al [18]. The influence of the hydrophobic chain of quaternary dispersers on the properties of pigment dispersion was discussed by Fang et al and good dispersion effects were obtained when the hydrophobic chain were 14 or 16 [21]. The aqueous suspensions of organic pigment particles using cetyltrimethylammonium bromide (CTAB) and sodium dodecylbenzene sulfonate (SDBS) as additives were prepared by Wu et al and a uniform hydrous alumina film could be formed on the organic pigment particle surface with anion surfactant SDBS [22].

But the dispersion system only using low molecular weight dispersants is still suffered from the lack of stability for long-term storage or at high temperature [8], so additional process or dispersant are necessary.

2.1.2. Polymeric dispersant

Polymeric dispersants are a class of specially designed, structured materials and show good properties in the stabilization of organic UMP. In aqueous media, polymeric dispersants are adsorbed onto UMP surface via anchor groups to build voluminous shells or intensify the charges around UMP surface, thereby preventing flocculation and coagulation of the UMP, which can greatly improve the stability of the UMP dispersion [23-25].

The UMP particles are usually quite hydrophobic. In order to achieve a good stabilization in aqueous UMP dispersions, many formulations have been proposed. The application of polymer surfactants in combination with ultrasonic action can significantly improve the quality of dispersed systems. Some aspects concerning UMP-polymer interaction and formation of adsorption layers under mechanical action need additional elucidation. The colloid stabilization of aqueous dispersions with polymer surfactants is believed to be a consequence of adsorption of the amphiphilic macromolecules on the particle surface resulting in mono- or multi-layers of certain structure and thickness which provide certain sterical and/or electrostatic stabilization effects.

Polymer adsorption from aqueous solution on a particle surface is a result of specific interactions of various active sites on the particle surface with corresponding sites (groups) of the macromolecule. Therefore the chemical structures of the stabilizers are believed to be adjusted to the nature of each type of the particles [26]. Fu and his coworkers reported that pigment particles with the diameter of 20-120 nm were uniformly distributing in aqueous media. –COOH of PSMA which encapsulated onto the surface of pigment would build a voluminous shell and also intensify the charges of particles, which could effectively hinder the attraction among particles [27].

2.1.3. Copolymer dispersant

Copolymer dispersants are advantageous for providing multiple anchoring sites toward UMP surface as well as structurally more designable for solvating with the selected solvents. Polymeric structures of random, A-B block, comb-like copolymers prepared by various synthetic techniques have been employed as stabilizers against particle flocculation. However, the methods of anionic and group transfer polymerization are less appropriate since the synthesis of dispersants often involves the monomers with polar functionalities.

Copolymer dispersants are suitable for stabilizing the UMP particles against flocculation during the grinding disruption and storage. The principle for achieving a fine dispersion is a thermodynamically driven interaction among dispersant molecules, UMP particles, and solvents in a collective manner of mutual non-covalent bonding such as electrostatic charge attraction, hydrogen bonding, p-p stacking, dipole-dipole interaction, and van der Waals forces [28].

Copolymer dispersants of high molecular weight have been employed as dispersants to resolve some problems through the molecular designs with multiple anchoring functionalities for interacting with the UMP surface and simultaneously with the involved solvents. The inhomogeneity in geometric shapes of any two nanoparticles may also play an important role for excluding each other from [8]. Recent developments in living/controlled polymerization including nitroxide mediated polymerization (NMP), reversible addition-fragmentation chain transfer (RAFT), and atom transfer radical polymerization (ATRP) have been reported. The copolymers with specific functionalities can be prepared from the monomers with diversified functionalities such as C_1-C_{12} alkyl(meth)acrylate, amine-functionalized (meth)acrylate, and acid-functionalized (meth)acrylate. In addition, copolymer structures can be controlled for their molecular weight distribution and are tailored for specific UMP applications [28].

2.1.4. Siloxane dispersant

Siloxane dispersant can make UMP well dispersed in organic binders due to their hydrophilic/ hydrophobic nature. They convert the hydrophilic surface of UMPs into hydrophobic components which make UMPs compatible with hydrophobic organic resins. Siloxane dispersants can be incorporated very easily into liquids and showed an increase of storage stability, as they tend to deposit more slowly. The dispersing extent and flowability of UMP treated with different siloxane dispersants in water medium are excellent. And siloxane also shows better affinity to UMP than ammonium and nonionic polyether [30]. Siloxane with long alkane chain shows great potential to be a new type of high performance dispersant [29].

Moreover, when the pigment powder modified by siloxane dispersant (10%) is added into the water, the treated sample (b) is able to easily be wetted and enter into the water while the untreated one (a) still floats on the water surface. It is obvious that the siloxane dispersant brings good wettability to Pigment Red CI 170.

2.2. Grinding process

It is necessary to de-aggregate and de-agglomerate the UMP particles. This is usually accomplished by mechanical action provided by high impact mill equipment, such as the sand mill and ball mill. As the UMP powder is broken down to individual particles by mechanical shear, higher surface areas are exposed to the vehicle and larger amounts of additives are required to wet out newly formed surfaces.

The grinding process can be regarded as a de-flocculation process. In the absence of stabilizing agents, some changes such as reducing K/S value, decreasing gloss and altering rheology probably occur. Grinding process offers UMP in liquids and is suitable for discrete pass as well as for circulation operation. The product passes through a high energy grinding zone inside a grinding chamber and is reduced to the targeted particle size, down to the nanometer range if required. Using a higher specific weight grinding media is likely to reduce the bead size without losing milling energy by use of equal bead filling volumes. The milling process is improved and an optimum grinding result from both a quality and cost perspective can also be achieved by selecting the best bead material, bead size and mill speed from their dependence on the milling product properties [31].

The mechanical grinding of UMPs with the aid of a dispersant is the most convenient method used to produce UMP particles [1]. But mechanical grinding process inevitably has such defects as the relatively large particle size and broad size distribution. The typical size range for particles and/or aggregates produces by traditional mechanical grinding was from 200 nm to 1000 nm. This process often combines with the utilization of ceramic grinding media for particle size reduction. K. Hayashi et al investigated the dry grinding of UMPs with silica particles generates core-shell structures with an average size comparable to the parent silica particles (20 nm) [5].

2.3. Ultrasonic wave process

As an environmental and efficient approach, ultrasonic energy was firstly reported to assist various textile processes by Sokolov and Tumansky in 1941 [32]. Sound is transmitted through a medium by inducing vibrational motion of the molecules through which it is travelling. Power ultrasound induces cavitations in liquids, which is the origin of the sound effect in cleaning and in chemical processes [33]. Ultrasound has been widely applied in textiles, for example, preparing nano-scale pigments, fabric pretreatment, dyeing and finishing processes.

Power ultrasound can enhance a wide variety of chemical and physical processes, mainly due to the phenomenon known as cavitation in a liquid medium, which is the growth and explosive collapse of microscopic bubbles. Sudden and explosive collapse of these bubbles can generate hot spots, i.e. local high temperature, high pressure, shock waves, and severe shear force capable of breaking chemical bonds. Dispersion of UMPs involves the application of shear forces to the agglomerates so that they break down into primary particles [34]. If a proper amount of Ultramarine is added to water and stirred briskly, the agglomerates will simply be carried within the flow and hardly change their nature. Ultrafine modified pigment blue FFG was mixed with dispersant and dispersed with an ultrasonic homogenizer for 30 min, and the particle size was reduced to 91 nm from the original size 220 nm [35].

2.4. Microencapsulated process

Encapsulating UMP with various polymers is a promising approach for improving the quality of the UMP dispersion. In the last decade, many techniques have been developed for encapsulating UMP [36]. It must be noted that a successful encapsulation technique should not impair the original color appearance of UMP but enhance their dispersion stabilities. Polymeric resins in encapsulation work basically as an adsorbed surface layer around UMP particles and also a fixing agent upon being colored on a substrate. The microencapsulated UMP supplies dispersion stability, and the surface-modified UMP gives fixing on a hydrophilic body, such as the fiber in textiles [20].

The whole encapsulation process probably divides into four steps below [37]. Organic UMP are dispersed into solution of polymeric dispersant leading to the polymeric dispersant absorbing onto the UMP. The absorption auxiliary is added slowly and the copolymers precipitate and encapsulate onto the UMP surface by van der Waals forces. After the precipitate is filtered and dried, the copolymer is tightly encapsulated onto the UMP surface. The copolymer onto the encapsulated UMP is hydrolyzed, and then the UMP is dispersed uniformly in aqueous media.

Dispersion is controlled by attractive force between the UMP and the hydrophobic part of the polymer, and the stability of the particles is made both by electrostatic repulsive force between the particles in water and the polymer-polymer entropic effect. The UMP are dispersed in the solid phase, and then the encapsulation and flocculation are presented. Re-dispersion is necessary to separated microcapsules and finally the complete encapsulation in the medium occurs [20]. The main differences between the surface-modified and the microencapsulated UMP are that the hydrophilic moiety on the surface of particles, i.e. the carboxylic groups in the microencapsulated UMP and the sulfonated groups on the surface of the modified UMP, and additionally the existence of a polymer shell only in the microencapsulated UMP [20,36].

2.5. Microfluidics process

Microfluidics with excellent dispersing and smashing effects is in favor of leading particles in liquid to reducing particle size to a submicron level to create pure and stable nano-emulsions and suspensions [38]. With this process, the effectiveness of high-performance materials increases, because particles are more uniformly and stably dispersed. Through microfluidizer high shear fluid processors, the fabrics with UMP easily achieve high K/S value and gloss, and the amount of volatile organic compounds by increasing water content is declined. Microfluidizer high shear fluid processors are frequently used to reduce particle size to less than 150 nm.

For example, Pigment Red CI 22 was stirred with an aqueous solution of an anionic polymeric dispersant for 30 min at 10,000 rpm, and the average particle diameter of the obtained UMP dispersion was 1, 500 nm. Whereas, the average particle diameter of 128 nm was available as the dispersion was converted to modified dispersion in a microfluidizer at pressure 22,000 Pa for 2.5 h [13]. Also, a UMP suspension system was prepared by adding the Gemini dispersant and water. Then it was mixed and dissolved and the deformer was added in the mixing process. The Pigment Red CI 22 was added into the above dispersant solution stirring at 600 rpm for 10 min and 9000 rpm for 30 min. The UMP system was treated using a microfluidizer at 172, 368 kPa for 50 min to prepare UMP suspension system. And particle size of the UMPs was the 211.9 nm, the Zeta potential was 32.4 mV and the viscosity was 1.33 Pa•s [16].

2.6. Combined process

It is exceedingly difficult to handle a dry powder comprising of UMP particles as the UMP is in the form of larger soft agglomerates. These agglomerates must be broken down into the medium in a process called dispersion. The UMP dispersion can be stabilized by dispersing agents in order to prevent the formation of uncontrolled flocculates. Combining two or more dispersing methods, for example, grinding/ultrasonic wave, microencapsulation/ grinding and dispersant/microfluidizer, the homogeneity of making UMP dispersion along with the use of organic dispersants for viscosity control and the prevention of particle from further agglomeration can be realized [8].

Dispersions are usually prepared by ball grinding a mixture of the pigment, dispersant and solvent. This is a simple combined process in which grinding process is used together with dispersant. The ultrasonic process can also disperse the UMP and this method can endow a high-throughput approach. The methacrylic copolymer was used to disperse carbon black in water using both ball mill and ultrasonic approaches. The particle size distribution was determined by laser diffraction. After 16 min both ultrasonification and ball grinding achieve the same particle size distribution (Figure 1). The ultrasonic approach could be used to obtain much smaller sample volumes than the ball mill approach [39].

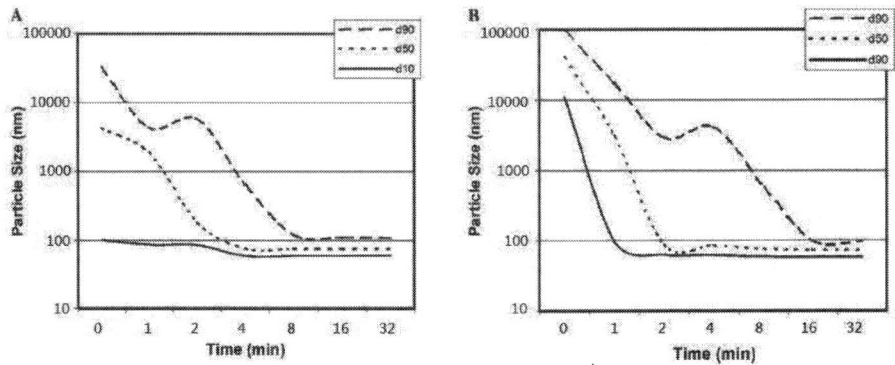

Figure 1. (A) Particle size distribution data from ball grinding; (B) Particle size distribution data from ultrasonication

3. Characterization of UMP

The properties of UMP dispersion can be described by the stability, the adsorbed layer thickness and the Zeta potentials, which are greatly affected by ionic strength, dispersants and dispersing methods. The adsorbed layer thickness and the Zeta potentials can reflect the forces among polymeric dispersants, media, and UMP particles. The characteristics of UMP dispersion affect the applications of the dyed fabric on K/S value, gloss, brightness, and transparency. Some key dispersion characteristics, such as particle size, disperse stability, color properties and Zeta potential properties are investigated as follows.

3.1. Particle size

UMPs are composed of fine particles which are normally in the submicrometer size range. The UMPs particle size has influence on the color, hide and settling characteristics. Large particles usually settle faster than smaller ones. UMP size and its distribution also influence the light scattering, colloidal stability, appearance and the color properties of the dyed fabric. Therefore, an assessment on the degree of dispersion is necessary to be considered in terms of these critical measurements. In general, color properties, such as strength, transparency, gloss and light fastness of all UMP systems, are affected to a greater or lesser extent by the size and distribution of the UMP particles in the dispersion [1].

UMP particle size and its distribution are greatly influenced by the dosage of dispersants. The particle size decreases at first and then increases with the enhancement of the dispersant, and the particle size reaches its minimum when the ratio is about 10% for phase separation method. The smaller the particle size is, the larger the UMP surface is, thus the more amount of dispersant is needed. However, excessive dispersants, higher than 10%, will dissolve in media instead of attaching onto UMP surface, resulting in increasing viscosity of dispersing media and poor wetting performance in that excess dispersants (Figure 2) [30].

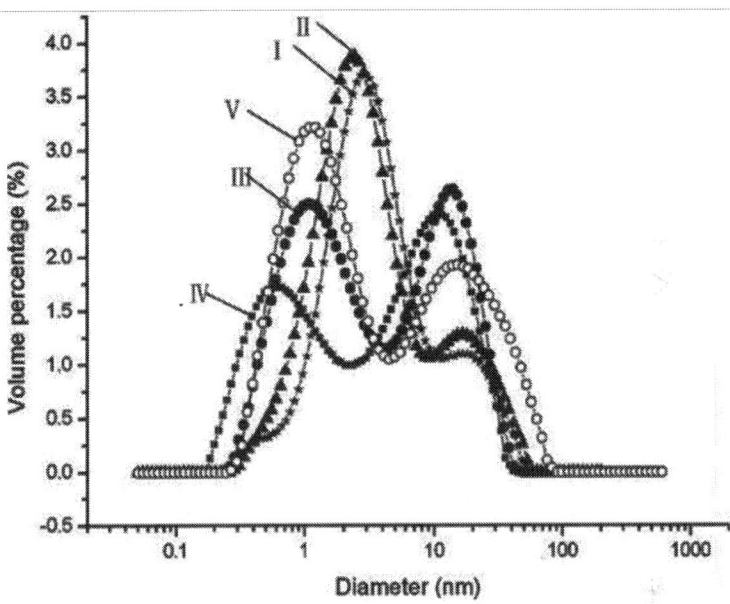

Figure 2. Particle size distribution of treated and untreated pigment red 170. (I) 1wt%pigment, 0 wt% dispersant; (II) 1wt% pigment, 0.01wt% dispersant; (III) 1wt% pigment, 0.02wt% dispersant; (IV) 1wt% pigment, 0.1wt% dispersant; (V) 1wt% pigment,10wt% dispersant.

The -COOH group of copolymers encapsulated onto the surface of UMPs would form a polymeric shell around the particles and also increase the surface charges of UMP particles. The interaction between UMP particles is weakened due to the existence of these charges and polymeric layer on the UMP particle surface, so that the UMP particles can be finely dispersed in aqueous media [37]. With the dosage of siloxane dispersant descending, the percentage of big particle ascends at first. However, while the dispersant is greatly excessive (10 wt%), the particle size increases.

Initial experiments, using dilute aqueous dispersions of carbon black of known average-particle size distribution, were performed using both a standard UV/Vis spectrometer and a digital camera to estimate light-transmission/attenuance. The results, summarized in Figure 3(A) and

(B), showed a strong relationship between average particle size and light-transmission/ attenuance over the size range examined by both these approaches, as demonstrated by the high R^2 values of the polynomial trend-line. Because the limited nature of the particle size distributions obtained during these initial experiments, we were only able to demonstrate a correlation over a small size. The digital image method was developed further on account of its potential to allow parallel analysis of multiple samples from the same image [39].

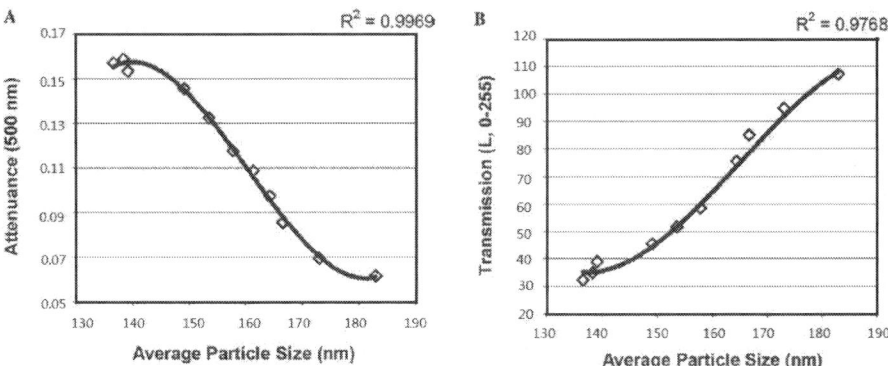

Figure 3. (A) Pigment average particle size versus attenuance (D) using UV/Vis analysis. (B) Pigment average particle size versus light-transmission using image analysis

There was a very interesting phenomenon of the particle size distribution of Pigment Red CI 170 with the increase of the cationic siloxane dispersant. In order to clearly describe the distribution process of UMP in water, a experiment was proposed according to the principle of physical chemistry. UMP could not be well dispersed in water with the absence of dispersant. Then, a small amount of dispersant was added into the system. Since the siloxane group of dispersant showed great affinity to the UMP, it tended to anchor on the surface of UMP particle.

3.2. Disperse stability

Stability of aqueous UMP dispersion is referred to that colloidal properties such as Zeta potentials, viscosity, and storage stability will not change in certain period. It's an important performance of UMP dispersion. Disperse stability can be supported by two generally mechanisms: steric stabilization and charge stabilization. Steric stabilization is due to steric hindrance resulting from the adsorbed dispersing agent, the chains of which become solvated in the liquid medium, thus creating an effective steric barrier that prevents the other particles from approaching too close. Charge stabilization is due to electrical repulsion forces, which are the result of a charged electrical double layer surrounding the particles. The charged electrical double layer developed around the particles extends well into the liquid medium, and since all the particles are surrounded by the same charge (positive or negative), they repel each other when they come into close proximity. The disperse stability usually includes deposited stability, heat stability, pH stability, electrolyte stability and centrifugal stability [35].

Repulsive and attractive forces among UMP particles determine the stability of the UMP dispersion. When the van der Waals forces between UMP particles are higher than steric and static repulsive forces, the UMP particles will combine each other and generate large particles. Instead, the dispersion will stably exist in aqueous media. In aqueous UMP dispersion, polymeric dispersants encapsulated onto the surface of UMPs are ionized in acid solution, and produce some negative charges onto the surface of UMPs to create the static forces between UMP particles. Further, the PSMA encapsulated onto the surface of UMPs can also generate steric repulsion [27]. The whole steric and static forces are larger than van der Waals force between UMP particles, the dispersion will exist in long time, even under centrifugal force or treated at high temperature. It was reported the effect of Triton X-100 on the colloidal dispersion stability of CuPc-U (unsulfonated and hydrophobic copper phthalocyanine) particles and was concluded that the stabilization mechanism for the CuPc-U is inferred to be primarily steric and adding $NaNO_3$ had no obvious effect on the dispersion stability [18].

Stabilization of UMP with polymeric dispersants has been proven to be a good way for the preparation of UMP dispersion with high stability, small particles size, low viscosities, and low moisture sensitivity. In aqueous media, the polymeric dispersant can build polymeric shell around the UMP particles and increase the surface charge on the UMP particles, so that the UMP particles are able to finely dispersed in the aqueous media.

3.3. Zeta potential

The Zeta potentials of UMP dispersion are analyzed to estimate their disperse stability. The stability of UMP dispersion determined by forces among the UMP particles. When the repulsive force is higher than attractive force, the UMP particles will stably disperse into the media. Otherwise, the UMP particles will arregate together. In the UMP dispersion with polymer dispersant, the repulsion forces among particles produced by the adsorbed polymer include static-electronic and steric repulsion [40]. The former can be measured by the Zeta potentials, and the latter is closely connected with the thickness of adsorbed polymer layer. The studies on the change of Zeta potentials and thickness of adsorbed polymer layer after the addition of solvent are important in understanding the effects of solvent on the stability of UMP dispersion [35,41].

The dispersion mechanism may be interpreted with a pair of electricity layers principle. Although many cationic dispersants gathered on the UMP surface neighbourhood, the quiet balance of positive negative charge didn't change. The cation on the UMP surface adsorbed a great many anions. With cation gathering gradually on the UMP surface, a pair of electricity layers formed. The particles stably existed in dispersion system for the electric charge rejected function mutually. An optimum dispersant dosage would exist, which promoted the formation of double electricity layers [16,42].

3.4. Color properties

UMP particles which are dispersed finely in dispersions enhance the color depth of UMP dispersions (e.g. transmittance, chroma, and lightness) and improve color display performance. The transmittances were enhanced with decreasing the mean size of dispersion (Figure 4). A color analysis program estimated the lightness (L*) and chroma (C*) from a spectrum of dispersion samples at a wavelength of 380 to 780 nm based on the L*a*b* Color System and D65 light source. The finer sizes of UMP particles in dispersion resulted in the higher lightness (L*) and chroma (C*). Serious aggregation of UMP particles without a dispersant in the dispersion would be performed. Holding the supercritical fluid-assisted dispersion process at the supercritical region conduced to better dispersion [43].

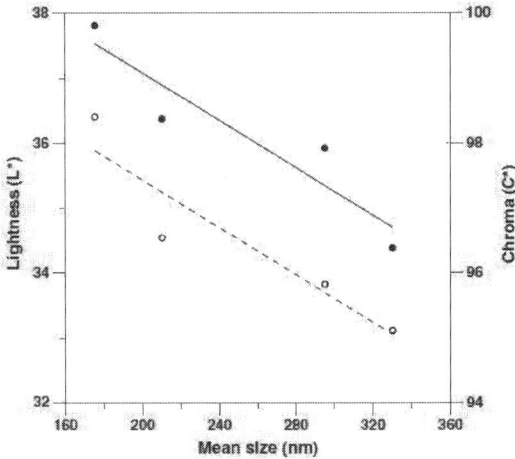

Figure 4. The color analyses from the four representative samples of dispersions. (o), Lightness; (•), chroma

Color properties are presented the color of the UMP dispersion. The absorbance of UMP hydrous dispersion is slightly larger than that of containing organic solvents. The change is mainly attributed to the disparate polarities of the solutions. In these two solutions, while the polarity of solvent UMP dispersion, which contains more ethanol, is weaker. The color of UMP is aroused by the $\pi \rightarrow \pi^*$ energy transition when the chromophore groups are irradiated. The weak ionization of the UMP increases in a solution with high polarity (such as H_2O), and the electric charge in the conjugated system can transport more easily, which leads to an increases in the absorbance. However, the maximum absorption wavelengths for both systems with or without solvent nearly remain the same. This indicates that the chromophore groups of the UMP are not damaged in the UMP dispersion containing organic solvent and the conjugated systems are essentially not altered, therefore, the color hue of the solvent UMP dispersion remains identical to that of the UMP hydrous dispersion [35].

4. Dyeing properties

4.1. Dyeing cotton fabric with UMP

In order to improve the dyeability with UMP, the cationic pretreatment of fabric is used. It can introduce positively charged sites on cotton fabric. Without pigment modification, there are two inhibiting factors. Firstly, pigments are insoluble and have no affinity for cotton and, secondly, the surfaces of pigment particles are usually negatively charged. UMP dyeing has several noticeable advantages compared to dyes. This method is a short process, simple operation, saving energy consumption and low costs, and matching intuitive easy color imitation, steady color hue, hiding power. Because UMP particles have no affinity for fibers, the UMP dyeing fibers does not exist the selective problem and is appropriate for all spices.

There are many reports on cationic modification of cotton. Karrer researched the cationic pretreatment to the cellulose fiber [44]. Guthrie studied the cationic pretreatment of cellulose fiber and dyed it with acidic dye [45]. The chemical modification of cotton to promote dyeability, light fastness and washing fastness has been researched by Cai [12]. Wang investigated chemical modification of cotton to promote fiber dyeability [46]. Burkinshaw analysized cationic pretreatment of cellulose fiber to improve the dye reactivity [47]. Hauser examined the dyeing behavior of cotton that had been rendered cationic by reaction with 2,3-epoxy propyltrimethylammonium chloride and the result showed that excellent dye yields and color fastness properties were obtained without the use of electrolytes, multiple rinsings or fixation agents [48].

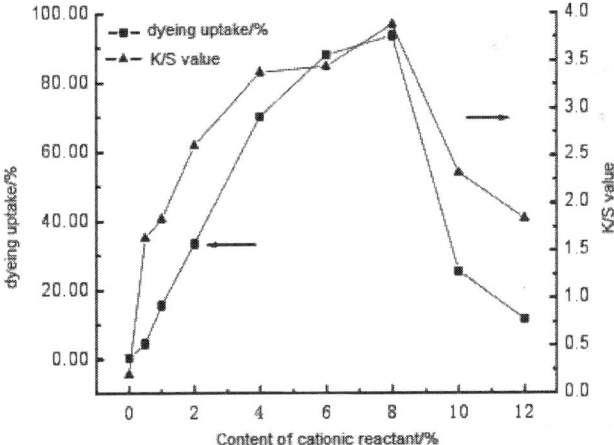

Figure 5. Effect of cationic reactant concentration on pigment exhaustion and the K/S value

During exhaustion dyeing process, the dyeing uptake of cotton fabric without modification was very low. Via cationic reagent modification, the uptake of the cationic fabric was increased significantly. Cationic reagent with higher proportions increased the number of positive charges on the cotton fiber, in turn increased pigment uptake.

For example, the cationic polymerization modifier was used to modify the charge of cotton fiber, and this could enhance the uptake of UMP and K/S value with different cationic reagent concentration (Figure 5). The application of cationic polymerization modifier on fibers for cationic modification could reduce the production cost, reduce dyeing temperature, shorten the dyeing process and improve production efficiency [49].

4.2. Dyeing silk fabric with UMP

Silk is a natural protein fiber, and the shimmering appearance of silk is due to the triangular prism-like structure of the silk fiber, which allows silk cloth to refract incoming light at different. But the silk fiber also has some defects, such as the problem of bleaching, the loss of pupa protein and the yellowing [50,51].

In order to improve the performance of the silk fiber, silk was modified with cationic dispersant [52]. This improved substantivity of UMP due to the introduction of cationic groups into the silk fabric, and the balance might be due to the adsorption of cationic reagent on surface of silk fabric reaching saturation values. In this situation, more cationic reagent molecules combined with UMP. The bounds between UMP and silk fabric led to a higher K/S values. The different amount of the cationic reagent in pretreatment could impart different amount of positive charge to silk surface, which could affect the K/S values. The UMP was not substantive to silk, the K/S values were low on silk without cationization pretreatment. The K/S values greatly increased and then kept in a balance with the increase of the cationization concentration [53].

Furthermore, penetration property and content of ammonium of silk would also restrict to produce the stronger ionic attraction between the cationic fiber and the anionic UMP. As a result, the adsorption of UMP on silk fabric did not change any more accordingly [54]. Wei reported that after modifying with glycidyltrimethylammonium chloride (EPTAC), the reactions of some amino acids were more rapid than the reaction of glucose. The dyeing ability of modified silk fiber had been improved remarkably. A theory which was called "fixed points" was given to this phenomenon [51]. Wang found that pH in pretreatment affected significantly on the dyeing properties of silk fabric with UMP, because in alkali condition, the chlorohydroxypropyl group of 3-chloro-2-hydroxypropyltrimrthylammonium chloride (CHTAC) converted to an epoxy group and then formed 2, 3-epoxypropyltrimethylammonium chloride which reacted with the nucleophilic amine group in the silk. And The highest K/S values could achieve at pH 8 [53].

With the increase of cationic reagent concentration, the rubbing fastness values appeared increasing trend (Table 1). The washing stain fastness of treated fabric was similarly comparable to the results without fastness improving reagent treatment because the UMP had no affinity with silk without treatment. Compared with untreated silk, the washing change fast-

nesses of cationic silk increased greatly. This was probably due to the fastness improving reagent acting as the binder between UMP and silk fabric [53].

Fastness improving reagent conc. (g/L)	Rubbing fastness (Grade)		Washing fastness (Grade)		
	Dry	Wet	Color change	Stain cotton	Stain silk
0	2	1-2	2	4	4
10	2-3	2	3-4	4-5	4
20	3	3	4	4-5	4
40	3-4	3	4	4-5	4-5
80	4	3-4	4	4-5	4-5

Table 1. Color fastness at different cationic reagent concentration

Cationic pretreatment depended on the extents of reaction between cationic reagent and
silk. This might affect the physical properties of the treated fabrics. The tensile strength and
elongation at break of fabrics changed with the increase of pH. It was clear that the reaction
between cationic reagent and silk took place significantly when pH of the pretreatment solution increased. The tensile strength decreased with the peptide chain hydrolyzation when
the pH increased. Also the elongation at break kept slightly decreasing. This revealed that
the cationic pretreatment has little effect on the physical properties. The bending rigidity
and hysteresis of cationization silk had no change compared with untreated silk. It revealed
that the cationization treatment had no impact on the soft properties of fabrics. The handle
of fabric decreased a little after dyeing, it might be due to the action of fastness improving
reagent.

4.3. Dyeing wool fabric with UMP

The wool fabric was modified by cationic reagent and dyed with UMP which was prepared with anionic polymer disperser via exhaust method [55]. The influence of pretreatment conditions such as concentration of cationic reagent, pH value of the bath,
temperature and duration of treatment on the dyeing property of the fabric was of importance to the drying properties. With the increase of the cationic content, the K/S value enhanced sharply and then kept a constant value when the content on cationic reagent was
higher than 10% (Figure 6). The dry and wet rubbing fastness of the wool fabric dyed
with UMP via exhaustion was gray scale ratings of 3-4 and 4, respectively. The tensile
strength and elongation at break decreased slightly with the increase of bath pH value.
The test results of bending rigidity and hysteresis of bending revealed that the wool fabric had soft handle and good elasticity.

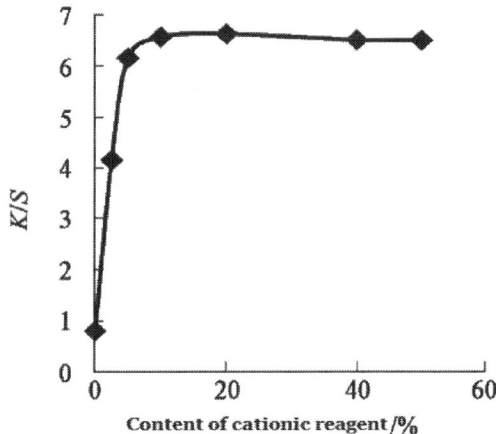

Figure 6. The contents of cationic reagent on the K/S value of dyed fabric

4.4. Dyeing cotton fabric with carbon black UMP

The excellent color and fastness properties make carbon black (CB) particles possible application in textile dyeing. Compared with common sulfur dye, the dyeing process of carbon
black UMP can save water and energy due to the shorter dyeing time. However, the dispersion of carbon black UMP in aqueous solutions and adsorption of the particles to cotton fibers are critical in exhaustion dyeing process, because carbon black UMP is hydrophobic
and easily aggregate in aqueous solutions.

On the basis of cotton fabrics modified with cationic reagent, the dyeing process and properties of CB dispersions were reviewed (Figure 7). The mechanism of cationic cotton dyed by
an exhaustion process using aqueous carbon black UMP dispersions was investigated. Cotton modified with a cationic reagent enhanced the dyeing properties of aqueous carbon
black UMP dispersions. There were higher affinity between carbon black UMP and cationic
cotton. CB particles could quickly adsorb on the surface of cationic cotton fibers within 5
min and diffuse into the inner pores and cracks on the surface layer of cationic cotton step
by step [56]. The cross sections of cationic-modified cotton fibers dyed with different sizes of
CB particles were colorless region, and CB particles were only in the cotton fiber grooves,
just as "ring-dyeing". After cationic modification, the cotton fiber could have a dark color
for the adsorbability of the anionic CB [13].

The color assessment of CB dyed cellulose is correlative to the amount carbon black UMPs
on the cotton via electrostatic force. It made possible to compare the charge density difference between nanoparticle ionic materials (NIMs) deposition process and chemical modified
process.

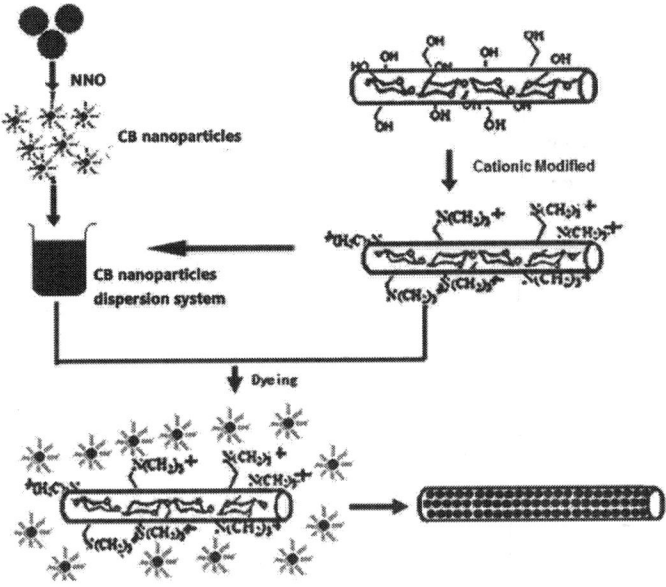

Figure 7. The dyeing process of carbon black UMP on modified cotton fabric

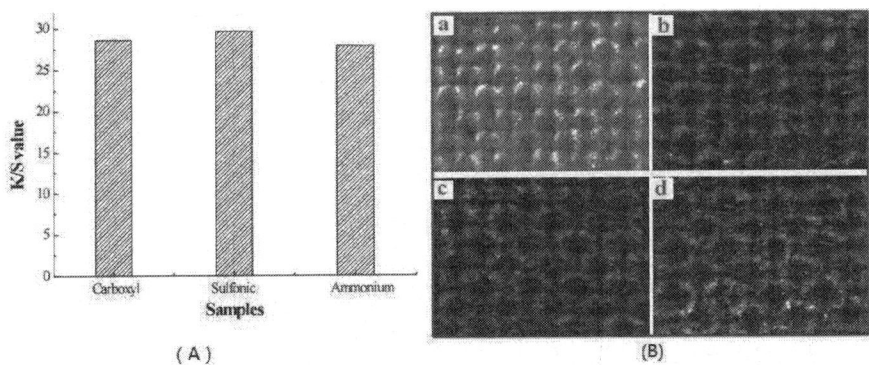

Figure 8. Color assessment of CB dyed cellulose matrixes: (A)K/S values; (B) Video microscope microscope:(a) Cotton fabrics without modification, (b) Carboxyl modified cotton fabric deposited with positive NIMs and carbon black UMPs, (c) Sulfonated modified cotton fabric deposited with positive NIMs and carbon black UMPs, (d) Ammonium modified cotton fabric deposited with carbon black UMPs

The results of the K/S presented that the sulfonated cellulose had the higher K/S value of 29.7 than the carboxyl of 28.6 and aminocellulose of 27.9 (Figure 8(A)). The results proved the conclusion about the color depth of the decorated samples. The images (Figure 8(B)) of the samples before and after the dyeing showed that the color depths of the dyed carboxyl and ammonium fabrics were not as deep as the sulfonated cellulose fabric, which further confirmed the color depth conclusion. Especially for the image of the ammonium sample, the fabric surface was uncoated with carbon black UMP, and there were some starred white flake on the fabric.

4.5. Dyeing cotton fabric with cationic UMP

Surface treatments are effective for UMP because its surface contain polar or polarized functional groups, which can serve as adsorption sites for the hydrophilic or lipophilic groups of the surfactants [1]. The cationic UMP dispersion has great prospects in cellulose dyeing, inkjet printing, and so on. Compared with the anionic and non-ionic UMP, the cationic UMP has stronger combining force with cellulose substrate, better K/S value, and vividness, thus it can increase the uptake of the UMP, reduce the environment pollution and improve the quality of final products.

For the specific structure, there are a lot of the characteristics on Gemini cationic dispersant. The Zeta potential of the Gemini cationic UMP is higher than that of the ordinary cationic UMP. Therefore, it is easier to color. The dry and wet rubbing fastness is good with both Gemini cationic UMPs after the film fixation (Table 2). With the dispersant dodecyltrimethylammonium, the dyeing of Gemini cationic UMPs exhibits deeper K/S value and the fabric is easier to be colored. The adsorption capacity of the Gemini cationic UMPs to cotton fabric is stronger, and can cover more fiber surface. Cotton fabric shows negative in a water solution, which has a weak charge bonding force with cationic UMPs. The more cationic UMPs containing electropositive, the easier the cotton fabric adsorbing the UMPs [16].

Sample	K/S value	Dye rubbing fastness	Wet rubbing fastness
Traditional pigment	42.64	3-4	2
UMP	51.99	4	2

Table 2. K/S value and fastness of Gemini cationic UMP dispersion on cotton

4.6. Dyeing acrylic yarns with cationic UMP

Acrylic fibers can be dyed by cationic UMPs, and the electrostatic attraction between UMPs and acrylic fibers is beneficial to the dyeing properties, especially its light fastness property. Compared with anionic and nonionic UMP systems, the cationic UMPs could bond with the acrylic fibers by electrostatic attraction. Therefore, the UMP presented a better dyeing result and a higher utilization ratio [57].

The dye bath pH presented great influences on cationic UMP dispersion stability during dyeing acrylic fiber with cationic UMP. The increase of UMP dosage resulted in lower dyeuptake and more brilliant colors, and dyeing acrylic yarns with UMP (40%, o.w.f) could reach the maximum K/S value. The K/S value of acrylic yarns reduced after adding adhesives, compared to the cationic dyes, the light fastness was improved significantly [58].

From Figure 9, when the temperature was lower than 65 °C, the dyeing rate was quite low, and the adsorption of pigments was unstable, while the temperature was higher than 65 °C, the dyeing rate increased significantly with the dyeing time prolonging. A higher dyeing temperature would shorten the dyeing time. As the temperature was higher than 95 °C, the dyeing time was only 15 min. The reason was that in a low temperature condition, the acrylic molecular chains were not moving and the adsorption only took place in the surface of fiber. The pigments adsorbed to the yarns were easily re-dissolved into the dye bath.

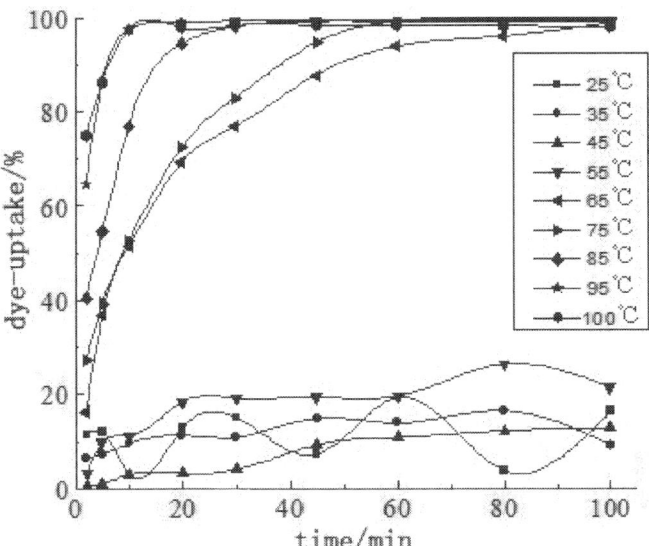

Figure 9. Effects of dyeing temperature on dye-uptake of acrylic yarns

In order to evaluate the light fastness of acrylic yarns dyed by cationic dye, the light fastness of cationic dye is attempted and assessed (Figure 10).

Cationic UMP presented better light fastness than that of the cationic dye. Acrylic yarns dyed by cationic UMP presented higher stability to light than that dyed by cationic dye. The main reason was that the chromophoric group of pigment was more difficult to destroy for its stable chemical structure.

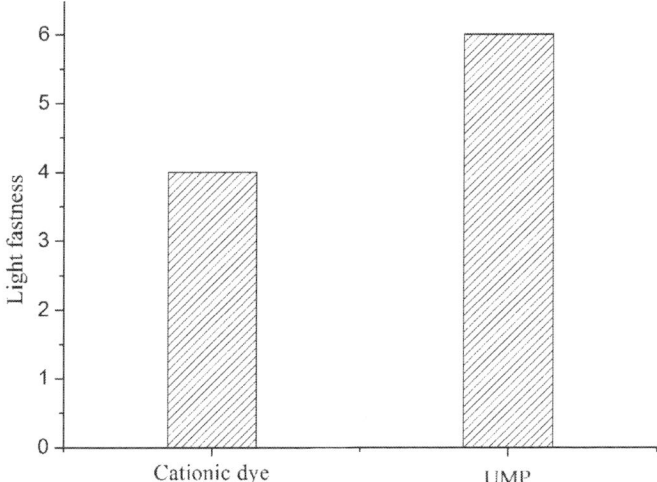

Figure 11. Light fastness of acrylic yarns dyed by cationic dye and ultra-fine pigment

5. Conclusion

A stable UMP dispersion is able to be prepared via grinding, ultrasonic wave, microencapsulation and microfluidics processes with different dispersants like low molecular weight dispersant, polymeric dispersant, copolymer dispersant and siloxane dispersant. It's more efficient to prepare the uniform and stable UMP dispersion via comprehensive utilization of these dispersing processes.

The key dispersion characteristics of UMP dispersion be evaluated by particle size, disperse stability, Zeta potential, color properties and dyeing properties, which are greatly affected by ionic strength, dispersants and dispersing methods. The disperse stability and color performance are promoted with the smaller particle size and higher charged UMP.

The dyeing properties of UMP on many fabrics such as cotton, silk and wool pretreated with cationic reagent is enhanced. The K/S value and rubbing fastness are able to be distinctly improved. Cationic cotton can also enhance the dyeing properties of aqueous carbon black UMP dispersions due to the higher affinity between carbon black UMP and cationic cotton. The cotton fabric dyed with cationic UMP can also promote the K/S value and color fastness. For acrylic yarns, the increase of UMP dosage results in lower dye-uptakes and more brilliant colors, and dyeing with UMP (40%, o.w.f) can reach the maximum K/S value on acrylic yarns.

As an advanced pigment dyeing method, the UMP takes many advantages of pigment. The specific characteristics of the UMP, including stability and high K/S value, reveal its significant potential on fabric dyeing, and especially for its environment-friendly characteristics, the development of UMP will present a more sharply speed than that of the dyes. Base on the textile application, the further research interests about UMP are focus on the novel dispersing process to obtain stable UMP system, modification with functional organic or inorganic materials, developing the multifunction and multipurpose of UMP.

Acknowledgements

The authors are grateful for the financial support of the National Natural Science Foundation of China (21174055) and the 333 Talent Project Foundation of Jiangsu Province (BRA2011184).

Author details

Chaoxia Wang* and Yunjie Yin

*Address all correspondence to: wangchaoxia@sohu.com

Key Laboratory of Eco-Textile, Ministry of Education, School of Textiles and Clothing, Jiangnan University, Wuxi, China

References

[1] Vernardakis, T., & Pigment, Dispersion. (2006). In: Tracton A A, editor., *Coatings Technology Handbook Third Edition.*, Taylor & Francis Group, LLC2006., 76-71.

[2] Gage, J. (1999). Color and Culture: Practice and Meaning from Antiquity to Abstraction. *University of California Press,*, 12-18.

[3] Velmurugan, P., Kim, M., Park, J., Karthikeyan, K., Lakshmanaperumalsamy, P., Lee, K., Park, Y., & Oh, B. (2010). Dyeing of Cotton Yarn with Five Water Soluble Fungal Pigments Obtained from Five Fungi[J]. *Fiber Polym,*, 11(4), 598-605.

[4] El -Shishtawy, R, & Nassar, S. Cationic pretreatment of cotton fabric for anionic dye and pigment printing with better fastness properties[J]. Color Technol, (2002).

[5] Hayashi, K., Morii, H., Iwasaki, K., Horie, S., Horiishi, N., & Ichimura, K. (2007). Uniformed nano-downsizing of organic pigments through core-shell structuring[J]. *Journal of Materials Chemistry,*, 17(6), 527-530.

[6] Kelley, A., Alessi, P., Fornalik, J., Minter, J., Bessey, P., Garno, J., & Royster, T.(2010). Investigation and Application of Nanoparticle Dispersions of Pigment Yellow 185 using Organic Solvents[J]. ACS Appl Mater Interfaces, , 2(1), 61-68.

[7] Kobayashi, Y., Chiba, H., Mizusawa, K., Suzuki, N., Cerda-Reverter, J. M., & Takaha-shi, A. (2011). Pigment-dispersing activities and cortisol-releasing activities of mela-nocortins and their receptors in xanthophores and head kidneys of the goldfish Carassius auratus[J]. *Gen Comp Endocr,,* 173(3), 438-446.

[8] Lan, Y., & Lin, J.(2011). Clay-assisted dispersion of organic pigments in water[J]. *Dyes Pigment,,* 90(1), 21-27.

[9] Ricaurte-Avella, G., Osorio, Velez. J., Nino, Cardona. J., Correa, J., & Alberto, Rodri-guez. H.(2010). Evaluation of the pigment dispersion in acrylic resins for the manu-facture of artificial teeth[J]. *Revista Facultad De Ingenieria-Universidad De Antioquia,,* 53, 54-63.

[10] Ahmed, N., & El -Shishtawy, R.(2010). The use of new technologies in coloration of textile fibers[J]. *Journal of Materials Science,,* 45(5), 1143-1153.

[11] Wang, C., Zhang, X., Lv, F., & Peng, L.(2012). Using carbon black nanoparticles to dye the cationic-modified cotton fabrics[J]. *J Appl Polym Sci,,* 124(6), 5194-5199.

[12] Cai, Y., Pailthorpe, M., & David, S. (1999). A new method for improving the dyeabili-ty of cotton with reactive dyes[J]. *Textile Research Journal,,* 69(6), 440-446.

[13] Wang, C., & Zhang, Y.(2007). Effect of cationic pretreatment on modified pigment printing of cotton[J]. Materials Research Innovations, , 11(1), 27-30.

[14] Ishwarlal, R.(2000). Dyeing anionic cellulosic fabric materials with pigment colors having a net cationic charge using a padding process. US , 27.

[15] Koyanagi, T. (2002). Ink sets for cellulose fiber-based recording media, printing method therewith and pinted products therefrom. 23.

[16] Wang, C., Yin, Y., & Fang, K. (2008). Investigation of gemini cationic water based pig-ment dispersion[J]. *Materials Research Innovations,,* 12(1), 7-11.

[17] Diaz-Blanco, C., Diaz, Gonzalez. M., Daga, Monmany. J. M., & Tzanov, T. (2009). Dyeing properties, synthesis, isolation and characterization of an in situ generated phenolic pigment, covalently bound to cotton. *Enzyme and Microbial Technology,* 380 EOF-385 EOF.

[18] Dong, J., Chen, S., Corti, D. S., Franses, E. I., Zhao, Y., Ng, H. T., & Hanson, E. (2011). Effect of Triton X-100 on the stability of aqueous dispersions of copper phthalocya-nine pigment nanoparticles[J]. *J Colloid Interface Sci,,* 362(1), 33-41.

[19] Huhnerfuss, K., von, Bohlen. A., & Kurth, D. (2006). Characterization of pigments and colors used in ancient Egyptian boat models[J]. *Spectrochimica Acta Part B-Atomic Spectroscopy,,* 61(10-11), 1224-1228.

[20] Leelajariyakul, S., Noguchi, H., & Kiatkamjornwong, S. (2008). Surface-modified and micro-encapsulated pigmented inks for ink jet printing on textile fabrics[J]. *Progress in Organic Coatings,,* 62(2), 145-161.

[21] Fang, K., Jiang, X., Wang, C., Wu, M., & Yan, Y. (2008). Properties of the Nanoscale Hydrophilic Cationic Pigment Based on Quaternary Surfactant[J]. *J Disper Sci Technol,*, 29(1), 52-57.

[22] Wu, H., Gao, G., Zhang, Y., & Guo, S. (2012). Coating organic pigment particles with hydrous alumina through direct precipitation[J]. *Dyes and Pigments,*, 92(1), 548-553.

[23] Chang, C., Chang, S., Tsou, S., Chen, S., Wu, F., & Hsu, M.(2003). Effects of polymeric dispersants and surfactants on the dispersing stability and high-speed-jetting properties of aqueous-pigment-based ink-jet inks[J]. J Polym Sci Pt B-Polym Phys, , 41(16), 1909-1920.

[24] Wang, C., Yin, Y., Wang, X., & Bu, G. (2008). Improving the Color Yield of Ultra-fine Pigment Printing on Cotton Fabric[J]. *AATCC Review,* 8(12), 41-45.

[25] Yoon, C., & Choi, J. (2008). Syntheses of polymeric dispersants for pigmented ink-jet inks[J]. *Coloration Technology,*, 124(6), 355-363.

[26] Bulychev, N., Kisterev, E., Ioni, Y., Confortini, O., Du, Prez. F., Zubov, V., & Eisenbach, C. (2011). Sufface Modification in Aqueous Dispersions with Thermo-Responsive Poly (Methylvinylether) Copolymers in Combination with Ultrasonic Treatment[J]. *Chemistry & Chemical Technology,*, 5(1), 59-65.

[27] Fu, S., Du, C., Zhang, K., & Wang, C. (2011). Colloidal properties of copolymer-encapsulated and surface-modified pigment dispersion and its application in inkjet printing inks[J]. *J Appl Polym Sci,*, 119(1), 371-376.

[28] Chen, Y. M., Hsu, R. S., Lin, H. C., Chang, S. J., Chen, S. C., & Lin, J. J. (2009). Synthesis of acrylic copolymers consisting of multiple amine pendants for dispersing pigment[J]. *J Colloid Interface Sci,*, 334(1), 42-49.

[29] Zhang, W., Deodhar, S., & Yao, D. G. (2010). Processing Properties of Polypropylene With a Minor Addition of Silicone Oil[J]. *Polymer Engineering and Science,*, 50(7), 1340-1349.

[30] Wu, J., Wang, L. M., Zhao, P., Wang, F., & Wang, G. (2008). A new type of quaternary ammonium salt containing siloxane group and used as favorable dispersant in the surface treatment of CI pigment red 170[J]. *Progress in Organic Coatings,*, 63(2), 189-194.

[31] Weber, U. (2010). The Effect of Grinding Media Performance on Milling a Water-Based Color Pigment[J]. *Chem Eng Technol,*, 33(9), 1456-1463.

[32] De Vallejuelo, S., Barrena, A., Arana, G., De Diego, A., & Madariaga, J.(2009). Ultrasound energy focused in a glass probe: An approach to the simultaneous and fast extraction of trace elements from sediments[J]. Talanta, , 80(2), 434-439.

[33] Jia, X., Caroli, C., & Velicky, B. (1999). Ultrasound propagation in externally stressed granular media[J]. *Physical Review Letters,*, 82(9), 1863-1866.

[34] Johnson, J., Barbato, M., Hopkins, S., & O'Malley, M.(2003). Dispersion and film properties of carbon nanofiber pigmented conductive coatings[J]. Progress in Organic Coatings, , 47(3), 198-206.

[35] Yin, Y., Wang, C., & Wang, Y. (2012). Preparation and colloidal dispersion behaviors of silica sol doped with organic pigment[J]. Journal of Sol-Gel Science and Technology,, 62(2), 1-7.

[36] Fu, S., Xu, C., Du, C., Tian, A., & Zhang, M.(2011). Encapsulation of CI Pigment blue 15:3 using a polymerizable dispersant via emulsion polymerization[J]. Colloid Surf A-Physicochem Eng Asp, , 384(1-3), 68-74.

[37] Fu, S., & Fang, K.(2008). Properties of waterborne nanoscale pigment red 122 dispersion prepared by phase separation method[J]. J Appl Polym Sci, , 108(6), 3968-3972.

[38] Squires, T., & Quake, S.(2005). Microfluidics: Fluid physics at the nanoliter scale[J]. Reviews of modern physics, , 77(3), 977.

[39] Merrington, J., Hodge, P., & Yeates, S. (2006). A High-Throughput Method for Determining the Stability of Pigment Dispersions[J]. Macromolecular Rapid Communications,, 27(11), 835-840.

[40] Chang, C., Chang, S., Shih, K., & Pan, F.(2005). Improving mechanical properties and chemical resistance of ink-jet printer color filter by using diblock polymeric dispersants[J]. J Polym Sci Pt B-Polym Phys, , 43(22), 3337-3353.

[41] Nsib, F., Ayed, N., & Chevalier, Y. (2007). Selection of dispersants for the dispersion of CI Pigment Violet 23 in organic medium[J]. Dyes Pigment,, 74(1), 133-140.

[42] Xu, D., Fang, K., & Fu, S. (2009). Effects of Ethanol on the Stability of Pigment Colloidal Dispersion[J]. Journal of Dispersion Science and Technology,, 30(4), 510-513.

[43] Wu, H., Huang, K., & Lee, K.(2011). Supercritical fluid-assisted dispersion of CI pigment violet 23 in an organic medium[J]. Powder Technology, , 206(3), 322-326.

[44] Karrer, P. Improvements in or relating to the treatment of cotton fibers preparatory to dyeing. GB Patent 2498421926.

[45] Guthrie, J.(1947). Introduction of Amino Groups into Cotton Fabric by Use of 2-Aminoethylsulfuric Acid [J]. Textile Research Journal, , 17(11), 625-629.

[46] Wang, H., & Lewis, D. (2002). Chemical modification of cotton to improve fibre dyeability[J]. Coloration Technology,, 118(4), 159-168.

[47] Burkinshaw, S., Lei, X., & Lewis, D. (1989). Modification of cotton to improve its dyeability. Part 1-pretreating cotton with reactive polyamide-epichlorohydrin resin[J. Journal of the Society of Dyers and Colourists, 105(11), 391-398.

[48] Hauser, P., & Tabba, A.(2001). Improving the environmental and economic aspects of cotton dyeing using a cationised cotton[J]. Color Technol, , 117(5), 282-288.

[49] Jin, H. (2009). The Synthesis and Application of Cationic Polymer (PDA) on Fibers Modification[D]. Master thesis. Jiangnan University;.

[50] Dubas, S., Chutchawatkulchai, E., Egkasit, S., Lamsamai, C., & Potiyaraj, P.(2007). Deposition of polyelectrolyte multilayers to improve the color fastness of silk[J]. Textile Research Journal, , 77(6), 437-441.

[51] Wei, J. (2003). Studies on Weighting Methods and Properties of Silk Fiber[D]. Master thesis. Suzhou University;.

[52] Qin, C., Chen, D., Tang, R., Huang, Y., Wang, X., & Chen, G.(2010). Studies on silk fabric dyed with a hemicyanine dye[J]. Coloration Technology, , 126(5), 303-307.

[53] Wang, C., Fang, K., & Ji, W. (2007). Superfine pigment dyeing ot silk fabric by exhaust process[J]. Fiber Polym,, 8(2), 225-229.

[54] Murphy, A., John, P., & Kaplan, D.(2008). Modification of silk fibroin using diazonium coupling chemistry and the effects on hMSC proliferation and differentiation[J]. Biomaterials, , 29(19), 2829-2838.

[55] Wang, C., & Ji, W. (2006). Dyeing property of the modified wool fabric with superfine pigment[J]. Journal of Textile Research,, 27(6), 74-77.

[56] Yuan, X., & Fang, K. J. (2006). Preparation of Waterborne Nanoscale Carbon Black Dispersion with Sodium Carboxymethyl Cellulose[J]. Journal Of Donghua University(English Edition), 23(5), 119-121.

[57] Yang, R., & Dai, M. (1996). Study on Properties of Acid-Modified Multicomponent Copolyesterification Fibre[J]. Journal Of Textile Research,, 17(1), 27-31.

[58] Zhou, Y., Yu, D., Xi, P., & Chen, S. L. (2003). Influence of styrene-maleic anhydride copolymers on the stability of quinacridone red pigment suspensions[J]. Journal of Dispersion Science and Technology,, 24(5), 731-737.

Chapter 5

The Use of New Technologies in Dyeing of Proteinous Fibers

Rıza Atav

Additional information is available at the end of the chapter

http://dx.doi.org/10.5772/53912

1. Introduction

Wool consists principally of one member of a group of proteins called keratins; other members of this group include the proteins of hair, feathers, beaks, claws, hooves, horn and even certain types of skin tumour [1]. The main substance of wool is a keratin. Keratin macromolecules are crosslinked with cystine residues and contain a variety of side chains, some basic and some acidic [2]. Wool fibers consist of cortical cells and cuticular cells, which are located in the outermost part of the fiber surrounding the cortical cells. They consist of endocuticle, A and B exocuticle, and an exterior hydrophobic thin membrane called the epicuticle [3]. Both layers have a tremendous influence on dyeing because of their hydrophobic characteristics. The cuticle is separated from the underlying cortex by the intercellular material, which is called the cell membrane complex (CMC) and consists of non-keratinous proteins and lipids [2].

The morphology of the wool fiber surface plays an important role in textile finishing processes. The covalently bound fatty acids and the high amount of disulphide bridges make the outer wool surface highly hydrophobic. Especially in the printing and dyeing of wool, the hydrophobic character of the wool surface is disturbing. Diffusion of the hydrophilic dyes at and into the fibers is hindered. For this reason, the hydrophilicity and dyeability properties of the wool fiber should be developed [4]. Wool dyeing is a degradative process involving high temperature for long periods in acidic to neutral pH medium to achieve good penetration, optimum fastness, and dye uptake. The results can be harsh handle, discomfort, and a deterioration of properties that impact consumer wear, care, and aesthetic appreciation [5].

Silk fiber is protein fiber that is produced from silk worms [6]. Silk has been called "the queen of fibers", its natural luster, handle and draping properties being superior to those of many other textile fibers [7]. It is composed of different alpha amino acids orienting to form

long chain polymer by condensation and polymerization. Silk fiber consists of 97% protein and the others are wax, carbohydrate, pigments, and inorganic compounds. The proteins in silk fiber are 75% fibroin and 25% sericin by weight, approximately. The sericin makes silk fiber to be strong and lackluster; therefore, it must be degummed before dyeing [6]. Silk fibroin, like wool keratin, is formed by the condensation of α-amino acids into polypeptide chains, but the long-chain molecules of silk fibroin are not linked together by disulfide bridges as they are in wool. Chemical treatments can cause modification of main peptide chains, and side chains of amino acids, which in turn influence the fiber's chemical, physical, and mechanical properties [8]. Silk fiber is easily damaged when dyeing at the boil, so low-temperature dyeing is usualy preferred [7]. Because the brilliancy of dyed and printed silk fabrics is a decisive factor for evaluating the quality of silk fabrics, dyeability of silk fibers is one of the most attractive topics for applied and basic research [9].

In recent years, many attempts have been made to improve various aspects of dyeing, and new technologies have been, and are being developed to reduce fiber damage, decrease energy consumption and increase productivity [10]. In this chapter, new technologies that improve the dyeability of proteinous fibers such as ultrasound, ultraviolet, ozone, plasma, gamma irradiation, laser, microwave, e-beam irradiation, ion implantation, and supercritical carbondioxide will be overviewed.

2. Ultrasound technology

Sound is transmitted through a medium by inducing vibrational motion of the molecules through which it is traveling. This vibrational motion represents the sound frequency [11]. Ultrasound is sound of a frequency that is above the threshold of human hearing [12]. The lowest audible frequency for humans is about 18Hz and the highest is normally around 18-20 kHz for adults, above which it becomes inaudible and is defined as ultrasound [11]. In recent decades the use of ultrasound technology has established an important place in different industrial processes such as the medical field, and has started to revolutionize environmental protection. The idea of using ultrasound in textile wet processes is not a new one. On the contrary there are many reports from the 1950s and 1960s describing the beneficial effects of ultrasound in textile wet processes. In spite of encouraging results from laboratory-scale studies, the ultrasound-assisted wet textile processes have not been implemented on an industrial scale as yet [13].

In practice, three ranges of frequencies (Fig. 1) are reported for three distinct uses of ultrasound: low frequency or conventional power ultrasound (20-100 kHz), medium frequency ultrasound and diagnostic or high frequency ultrasound (2-10 MHz) [13].

As the sound wave passes through water in the form of compression and rarefaction cycles, the average distance between the water molecules varies. If the pressure amplitude of the sound is sufficiently large, then the distance between the adjacent molecules can exceed the critical molecular distance during the rarefaction cycle. At that moment a new liquid surface is created in the form of voids. This phenomenon is called acoustic cavitation [15].

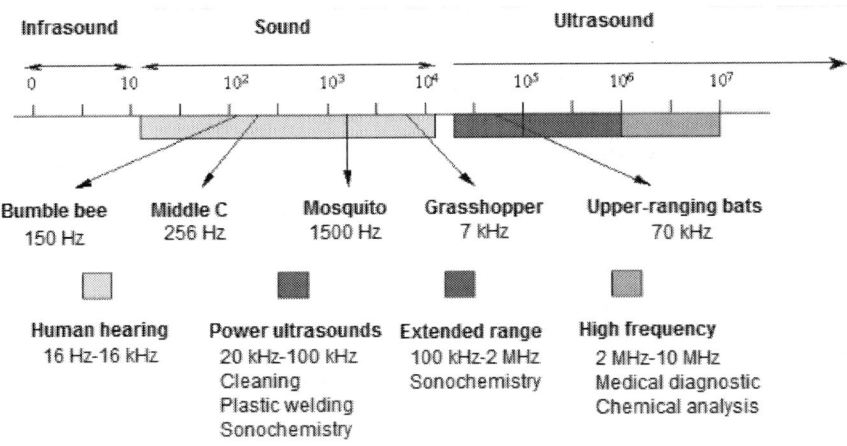

Figure 1. Classification of sound according to the frequency [14]

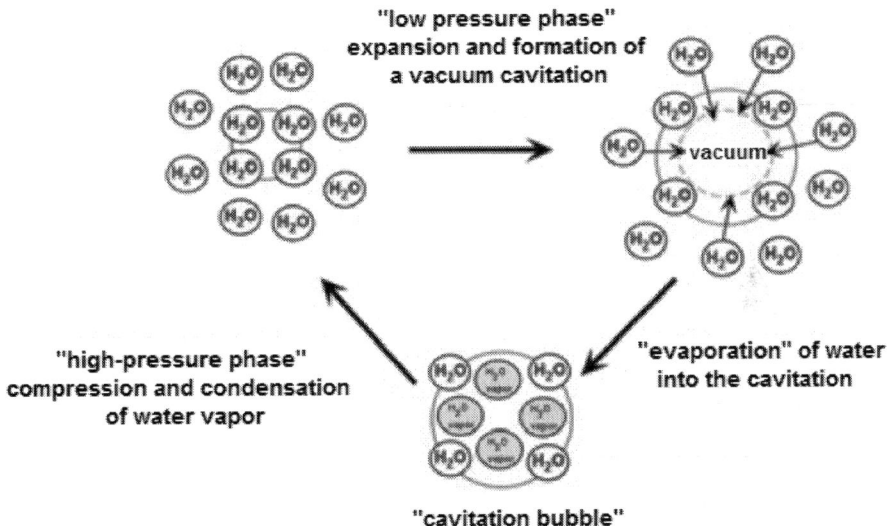

Figure 2. Formation of a cavitation buble [16]

Power ultrasound can enhance a wide variety of chemical and physical processes, mainly due to the phenomenon known as cavitation in a liquid medium that is the growth and explosive collapse of microscopic bubbles. Sudden and explosive collapse of these bubbles can generate "hot spots", i.e., localized high temperature, high pressure, shock waves and severe shear force capable of breaking chemical bonds [17]. High temperature and pressures resulting from the collapse of the transient cavitation bubbles are responsible for all the observed effects of ultrasound [18]. Parameters which affect cavitation and bubble collapse are:

• Properties of the solvent: The solvent used to perform sample treatment with ultrasonication must be carefully chosen. As a general rule, most applications are performed in water. However, other less polar liquids, such as some organics, can be also used, depending on the intended purpose [19]. Cavities are more readily formed when using a solvent with high vapor pressure, low viscosity and low surface tension. But at high vapor pressure more vapor enters the cavitation bubble during its formation and the bubble collapse is cushioned and less violent [13].

• Properties of gases: Soluble gases should result in the formation of a larger number of cavitation nuclei, but the greater the solubility of the gas is the more gas molecules should penetrate the cavity. Therefore, a less violent and intense shock wave is created on bubble collapse [13].

• External pressure: With increasing external pressure, the vapor pressure of the liquid decreases and higher intensity is necessary to induce cavitation [13]. In addition, there is an increment in the intensity of the cavitational bubble collapse and, consequently, an enhancement in sonochemical effects is obtained. For a specific frequency there is a particular external pressure that will provide an optimum sonochemical reaction [19].

• External temperature: Higher external temperature reduces the intensity necessary to induce cavitation due to the increased vapor pressure of the liquid. At higher external temperatures more vapor diffuses into the cavity, and the cavity collapse is cushioned and less violent [13].

• Frequency of the sound wave: At high sonic frequencies, on the order of the MHz, the production of cavitation bubbles becomes more difficult than at low sonic frequencies, of the order of the kHz. To achieve cavitation, as the sonic frequency increases, so the intensity of the applied sound must be increased, to ensure that the cohesive forces of the liquid media are overcome and voids are created [19]. Lower frequency produces more violent cavitation and, as a consequence, higher localized temperatures and pressures. At very high frequency, the expansion part of the sound wave is too short to permit molecules to be pulled apart sufficiently to generate a bubble [13].

Cavitation induced by ultrasound will allow accelerating processes and obtaining the same results as existing techniques but with a lower temperature and low dye and chemical concentrations [20]. For this reason textile wet processes assisted by ultrasound are of high interest for the textile industry. A review of earlier studies using ultrasound in textile wet processes was compiled by Thakore et al. Ultrasound-assisted textile dyeing was first reported by Sokolov and Tumansky in 1941 [13]. Some of the benefits of using of ultrasonics in dyeing can be listed as below;

- energy savings by dyeing at lower temperatures and reduced processing times,

- environmental improvements by reduced consumption of auxiliary chemicals,

- lower overall processing costs (due to less energy and chemical consumption), thereby increasing industry competitiveness [21]

Improvements observed in ultrasound-assisted dyeing processes are generally attributed to 3 main phenomenons,

- Dispersion: breaking up of micelles and high molecular weight aggregates into uniform dispersions in the dye bath,

- Degassing: expulsion of dissolved or entrapped gas or air molecules from fiber into liquid and removal by cavitation, thus facilitating dye-fiber contact, and

- Diffusion: accelerating the rate of dye diffusion inside the fiber by piercing the insulating layer covering the fiber and accelerating the interaction or chemical reaction, if any, between dye and fiber [11].

A good wool dyeing process must provide a satisfactory uptake of dye bath and an adequate penetration of dye into the fiber, with the practical advantages of good wet fastness and uniform coloration. The conventional methods for wool dyeing are based on long times at temperature close to the boiling point, in order to ensure good results of dye penetration and leveling. These conditions can damage the fibers, with negative effects on the characteristics of the finished material. Such damage can be minimized by reducing the operation time or, better yet, by reducing the dyeing temperature. Recently, ultrasound assisted wool dyeing was studied with the aim to reduce temperature or dyeing time with respect to the conventional dyeing technique [22]. Some literature related to the use of ultrasound technology in dyeing of proteinous fibers is summarized below.

Shukla and Mathur (1995) studied the dyeing process of silk using cationic, acid and metal-complex dyes at low temperatures, assisted by a low frequency ultrasound of 26 kHz and compared the results of dye uptake with those obtained by conventional processes. Their results show that silk dyeing in the presence of ultrasound increases the dye uptake for all classes of dyes at lower dyeing temperatures (45°C and 50°C) and a shorter dyeing time (15 min.), as compared with conventional dyeing at 85°C for 60 min. Furthermore, there was no apparent fiber damage caused by cavitation [23].

Kamel et al. (2005) have investigated the dyeing of wool fabrics with lac as a natural dye in both conventional and ultrasonic techniques. The extractability of lac dye from natural origin using power ultrasonic was also evaluated in comparison with conventional heating. The results of dye extraction indicate that power ultrasonic is rather effective than conventional heating at low temperature and short time. Color strength values obtained were found to be higher with ultrasonic than with conventional heating. The results of fastness properties of the dyed fabrics were fair to good [17].

Vankar and Shanker (2008) have extracted coloring pigment from Hollyhock (Alcea rosea) flower and used for dyeing wool yarn, silk and cotton fabrics. It is observed that the dyeing with hollyhock gives fair to good fastness properties in sonicator in 1 hour and shows good dye uptake as compared with conventional dyeing [24].

Battu et al. (2010) observed that in wool dyeing at 85°C with acid dyes, ultrasound caused an improvement of the dye uptake as much as 25%, or dyeing time would be nearly 20% shorter than conventional dyeing [25].

Yukseloğlu and Bolat (2010) stated that the wool fabrics have presented similar color yield (K/S) and acceptable color differences (ΔE) with the use of ultrasonic energy. Ultrasonic energy was found to be advantageous to be used for wool dyeing at lower temperatures (such as 80°C and 90°C) and lower dyeing times (i.e. 80 min. or 90 min.) as an alternative process for conventional dyeing (100°C and 144 min) [26].

McNeil and McCall (2011) investigated the effects of ultrasound at 35-39 kHz on several wool dyeing and finishing processes. Ultrasound pre-treatment increased the effectiveness of subsequent oxidative-reductive bleaching, but had no effect on the uptake of acid leveling and acid milling dyes. The pre-treatment retarded the uptake of reactive dye, possibly by increasing the crystallinity of the fiber or removing surface bound lipids. Ultrasound did not improve dyeing under conditions that are currently used in industry, but did show potential to reduce the chemical and energy requirements of dyeing wool with reactive and acid milling dyes, but not acid leveling dyes [27].

Atav and Yurdakul (2011) investigated the effect of ultrasound usage on the color yield in dyeing of mohair fibers. They found that dyeing in the presence of ultrasound energy increases the dye-uptake of mohair fibers and hence higher color yield values are obtained. The difference between the samples dyed in the presence and absence of ultrasound, was greater for darker shades and for dyeing carried out in acidic medium (pH 5), and also for shorter dyeing periods. Furthermore there is no important difference between washing fastness and alkali solubility values of fibers dyed in the presence and absence of ultrasound [28].

Ferrero and Periolatto (2012) studied the possibility of reducing the temperature of conventional wool dyeing with an acid leveling dye using ultrasound in order to reach dye uptake values comparable to those obtained with the standard procedure at 98°C. Dyeings of wool fabrics were carried out in the temperature range between 60°C and 80°C using either mechanical or ultrasound agitation of the bath and coupling the two methods to compare the results. For each dyeing, the dye uptake curves of the dye bath were determined and the better results of dyeing kinetics were obtained with ultrasound coupled with mechanical stirring. Finally, fastness tests to rubbing and domestic laundering yielded good values for samples dyed in ultrasound assisted process even at the lower temperature [22].

3. Ultraviolet technology

Light is electromagnetic radiation or radiant energy traveling in the form of waves [29]. The electromagnetic spectrum is the range of all possible frequencies of electromagnetic radiation. The "electromagnetic spectrum" of an object is the characteristic distribution of electromagnetic radiation emitted or absorbed by that particular object. The electromagnetic spectrum extends from low frequencies used for modern radio communication to gamma radiation at the short-wavelength (high-frequency) end, thereby covering wavelengths from thousands of kilometers down to a fraction of the size of an atom [30]. UV energy is found in the electromagnetic spectrum between visible light and x-rays [29].

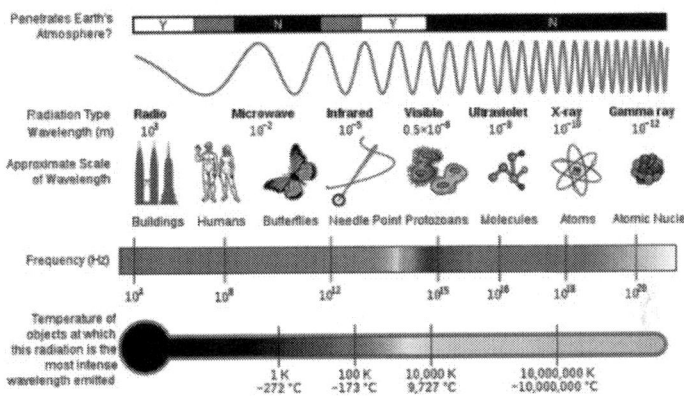

Figure 3. A diagram of the electromagnetic spectrum [30]

Ultraviolet or UV radiation is part of the electromagnetic (light) spectrum that reaches the earth from the sun. It has wavelengths shorter than visible light, making it invisible to the naked eye [31]. Ultraviolet radiation constitutes to 5% of the total incident sunlight on earth surface (visible light 50% and IR radiation 45%). Even though, its proportion is quite less, it has the highest quantum energy compared to other radiations [32]. Scientists classify UV radiation into three types or bands: UVA, UVB, and UVC (Fig. 4). The ozone layer absorbs some, but not all, of these types of UV radiation [33].

- UVA: Long-wavelength UVA covers the range 315-400 nm. Not significantly filtered by the atmosphere. Approximately 90% of UV radiation reaching the Earth's surface. UVA is again divided into UVA-I (340 nm - 400 nm) and UVA-II (315 nm - 340 nm) [34].

- UVB: Medium-wavelength UVB covers the range 280-315 nm. Approximately 10% of UV radiation reaching the Earth's surface [34].

- UVC: Short-wavelength UVC covers the range 100-280 nm [34]. They are the most dangerous among all the rays. However, these rays do not reach the earth's surface as they are completely absorbed by the ozone layer [31].

The deleterious effects of solar irradiation are perceived as changes in texture and color, dryness, etc., and can be evaluated in terms of reduced elasticity, increased porosity or swelling properties, altered

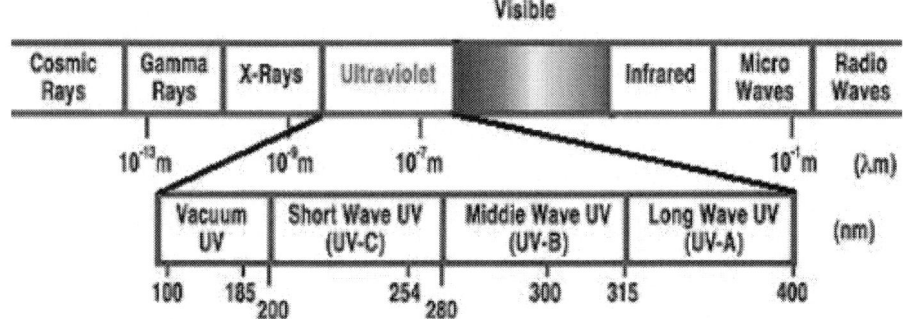

Figure 4. Classification of ultraviolet radiation [35]

dye sorption characteristics, and photofading of natural or artificial hair color [36]. The chemical changes caused by short-term UV irradiation of wool are confined to fibers at the fabric surface and UV is unable to penetrate beyond the surface to weaken the bulk fibers responsible for the mechanical strength. This has enabled the potential application of UV technology as a surface-specific treatment in several areas of wool processing [37]. For wool the UV-absorbing species are aromatic amino acid and cystine residues in the protein structure which absorb strongly below 350 nm. UVC radiation (200-280 nm) is the most effective range for modifying wool fiber surfaces [38].

The most commercially used process for antifelting and antipilling of wool is based on chlorination. However, recent concern over the release into the environment of adsorbable organohalogens (AOX) in process effluents has prompted the development of alternative, AOX-free processes. Different types of radiation techniques, such as ultraviolet radiation, are utilized as alternatives to chlorination in wool processing [39]. UV treatment can add value in coloration (dyeing and printing), since it is predominantly surface fibers in a fabric that absorb, reflect and scatter light. Photomodification of the surface fibers can allow:

- more dye to become fixed, producing deeper shades

- more rapid fixation of dyes dye fixation under less severe conditions (e.g. lower temperature) [38]

Modification of the dye uptake by exposure of wool fabric to UV radiation before dyeing has been known since the early 1960s. For most dye classes, UV-irradiated fabric takes up significantly more dye than untreated and when fabrics are irradiated through stencils, intricate tone-ontone effects can be produced [37]. Some literature related to the use of ultraviolet technology in dyeing of proteinous fibers is summarized below.

Millington (1998) stated that UV irradiation of wool can significantly increase dyeing color yields. The use of 1:1 metal-complex dyes was found to be particularly effective, and a 3% o.w.f. dyeing on UV-treated fabric could produce a better depth of shade than a 5% dyeing on untreated fabric [40].

Millington (1998) found that UV radiation of wool fabric exhibits some physical and chemical changes on its surface. This interaction not only modifies the fabric of wool but also improve the shades particularly grey and black. It also helps in even dyeing, deeper shades, chlorine free printing and improve the photo bleaching of wool [41].

Xin et.al (2002) exposed wool samples to UV radiation for 60 min. and investigated the surface modification of the wool fiber by X-ray photoelectron spectroscopy. The chemical change caused by the UV treatment was identified as surface oxidation of cystine (disulphide bonds) and thereby induced changes in the dyeing properties of the wool. The dyeability of UV-treated and untreated wool samples was determined at temperatures of 45, 50, 55 and 60°C using C.I. Acid Blue 7. The UV-treated wool samples showed greater levels of dye uptake compared with those of the untreated samples. The adsorption behavior and diffusion coefficients were also studied. The dyeing properties of wool were enhanced by UV radiation due to the increased diffusion coefficient of the dyes in the treated wool fibers [42].

4. Ozone technology

Ozone is a natural occurring gas that can be both beneficial and detrimental to organisms on Earth. It is important that sufficient amount of this pale blue gas is present in the stratosphere, where O_3 molecules would shield most of the UV radiation from reaching Earth [43]. Although ozone is a blue colored gas at normal temperatures and pressures; because of its low concentrations in its applications the observation of this blue color of ozone is impossible [44]. Ozone gas has a pungent odor readily detectable at concentrations as low as 0.02 to 0.05 ppm (by volume), which is below concentrations of health concern [45].

Ozone was first generated and characterized by a German scientist named Schonbein in 1840 [46]. Ozone is a nonlinear triatomic molecule possessing two interoxygen bonds of equal length (1.278 A) and an average bond angle of 116°49' [47].

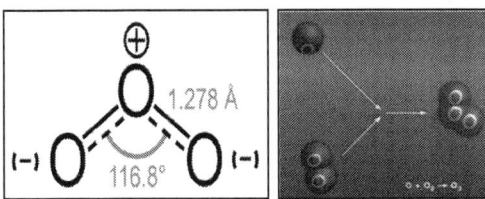

Figure 5. Ozone molecule [48, 49]

Ozone is formed naturally in the atmosphere by photochemical reaction with solar UV radiation and by lightening. It can also be generated artificially. Three most common ways of generating ozone artificially are:

- Corona discharge: In this method, ozone is generated when free, energetic electrons in the corona dissociate oxygen molecules in oxygen-containing feed gas that passes through the discharge gap of the ozone generator.

- UV light: Ozone can also be generated by UV light. The high energy UV light ruptures the oxygen molecules into oxygen atoms, and the subsequent combination of an oxygen atom with an oxygen molecule produces ozone (O_3).

- Electrolysis: A third method for generating ozone is electrolysis, which uses an electrolytic cell. Specifically, electrolysis involves converting oxygen in the water to ozone by passing the water through positively and negatively charged surfaces [46].

Ozone is a very powerful oxidizing agent, which is able to participate in a great number of reactions with organic and inorganic compounds. Among the most common oxidizing agents, it is only surpassed in oxidant power by fluorine and hydroxyl radicals (see Table 1) [50]. Ozone has strong tendency to react with almost any organic substance as well as with water. The reaction proceeds via several intermediates such as peroxides, epoxides and perhydroxyl and hydroxyl radicals [51].

Oxidation species	Oxidation power (V)	Oxidation species	Oxidation power (V)
Fluorine	3.03	Chlorine dioxide	1.50
Hydroxyl radical	2.80	Hypochlorous acid	1.49
Atomic oxygen	2.42	Hypoiodous acid	1.45
Ozone	2.07	Chlorine	1.36
Hydrogen peroxide	1.77	Bromide	1.09
Permanganate	1.67	Iodine	0.54
Hypobromous acid	1.59		

Table 1. Oxidation power of selected oxidizing species [50]

In an ozonation process two possible pathways have to be considered [50]: direct oxidation with ozone molecules or the generation of free-radical intermediates, such as the OH radical, which is a powerful, effective, and non-selective oxidizing agent [52].

$$O_3 + H_2O \rightarrow 2HO_2 \tag{1}$$

$$O_3 + HO_2 \bullet \rightarrow HO \bullet + 2O_2 \tag{2}$$

Molecular ozone can directly react with dissolved pollutants by electrophilic attack of the major electronic density positions of the molecule. This mechanism will take place with pollutants such as phenols, phenolates or tiocompounds. The radical mechanism predominates in less reactive molecules, such as aliphatic hydrocarbons, carboxylic acids, benzenes or chlorobenzenes [50].

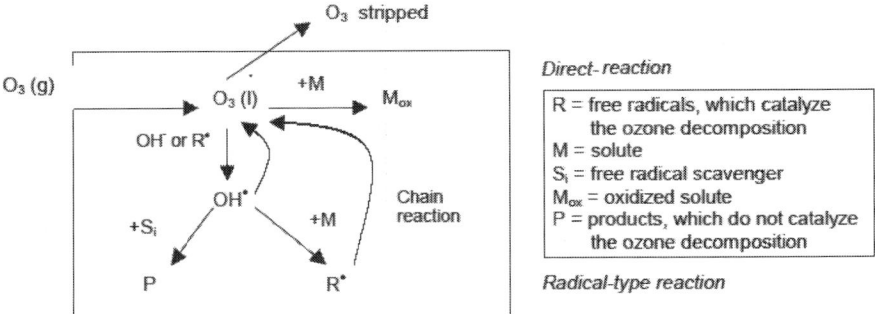

Figure 6. Scheme of reactions of ozone added to an aqueous solution [50]

The chlorine/ Hercosett process, the most widely used treatment for the wool dyeing process, causes dangerous ecological problems due to the contamination of waste water with absorbable organic halides (AOX). Because of legal restrictions and national and international awareness of ecology and pollution control, an AOX-free pretreatment is required to offer environmental advantages [53]. Alternative surface modifications for improving wool dyeability are therefore being explored. One of them is ozonation process [54]. Ozone treatments of proteinous fibers such as wool, mohair, angora and silk have been investigated by many authors. When literature is examined, it can be well understood that, increase in dyeability of protenious fibers caused by ozonation process depends on the following parameters;

- **pH:** Typically, at pH<4 direct ozonation dominates and above pH>9 the indirect pathway dominates. In the range of pH 4-9, both of them are important. The pH influences the generation of hydroxyl radicals [55].

$$O_3 + OH^- \rightarrow O_3^- \bullet + OH\bullet \qquad (3)$$

$$O_3^- \bullet \rightarrow O^- \bullet + O_2 \qquad (4)$$

$$O^- \bullet + H^+ \rightarrow OH\bullet \qquad (5)$$

Generally at neutral medium reaction rate of ozone gas is slow due to its low solubility. While molecular ozone reacts at low pH values, at high pH values radicals react. Since the oxidation potential of hydroxyl radicals exceeds that of ozone molecules, oxidation is faster in indirect reactions. Additionally HO• is not the only radical that is formed. Even though the HO• radical is the most powerful radical with a 2.8 V oxidation potential, HO_2•, HO_3• and HO_4• radicals are also formed as shown in Eq. 6-11 [56].

$$O_3 + OH^- \rightarrow HO_2 \bullet + O_2^- \tag{6}$$

$$HO_2 \bullet \leftrightarrow O_2^- \bullet + H^+ \tag{7}$$

$$O_3 + O_2^- \bullet \rightarrow O_3^- + O_2 \tag{8}$$

$$O_3^- + H^+ \rightarrow HO_3 \tag{9}$$

$$HO_3 \bullet \rightarrow HO \bullet + O_2 \tag{10}$$

$$HO \bullet + O_3 \rightarrow HO_4 \tag{11}$$

At pH < 9, the highly selective ozone molecules react rapidly at sites of high electron density (such as aliphatic and aromatic double bonds) and slowly at less reactive sites (such as C-H bonds of saturated hydrocarbons). The presence of the •OH radicals above pH 9 have less selectivity and high oxidation potential (2.8 V) [51]. Ozonation efficiency decreases in the basic pH values when compared to acidic and neutral pH. Because in a basic solution, more hydroxide ions are present and these hydroxide ions act as an initiator for the decay of ozone [57]. Moreover, it is reported that the dissolved ozone concentration in water decreases from 4.3×10^{-4} mol/L at pH 4 to 1.5×10^{-4} mol/L at pH 10 [58].

• **Temperature:** As temperature increases, ozone becomes less soluble (see Table 2) [45]; however, it can not be said that ozonation efficiency reduces with the decrease in the solubility of ozone, because temperature rise also increases the reaction rate [56].

Temperature (°C)	Solubility (kg.m⁻³)
0	1,09
10	0,78
20	0,57
30	0,40
40	0,27
50	0,19
60	0,14

Table 2. Water solubility of ozone [59]

- **Ozone dose:** Because oxidation reactions are caused by molecular ozone or radical species formed by the reactions of ozone, with the increase in ozone dose ozonation efficiency increases [56]

- **Water content of the fiber:** In literature it was stated that the rate of oxidation is accelerated by the hydration of hydrophilic groups in the fiber and above a critical level of moisture content (at which fiber is completely hydrated) it is retarded by the water which enters the intermicellar and interfibrillar space. In many previous studies, the importance of water during ozonation process was reported [51, 57, 58]. Prabaharan and Rao have suggested a model in order to explain this phenomenon. According to this model it can be said that for ozonation to take place, water should be present inside of the fiber. However the quantity of water present has a definite effect on the rate of reaction [51].

Fig. 7 gives the schematic representation of ozone path from gas phase to reaction site in the fiber. When sufficient water is not present, d_1 and d_2 are absent and hence O is transported by convection across the distances d_1 and d_2 and then by diffusion across d_3 and d_4. Since sufficient water is not supplied, entire hydrophilic group in the fiber is not hydrated and hence the extent of attack at R is low. When sufficient water is supplied, d_1 is absent and d_2 is either fully or partially absent depending on the quantity of water available and hence O is transported by convection across d_1 and convection or diffusion across d_2 followed by diffusion across d_3 and d_4. Since sufficient water is supplied, entire hydrophilic group in the fiber is hydrated and hence the extent of attack at R is maximum. When excess water is present, d_1 and d_2 are present and hence O is transported by diffusion across d_1, d_2, d_3, and d_4. Since excess water is present at d_1 and d_2, dilution of ozone takes place and hence ozone attack at R is lower in spite of complete hydration at d_3 [58].

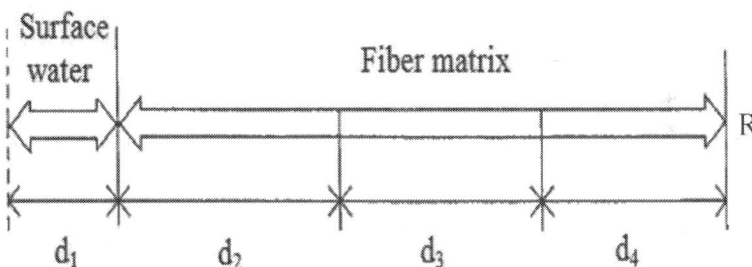

Figure 7. Schematic representation of ozone path from gas phase to reaction site [58]; (d_1: the distance occupied by surface water, d_2: the distance occupied by mobile water phase, d_3: immobile water phase, d_4: the distance between the immobile water phase and R)

Some literature related to the effect of ozonation on dyeing properties of various proteinous fibers is summarized below.

Micheal and El-Zaher (2003) has evaluated the effect of ultraviolet/ozone treatments for different times on the characteristics of wool fabrics with respect to wettability, permeability, yellowness index, and weight loss. The beneficial effects of this treatment on dyeability, color parameters, light fastness characteristics, and the change in color difference after exposure of the treated dyed samples to artificial daylight for about 150 hours were investigated. The results indicated that the improvement in wetting processes may have been due to surface modifications; this meant that an increase in the amorphousity of the treated samples, the oxidation of the cystine linkage on the surface of the fabrics, and the formation of free-radical species encouraged dye uptake [60].

Sargunamani and Selvakumar (2007) investigated the effects of process parameters (pick-up value, pH and time) in the ozone treatment of raw and degummed tassar silk fabrics on their properties such as yellowness index, breaking strength, breaking elongation, weight, amino group content. Decrease in yellowness index, breaking strength, breaking elongation, and weight as well an increase in amino group content was observed [58].

Perincek et.al (2008) investigated a novel bleaching technique for Angora rabbit fiber. For this purpose, a detailed investigation on the role of the fiber moisture, pH, and treatment time during ozonation was carried out. Also, the effect of ozonation on the dyeing properties of Angora rabbit fibers was researched. Consequently, it was found that ozonation improved the degree of whiteness and dyeability of Angora rabbit fiber [57].

Atav and Yurdakul (2010) investigated the use of ozonation to achieve dyeability of the angora fibers at lower temperatures without causing any decrease in dye uptake by modifying the fiber surfaces. The study was carried out with known concentration of ozone, involving process parameters such as wet pick-up (WP), pH, and treatment time. The effect of fiber ozonation was assessed in terms of color, and test samples were also evaluated using scanning electron microscopy (SEM). The optimum conditions of ozonation process were determined as WP 60%, pH 7 and 40 min. According to the experimental results it can be concluded that, ozonated angora fibers can be dyed at 90°C with acid and reactive dye classes without causing any decrease in color yield [61]. In an other study on ozonation process carried out by *Atav and Yurdakul (2011)* the optimum conditions for mohair fiber were determined as WP 60%, pH 7 and 30 min. Dyeing kinetics also studied and it was demonstrated that the rate constant and the standard affinity of ozonated sample increased [54].

5. Plasma technology

Faraday proposed to classify the matter in four states: solid, liquid, gas and radiant. Researches on the last form of matter started with the studies of Heinrich Geissler (1814-1879): the new discovered phenomena, different from anything previously observed, persuaded the scientists that they were facing with matter in a different state. Crookes took again the term "radiant matter" coined by Faraday to connect the radiant matter with residual molecules of gas in a low-pressure tube. Sufficient additional energy, supplied to gases by an electric field, creates plasma [62]. The plasma is referred to as the fourth state of matter (in addition to solid, liquid, gaseous) [63].

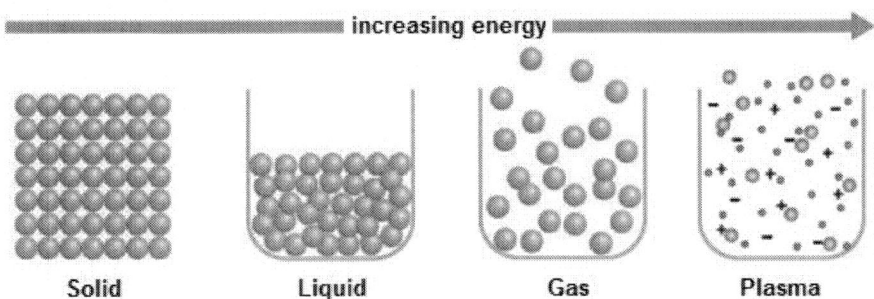

Figure 8. States of matter [64]

The plasma is an ionized gas with equal density of positive and negative charges which exist over an extremely wide range of temperature and pressure [65]. As shown in Fig. 9, the plasma atmosphere consists of free electrons, radicals, ions, UV-radiation and a lot of different excited particles in dependence of the used gas [66].

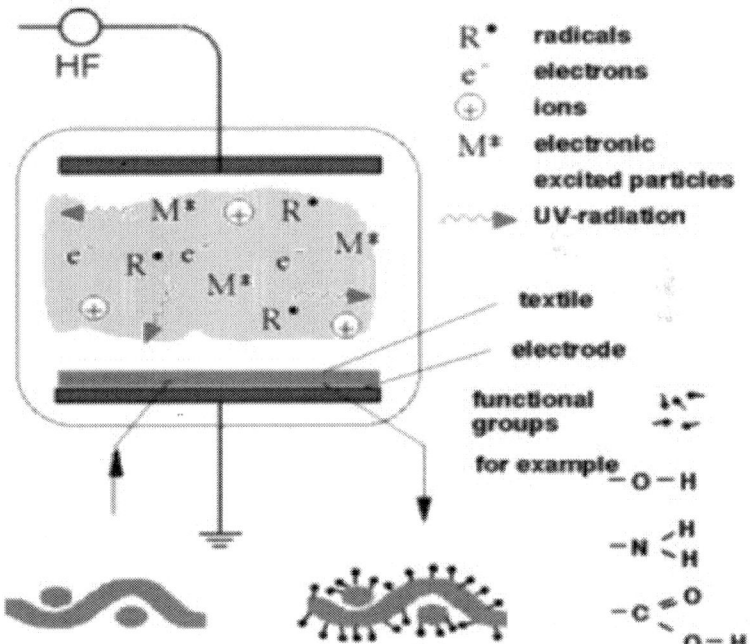

Figure 9. The principle of the plasma treatment [66]

There are different methodologies to induce the ionization of plasma gas for textile treatment [65]:

- Glow Discharge: It is the oldest type of plasma; it is produced at reduced pressure and assures the highest possible uniformity and flexibility of any plasma treatment [67]. The methodology applies direct electric current, low frequency over a pair of electrodes [65]. Alternatively, a vacuum glow discharge can be made by using microwave (GHz) power supply [68].

- Corona Discharge: It is formed at atmospheric pressure by applying a low frequency or pulsed high voltage over an electrode pair [65]. Typically, both electrodes have a large difference in size. The corona consists of a series of small lightning-type discharges. High local energy levels and problems related to the homogeneity of the classical corona treatment of textiles make it problematic in many cases [68].

- Dielectric-Barrier Discharge: DBD is produced by applying a pulsed voltage over an electrode pair of which at least one is covered by a dielectric material [65]. Although lightning-type discharges are created, a major advantage over corona discharges is the improved textile treatment uniformity [68].

Practically, one generates the plasma by applying an electrical field over two electrodes with a gas in between. This can be carried out at atmospheric pressure or in a closed vessel under reduced pressure. In both cases, the properties of the plasma will be determined by the gasses used to generate the plasma, as well as by the applied electrical power and the electrodes (material, geometry, size, etc.). The pressure of the gas will have a large influence on the plasma properties but also on the type of equipment needed to generate the plasma [69].

The plasmas can be classified as being of the low pressure and atmospheric type. Both plasmas can be used for the surface cleaning, surface activation, surface etching, cross linking, chain scission, oxidation, grafting, and depositing of materials, and generally similar effects are obtained; however, atmospheric plasma has many advantages when compared with vacuum plasma [70]. *Low pressure plasmas* are typically in the pressure range of 0.01 kPa. A vacuum chamber and the necessary vacuum pumps are required, which means that the investment cost for such a piece of equipment can be high. These plasmas are characterized by their good uniformity over a large volume. *Atmospheric plasmas* operate at standard atmospheric pressure (~ 100 kPa). Open systems using the surrounding air exist. The range of processes is not as wide as for low pressure plasmas. On the other hand, these systems are easily integrated in existing finishing lines, a major advantage from industrial view point. Of course, for an inline process to be feasible, the plasma treatment has to be done at sufficiently high line speeds, which is not evident for textile materials [69].

Due to increasing requirements on the finishing of textile fabrics, increasing use of technical textiles with synthetic fibers, as well as the market and society demand for textiles that have been processed by environmentally sound methods, new innovative production techniques are demanded [66]. Plasma technology is an important alternative to wet treatments, because there is no water usage, treatment is carried out in gas phase, short treatment time is enough, it does not cause industrial waste, and it provides energy saving [71]. In Cold plasma, that is, low-temperature plasma (LTP) treatment is the most commonly used physical method for a surface specific fiber modification, as it affects the surface both physically and chemically [72]. The advantage of such plasma treatments is that the modification turns out to be restricted in the uppermost layers of the substrate, thus not affecting the overall desirable bulk properties [62].

The plasma gas particles etch on the fabric surface in nano scale so as to modify the functional properties of the fabric. Unlike conventional wet processes, which penetrate deeply into fibers, plasma only reacts with the fabric surface and does not affect the internal structure of the fibers. Plasma technology modifies the chemical structure as well as the topography of the textile material surface. In conclusion, plasma can modify the surface properties of textile materials, deposit chemical materials (plasma polymerization) to add functionality, or remove substances (plasma etching) from the textile materials [65]. Essentially, four main effects can be obtained depending on the treatment conditions;

- The cleaning effect: It means the removal of impurities or substrate material from the exposed surface [73]. It is mostly combined with changes in the wettability and the surface texture. This leads for example to an increase in dye-uptake.

- Increase of microroughness: This affects, for example, an anti-pilling finishing of wool.

- Generation of radicals: Presence of free radicals induces secondary reactions like cross-linking. Furthermore, graft polymerization can be carried out as well as reaction with oxygen to generate hydrophilic surfaces [66].

- Plasma polymerization: In plasma polymerization, a monomer is introduced directly into the plasma and the polymerization occurs in the plasma itself [73]. It enables the deposition of solid polymeric materials with desired properties onto the substrates [66].

When a surface is exposed to plasma a mutual interaction between the gas and the substrate takes place. The surface of the substrate is bombarded with ions, electrons, radicals, neutrals and UV radiation from the plasma while volatile components from the surface contaminate the plasma and become a part of it. Whatever may be the final outcome on the surface, the basic effect that causes modification is based on the *radical formation* (attachment of functional group and deposition/polymerization) and *etching* phenomena. Fig. 10 illustrates the mechanism of plasma modification [74].

Low temperature plasma treatment of wool has emerged as one of the environmental friendly surface modification method for wool substrate. The efficiency of the low temperature plasma treatment is governed by several operational parameters like;

- Nature of the gas used

- System pressure

- Discharge power

- Duration of treatment

Plasma treatment can impart anti-felting effect, degreasing, improved dyestuff absorption and increased wetting properties to wool fibers [68]. These effects of the plasma process are attributed to several changes in the wool surface, such as;

- the formation of new hydrophilic groups (sulphonate and carboxylate),

- partial removal of covalently bonded fatty acids belonging to the outermost surface of the fiber, and

- the etching effect [75].

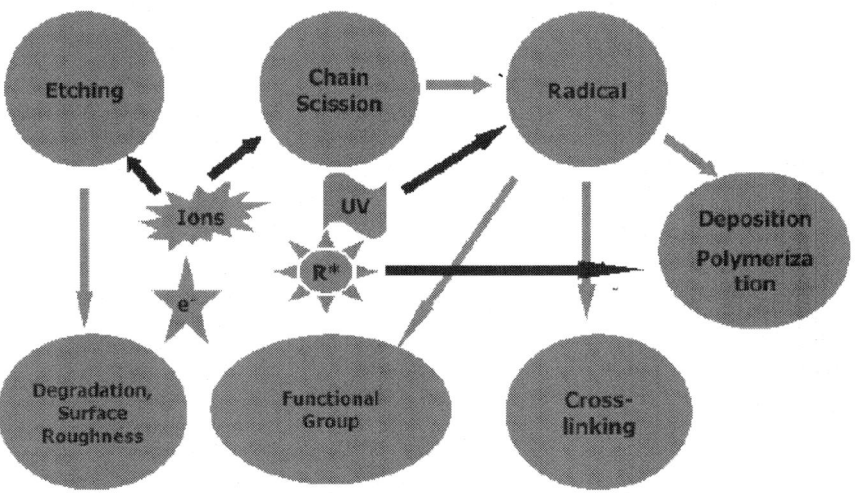

Figure 10. Mechanism of plasma-substrate interaction [74]

Plasma treatments modify the fatty acid monolayer present in the outermost part of the wool fiber, generating new hydrophilic groups as a result of the hydrocarbon chain oxidation and reducing the fatty acid chain length. The oxidation process also promotes the formation of Bunte salt and cysteic acid residues on the polypeptide chain. Particularly when oxidizing gasses are used, plasma induces cystine oxidation in the A-layer of the exocuticle, converting it into cysteic acid and thus reducing the number of crosslinkages in the fiber surface [72]. As the surface is oxidized, the hydrophobic character is changed to become increasingly hydrophilic [76]. The etching of the hydrophobic epicuticle and increase in surface area also contributes towards the improvement in the ability of the fibers to wet more easily [74].

Plasma treatment of wool fibers causes improvement in dyeability of wool fibers due to the changes occurred on fiber surface. For this reason dyeing kinetics, dye uptake and hence depth of shade are increased. In literature there are many studies related to the effect of plasma treatment on dyeability of proteinous fibers. Some of them are summarized below.

Wakida et al. (1993) treated the merino wool top with low-temperature plasmas of helium/argon and acetone/argon under atmospheric pressure for 30 seconds and then dyed with two leveling acid dyes, and two milling acid dyes. Dyeing rate and final dye uptake increased with the atmospheric low-temperature plasma treatments. In particular, helium/argon plasma was found to be much more effective than acetone/argon plasma at improving dyeing properties [77].

Yoon et al. (1996) treated the wool fabrics with low temperature oxygen plasma and examined their mechanical and dyeing properties. Plasma pretreatment caused an increase in strength. Furthermore, it was observed that when wool was dyed with a leveling acid dye, equilibrium dye uptake did not change, but the dyeing rate increased with a milling acid dye [78].

Jing (1996) investigated the surface modification of silk fabric by plasma graft copolymerization with acrylamide and acrylic acid. The dependence of graft degree was examined on the conditions of plasma grafting. The relationships were discussed between graft degree and factors such as crease recovery, dyeability, colour fastness and mechanical properties. It was shown that the dyeability and color fastness have been improved for samples grafted with acrylic acid [79].

Wakida et al. (1996) treated wool fibers with oxygen low-temperature plasma and then dyed with acid and basic dyes. Despite the increase of electronegativity of the fiber surface caused by the plasma treatment, the rate of dyeing of wool was increased with both dyes [80].

Wakida et al. (1998) treated wool fabrics with oxygen, carbon tetrafluoride, and ammonia low temperature plasmas and then dyed with several natural dyes. The dyeing rate of the plasma-treated wool increased considerably with cochineal, Chinese cork tree, and madder, but not with gromwell. Furthermore, plasma-treated wool fabrics dyed with cochineal and Chinese cork tree have increased brightness compared with untreated wool [81].

Kan et al. (1998) investigated the induced surface properties of wool fabrics created by the sputtering of low-temperature plasma treatment, such as surface luster, wettability, surface electrostatic and dyeability. After the low-temperature plasma treatment, the treated wool fabric specimens exhibited better hydrophilicity and surface electrostatic properties at room temperature, together with improved dyeing rate. The occurrence of some grooves on the fiber specimens was determined by scanning electron microscope and it was stated that these grooves might possibly provide a pathway for a faster dyeing rate [82].

Kan et al. (1998) treated the wool fiber with low-temperature plasma (LTP) using different non-polymerising gases and dyed with chrome dye. The rate of dyeing, the time of half-dyeing ($t_{1/2}$), the final dye uptake, the chromium exhaustion and the chromium fixation were studied. The results showed that LTP treatments can alter the dyeing properties to various degrees. The nature of the LTP gases plays an important role in affecting the behavior of chrome dyeing [83].

Kan et al. (1999), have studied the influence of the nature of gas (oxygen, nitrogen, and a 25% hydrogen/75% nitrogen gas mixture) in plasma treatments on the fiber-to-fiber friction, feltability, fabric shrinkage, surface structure, dyeability, alkali solubility, and surface chemical composition properties of wool substrates. After the low temperature plasma (LTP) treatment, those properties of the LTP-treated substrates changed, and the changes depended on the nature of the plasma gas used [84]. *Kan et al. (1999),* have also searched the surface characteristics of wool fibers treated with LTP with different gases, namely, oxygen, nitrogen and gas mixture (25% hydrogen / 75% nitrogen). Investigations showed that chemical composition of wool fiber surface varied differently with the different plasma gas used. The surface chemical composition of the different LTP-treated wool fibers was evaluated with different characterization methods, namely FTIR-ATR, XPS and saturated adsorption value [85].

Iriyama et al. (2002) treated the silk fabrics with O_2, N_2, and H_2 plasmas for deep dyeing and good color fastness to rubbing. C.I. Reactive Black 5 was used as a dye, and color was evaluated by total K/S. All plasma-treated silk fabrics showed weight loss, especially by O_2 plasma. Total K/S of dyed silk fabrics treated at 60 Pa of all plasmas was improved greatly. Total K/S increased with increasing plasma treatment time, weight loss of the fabrics in the treatment, and dye concentration in dyeing. They gained greater total K/S even dyed in 6% of dye concentration compared with untreated one dyed in 10%. Color fastness to wet rubbing of silk fabrics was not improved by plasma treatment. However, most of them were still within the level for commercial use [86].

Sun and Stylios (2004), have investigated the effects of LTP on pre-treatment and dyeing processes of cotton and wool. The contact angles, wicking properties, scourability, and dyeability of fabrics were affected by low-temperature plasma treatments. After treatment, the dye uptake rate of plasma treated wool has been shown to increase. It has been shown that O_2 plasma treatment increases the wetability of wool fabric and also the disulphide linkages in the exocuticle oxidize to form sulphonate groups which also enhances the wetability [87].

Binias et al. (2004) have investigated the effect of low temperature plasma on some properties (surface characteristic, water absorption capacity, capillarity, dyeability) of wool fibers. Selected properties of wool textiles were changed by the influence of low-temperature plasma on the fiber's surface layers, acting only on a very small thickness. The level of changes was limited by parameters of the low-temperature plasma. Lowering of the dyeing temperature was achieved [88].

Jocic et al. (2005), have investigated the influence of low-temperature plasma and biopolymer chitosan treatments on wool dyeability. Wool knitted fabrics were treated and characterized by whiteness and shrink-resistance measurements. Surface modification was assessed by contact-angle measurements of human hair fibers. It was stated that after plasma treatment the whiteness degree and hydrophility of fibers increased and fiber dyeability was improved [71].

Sun and Stylios (2005), have determined the mechanical and surface properties and handle of wool and cotton fabrics treated with LTP. This investigation showed that the mechanical properties of wool changed remarkably after oxygen plasma treatment. There were no sig-nificantly observed differences between plasma treated and un-treated fabrics after scouring and dyeing [89].

Masukuni and Norihiro (2006), studied the dyeing properties of Argon (Ar)-plasma treated wool using the six classes of dyestuffs, i.e., acid, acid metal complex, acid mordant, reactive, basic and disperse dyes. Ar-plasma treatment greatly improved the color yield and level-ness, together with the decrease of tippy dyeing. A condition in the plasma treatment en-hanced not only the color yield but also the anti-felting performance. The relationship between the improvement of dyeing properties by the plasma treatment and the chemical structure of the dyes was also examined. In the case of the acid dyes, the effect of plasma treatment on color yield was more significant for the milling type dyes with large molecular weight than the leveling type dye with low molecular weight. Furthermore, the hot water and rubbing fastness were improved by Ar-plasma treatment [90].

Kan and Yuen (2006) treated the wool fibers with oxygen plasma and then dyed these fibers with acid, chrome and reactive dye. For acid dyeing, the dyeing rate of the LTP-treated wool fiber was greatly increased, but the final dye uptake equilibrium did not show any signifi-cant change. For the chrome dyeing, the dyeing rate of the LTP-treated wool fiber was also increased, but the final dye uptake equilibrium was only increased to a small extent. For the reactive dyeing, the dyeing rate of the LTP-treated wool fiber was greatly increased, and the final dye uptake equilibrium was also increased significantly [91].

El-Zawahry et al. (2007) investigated the impact of plasma-treatment parameters on the sur-face morphology, physical-chemical, and dyeing properties of wool using anionic dyes. The LTP-treatment resulted in a dramatic improvement in fabric hydrophilicity and wettability, the removal of fiber surface material, and creation of new active sites along with improved initial dyeing rate. The nature of the plasma gas governed the final uptake percentage of the used acid dyes according to the following descending order: nitrogen plasma > nitrogen/oxygen (50/50) plasma > oxygen plasma > argon plasma ≥ control. Prolonging the exposure time up to 20 minutes resulted in a gradual improvement in the extent of uptake [92].

Demir et al. (2008), were treated the knitted wool fabrics with atmospheric argon plasma, en-zyme (protease), chitosan and a combination of these processes. The treated fabrics were evaluated in terms of their dyeability, color fastness and shrinkage properties, as well as bursting strength. The surface morphology was characterized by SEM images. In order to show the changes in wool surface after plasma treatment, XPS analysis was done. According to the experimental results it was stated that atmospheric plasma has an etching effect and increases the functionality of a wool surface [93].

Chvalinova and Wiener (2008) investigated the effects of atmospheric pressure plasma treat-ment on dyeability of wool fabric. Untreated and plasma treated wool materials were dyed with acid dye in weak acid solution (pH 6). Experiments showed invasion of surface layer of cuticle by plasma and it was observed that the plasma treated wool fabric for 100 seconds, absorbed double more dye than untreated wool fabric [94].

Naebe et al. (2010) treated the wool fabric with atmospheric-pressure plasma with helium gas for 30 seconds. X-ray photoelectron spectroscopy and time-of-flight secondary ion mass spectrometry confirmed removal of the covalently-bound fatty acid layer (F-layer) from the surface of the wool fibers, resulting in exposure of the underlying, hydrophilic protein material. Dye uptake experiments were carried out at 50°C to evaluate the effects of plasma on the rate of dye uptake by the fiber surface, as well as give an indication of the adsorption characteristics in the early stages of a typical dyeing cycle. The dyes used were typical, sulfonated wool dyes with a range of hydrophobic characteristics, as determined by their partitioning behavior between water and n-butanol. No significant effects of plasma on the rate of dye adsorption were observed with relatively hydrophobic dyes. In contrast, the relatively hydrophilic dyes were adsorbed more rapidly (and uniformly) by the plasma-treated fabric [95].

Demir (2010) treated the mohair fibers by air and argon plasma for modifying their some properties such as hydrophilicity, grease content, fiber to fiber friction, shrinkage, dyeing, and color fastness. The results showed that the atmospheric plasma has an etching effect and increases the functionality of a fiber surface. The hydrophilicity, dyeability, fiber friction coefficient, and shrinkage properties of mohair fibers were improved by atmospheric plasma treatment [96].

Atav and Yurdakul (2011) investigated the use of plasma treatment for the modification of fiber surfaces to achieve dyeing of mohair fibers at lower temperatures without decreasing dye uptake. The study was carried out by using different gases under various powers and times. The effect was assessed in terms of color. Test samples were also evaluated using scanning electron microscopy (SEM). The optimum conditions of plasma treatment for improving mohair fiber dyeability, is treatments carried out by using Ar gas at 140 W for 60″. According to the experimental results it can be concluded that plasma treated mohair fibers can be dyed at lower temperatures (90°C) shorter times (1 h instead of 1.5 h) with reactive dyes without decreasing color yield. Dyeing kinetics was also searched in the study and it was demonstrated that the rate constant and the standard affinity of plasma treated sample was increased [97].

6. Gamma irradiation technology

Gamma radiation, also known as gamma rays or hyphenated as gamma-rays and denoted as γ, is electromagnetic radiation of high frequency and therefore energy. Gamma rays typically have frequencies above 10 exahertz (or >10^{19} Hz), and therefore have energies above 100 keV and wavelengths less than 10 picometers (less than the diameter of an atom) [98]. Gamma rays are identical in nature to other electromagnetic radiations such as light or microwaves but are of much higher energy. Examples of gamma emitters are cobalt-60, zinc-65, cesium-137, and radium-226. Like all forms of electromagnetic radiation, gamma rays have no mass or charge and interact less intensively with matter than ionizing particles [99].

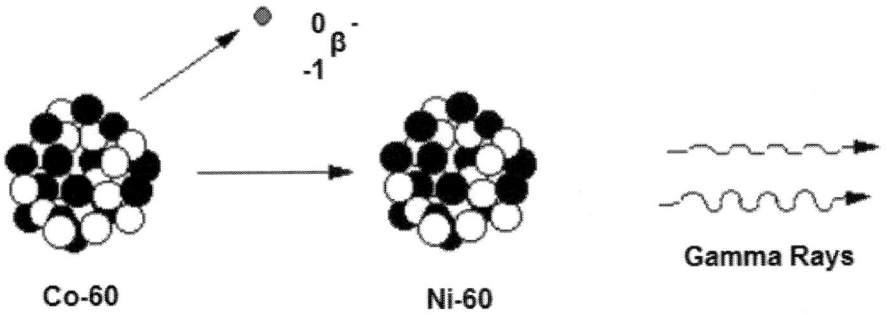

Figure 11. Gamma ray radiation [99]

Studies on the interaction of high energy radiation with polymers have attracted the attention of many researchers. This is due to the fact that high-energy radiation can induce both chain scission and/or crosslinking [100]. The efficiency of these two types of reactions depends mainly on polymer structure and irradiation atmosphere. However, dose rate, types of radiation source and temperature during irradiation can influence the reaction rates [101]. For some applications radiation degradation can be controlled and devoted to achieve a specific property [100]. Degradation, in broad terms, usually involves chemical modification of the polymer by its environment; modification that is often (but not always) detrimental to the performance of the polymeric material. Although the chemical change of a polymer is frequently destructive, for some applications degradation can be controlled and encouraged to achieve a specific property. In this regard, different vinyl monomers have been grafted onto gamma irradiated wool fabric to improve some favorable properties such as dyeability, and moisture regain. These studies have been based on the formation of stable peroxides on wool, upon irradiation, which are thermally decomposed to initiate polymerization [102].

Gamma rays are ionizing radiations that interact with the material by colliding with the electrons in the shells of atoms. They lose their energy slowly in material being able to travel through significant distances before stopping. The free radicals formed are extremely reactive, and they will combine with the material in their vicinity. The irradiated modified fabrics can allow: more dye or pigment to be fixed, producing deeper shades and more rapid fixation of dyes at low temperature [32]. In literature it is stated that two kinds of effects might occur in parallel in wool during the irradiation. The first effect as manifests as an evident decrease in dye accessibility at lower doses may not be altogether independent of crosslinking. On the other hand, the remarkable increase in the uptake at higher doses seems to be associated with strong structural damage of fibers. It is interesting to note that the increase in accessibility to dyes of the highly irradiated fibers is so great that the bilateral structure is hardly visualized by the partial staining. Thus the cross-sections of fibers irradiated with a dose of 10^8 roentgens are stained uniformly in dark tone even under the condition which does not give rise to the staining of unirradiated fibers [103]. In literature there are limited studies related to the effect of gamma irradiation treatment on dyeability of proteinous fibers. Some of them are summarized below.

Horio et al. (1962) investigated the effect of irradiation on dyeing property of wool fiber. It was found that the dyeing property of wool fiber was greatly affected by irradiation even at low doses where any changes in mechanical properties were not noticeable. The rate of dye absorption was strongly depressed by irradiation with Co-60 gamma radiation of low doses from 10^3 to 10^5 roentgens. Two dyestuffs, C.I. Acid Red 44 and C. I. Acid Green 28 were used. The dye absorption was strikingly suppressed at the range of doses from 10^3 to 10^5 roentgens, but fibers regain dye accessibility at higher doses [103].

Beevers and McLaren (1974) have been found that small doses of gamma radiation (0.5-10 Mrad) produce marked effects on some physical properties of wool. The results indicate that even small doses of gamma radiation break sufficient covalent bonds to make the cross-linked peptide chain structure more susceptible to the action of swelling and disordering agents. These small radiation-induced changes can be expected to affect properties of wool significantly in absorption and penetration processes, such as those involved in dyeing, chemical modification, and grafting treatments of wool [104].

Millington(2000) investigated the effects of γ-radiation (^{60}Co) on some chemical and physical properties of wool keratin and compared and contrasted with the effects of ultraviolet radiation in the UVC (200-280 nm) region. The effect of UVC doses up to 25 J/cm^2 on fabric strength was found to be small (5%), whereas γ-irradiated wool experienced strength reductions of 15% at doses over 100 kGy. Color changes following UVC and γ-irradiation were quite different: UVC wool was initially green changing to yellow under ambient conditions, γ-treated wool became pink-red at doses 25-250 kGy, and yellow at higher doses. The chromophores produced by UVC were easily removed by oxidative bleaching with hydrogen peroxide, whereas γ-treated wool remained yellow even at relatively low doses (25–50 kGy). This has implications for the use of γ-radiation as a means of sterilising wool for compliance with quarantine regulations. The effects of the two forms of radiation on the natural fluorescence of wool, permanent setting, printing properties and the epicuticle layer were also described [105].

7. Laser technology

In the last decade, considerable effort has been made in developing surface treatments such as UV irradiation, plasma, electron beam and ion beam to modify the properties of textile materials. Laser modification on material surface is one of the most studied technologies [106]. A laser is a device that emits light (electromagnetic radiation) through a process of optical amplification based on the stimulated emission of photons. The term "laser" originated as an acronym for "Light Amplification by Stimulated Emission of Radiation" [107]. Laser processing as a new processing method, with its processing of accurate, fast, easy, automatization, in leather, textile and garment industry increasingly widely used [108].

Laser technology has been widely used in surface modification of polymers [109]. Since the late 90s, different types of commercial lasers are available for surface modification of materials [110]. While most of the efforts in developing surface treatments have been made using UV laser, infra-red lasers, like CO_2 appear to be less concerning [109]. Adequate power levels for a specific application are very important in surface modification processes, because an excessive amount of energy can damage the polymers. Infra-red lasers like CO_2 are the most powerful lasers; with no suitable power level, severe thermal damage can be resulted. However, this short coming can be overcome by the use of pulsed-mode CO_2 lasers, which are easier to control than lasers operating in the continuous wave mode [110]. Excimer lasers, which are a form of ultraviolet lasers [111], are a special sort of gas laser powered by an electric discharge in which the lasing medium is an excimer, or more precisely an exciplex in existing designs. These are molecules which can only exist with one atom in an excited electronic state. Once the molecule transfers its excitation energy to a photon, therefore, its atoms are no longer bound to each other and the molecule disintegrates. This drastically reduces the population of the lower energy state thus greatly facilitating a population inversion. Excimers currently used are all noble gas compounds; noble gasses are chemically inert and can only form compounds while in an excited state. Excimer lasers typically operate at ultraviolet wavelengths [107].

It has been shown that materials like polymers, woods, metals, semiconductors, dielectrics and quartz modified by laser irradiation often exhibit physical and chemical changes in the material's surface [106]. Physical modifications occur in the form of a certain, regular surface structure of the irradiated sites. The high energy input of the excimer radiation into the polymer might also give rise to chemical changes of the surface [112]. In the case of polymers, some well-oriented structure of grooves or ripple structures with dimensions in the range of micrometer are developed on surface with irradiation fluence above the so-called ablation threshold. The laser irradiation of highly absorbing polymers can generate characteristic modifications of the surface morphology. The physical and chemical properties are also affected after laser irradiation. Hence, it is reasonable to believe that such surface modification of a polymer may have an important impact on its textile properties [106].

The possible applications of laser technology in the textile industry include removal of indigo dye of denim, heating threads, creating patterns on textiles to change their dyeability, producing surface roughness, welding, cutting textile webs. Laser irradiation on polymer surface is used to generate a modified surface morphology. The smooth surface of polymers is modified by this technique to a regular, roll-like structure that can cause adhesion of particles and coating, wetting properties and optical appearance [113].

Laser technology can also be used for improving dyeability since it is well known that the UV output from excimer lasers can modify the surface of synthetic fibers. But the use of laser energy in textile treatment is still uncommon and is generally limited with denim garment finishing [114]. Although there are many studies in the literature related to the investigation of the effect of laser treatment on dyeability of many fiber types, there is still no study carried out on wool. But, by taking into consideration that the chemical structure and dyeing property of proteinous fibers is similar to polyamide, in the light of the studies carried out on polyamide [114, 115], it can be said that dyeability of proteinous fibers with anionic dyes such as acid and reactive dyes may increase due to the decrease in the crystallinity and increase in free amino groups of the fiber.

8. Microwave technology

Microwaves (MW) which have broad frequency spectrum are electromagnetic waves that are used in radio, TV and radar technology [116]. MWs are radio waves with wavelengths ranging from as long as one meter to as short as one millimeter, or equivalently, with frequencies between 300 MHz (0.3 GHz) and 300 GHz [117].

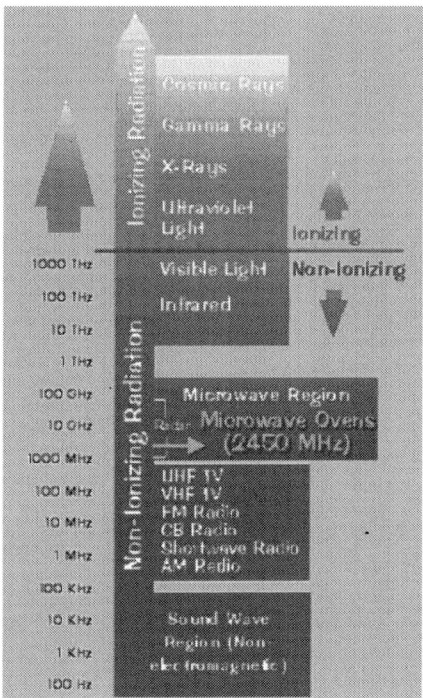

Figure 12. Frequency spectrum [118]

The term "microwave" denotes the techniques and concepts used as well as a range of frequencies. Microwaves travel in matter in the same manner as light waves: they are reflected by metals, absorbed by some dielectric materials and transmitted without significant losses through other materials. For example, water, carbon, foods with a high water, some organic solvents are good microwave absorbers whereas ceramics, quartz glass and most thermoplastic materials absorb microwaves slightly [119]. Electromagnetic waves can be absorbed and be left as energy units called photon. The energy carried by photon is depended on the wavelength and the frequency of radiation. Energy of MW photons is 0.125 kJ/mol. This value is very low considering the necessary energy for chemical bonds. Therefore MW rays can not affect the molecular structure of the material directly and change the electronic structures of atoms [116]. ·

Microwave-promoted organic reactions are known as environmentally benign methods that can accelerate a great number of chemical processes. In particular, the reaction time and energy input are supposed to be mostly reduced in the reactions that are run for a long time at high temperatures under conventional conditions [120]. Reactions conducted through microwaves are cleaner and more environmentally friendly than conventional heating methods [121].

The use of microwave radiation as a method of heating is over five decades old. Microwave technology originated in 1946, when Dr. Percy Le Baron Spencer, while conducting laboratory tests for a new vacuum tube called a magnetron, accidentally discovered that a candy bar in his pocket melted on exposure to microwave radiation. Dr. Spencer developed the idea further and established that microwaves could be used as a method of heating. Subsequently, he designed the first microwave oven for domestic use in 1947. Since then, the development of microwave radiation as a source of heating has been very gradual [121].

Microwave heating occurs on a molecular level as opposed to relying on convection currents and thermal conductivity when using conventional heating methods. This offers an explanation as to why microwave reactions are so much faster [122]. The fundamental mechanism of microwave heating involves agitation of polar molecules or ions that oscillate under the effect of an oscillating electric or magnetic field. In the presence of an oscillating field, particles try to orient themselves or be in phase with the field. However, the motion of these particles is restricted by resisting forces (inter-particle interaction and electric resistance), which restrict the motion of particles and generate random motion, producing heat [121]. Microwave (MW) heat systems consists of three main units; magnetron, waveguide and applicator. Magnetron is used as a microwave energy source in industrial and domestic type of microwave ovens. One of the oscillator tube, magnetron consists of two main parts as anode - cathode, and it converts the continuous current - electrical energy to MW energy. Circulator transmits approximately all of the waves that are sent from magnetron and shunts transmitted waves to water burden. Thus magnetron is protected. Electromagnetic waves are transmitting to the applicator by waveguides. Applicators are parts of the matter MW applied on. MW energy produced in generator is affected directly on the material in applicators in MW heating systems. Type of applicators used in practice can be divided into three groups as multi-mode (using 80% of the industrial systems), single-mode and near field MW applicators [116].

Since the beginning of the twentieth century MW technology has made significant contributions to scientific and technological developments. Also due to its initial intend to be used in telecommunications, very important progresses have been made in this area. Nevertheless from the second half of twentieth century, MW energy is finding increased number of application area in other industrial processes and these applications are surpassing telecommunication applications [116].

In textile processing it is necessary to apply heat as in dye fixation, heat setting or drying the product. Heat can be transferred to the material by radiation, conduction and convection. These three ways of transferring can be used either separately or in combination. The saving of time and energy is of immediate interest to the textile industry. The introduction of new techniques which will allow less energy to be used: is a highly important area of activity to consider. The textile industry has investigated many uses for microwave energy such as heating, drying, dye fixation, printing and curing of resin finished fabrics. In 1966, Ciba-Geigy obtained one of the earliest patents for using microwave heating in dyeing and printing fibrous material with reactive dyes. Since then many authors have investigated the feasibility of using microwaves for a variety of dyeing and finishing processes [123]. Although many studies have focused on investigating the feasibility of using microwaves to dye polyester fibers with disperse dyes, researches related to the use of microwave heating in dyeing of proteinous fibers are very limited.

Delaney and Seltzer (1972) used microwave heating for fixation of pad-dyeings on wool and they demonstrated the feasibility of applying certain reactive dyes to wool in fixation times of 30-60 s [124].

Zhao and He (2011) treated the wool fabric with microwave irradiation at different conditions and then studied for its physical and chemical properties using a variety of techniques, such as Fourier transform infrared spectroscopy, X-ray diffraction, and scanning electron. It was found that microwave irradiation of wool fabric significantly improved its dyeability. It was stated that this could be due to the change in wool surface morphological structure under microwave irradiation which implied that the barrier effect in wool dyeing was diminished. Although the breaking strength of the treated wool fabrics also improved with microwave irradiation, the chemical structure and crystallinity did not show any significant change [125].

In literature it was stated that dyeing time of mohair could be drastically reduced from the conventional 90 minutes to 35 minutes using the radio frequency technique, only 5 minutes of the 35 minutes representing actual exposure to the radio frequency field. An estimated saving of some 80% in dyeing energy costs could be achieved. Furthermore, dye fixation improved slightly from 93% to 96%. *Turpie* quoted unpublished data in which radio-frequency dyeing of tops produced better luster and enabled higher maximum spinning speeds than did conventional dyeing of mohair. *Smith,* claimed advantages of radio-frequency dyeing and bleaching, - for example, reduced chemical damage because of a shorter exposure to 100°C; and reduced energy, water and effluent costs [126].

9. E-beam irradiation technology

Radiation processing has been increasingly applied in industries to improve the quality of products, efficiency, energy saving and to manufacture products with unique properties

[127]. An irradiation can induce chemical reactions at any temperature in the solid, liquid and gas phase without any catalyst; it is a safe method that could protect the environment against pollution; radiation process could reduce curing time and energy saving; it could treat a large and thick three-dimensional fabrics that need not consider of the shape of the samples [128].

Physical techniques for activating fiber molecules in the absence of solvent for producing functional textiles are becoming increasingly attractive also from an ecological viewpoint. Among them, electron beam processing is particularly interesting as it offers the possibility to treat the materials without solvent, at normal temperature and pressure [129]. Whilst the energy of the electrons in gas discharge plasmas is typically in the range of 1-10 electron volt (eV), electron-beam (E-beam) accelerators generate electrons with a much higher energy, generally 300 keV to 12 MeV. These electrons may be used to modify polymer materials through direct electron-to-electron interactions. These interactions can create active species such as radicals, so there are different possible outcomes from the electron-beam irradiation of polymer materials, on the basis of the chosen operating conditions [130].

The formation of active sites on the polymer backbone can be carried out by several methods such as plasma treatment, ultraviolet (UV) light radiation, decomposition of chemical initiator and high-energy radiation [131]. At present, the most common radiation types in industrial use are gamma and e-beam. E-beam machines play a significant role in the processing of polymeric materials; a number of different machine designs and different energies are available. Industrial e-beam accelerators with energies in the 150-300 keV range are in use in applications where low penetration is needed, such as curing of surface coatings. Accelerators operating in the 1.5 MeV range are used where more penetration is needed. E-beam machines have high-dose rate and therefore short processing times. While they have limited penetration compared with gamma, they conversely have good utilization of energy because it can all be absorbed by the sample irradiated [132].

New radiation processing applications involving ion-beam treatment of polymers offer exciting prospects for future commercialization, and are under active investigation in many research laboratories. Much of the work has involved the irradiation of non-polar polyolefins (PE, PS, PP, etc.) as a means of inducing polar groups at surfaces, in order to enhance such properties as printability, wetability and exc. [132]. Studies related to the e-beam irradiation of proteinous fibers are limited.

Fatarella et al.(2010) wanted to clarify whether fiber surface treatments such as plasma and e-beam could affect the effectiveness of a TGase treatment by improving the accessibility of target amino acids in wool to the enzyme or improving the penetration of the enzyme into the wool fiber cortex, thus accessing more sites for reaction. Plasma treatments with different non-polymerizing gases (oxygen, air and nitrogen) and e-beam irradiation in air or nitrogen atmosphere were assessed as possible pretreatments to non-proteolytic enzymatic processes (such as TGase) to improve the accessibility of target groups in the wool proteins to the enzymes. Only limited inhibition and/or inactivation of the transglutaminase enzyme was found after treatment with plasma and e-beam, suggesting such treatments could be used as a preparative treatment prior to the application of the enzyme. In contrast, by increasing the energy of the electrons in e-beam treatments no significant superficial modifications were observed. In fact, they promoted the cleavage of high-energy bond, such as S-S linkage, by enhancing depolymerization reaction [130].

10. Ion implantation technology

It has been approximately 50 years since researchers first began exposing polymeric materials to ionizing radiation, and reporting the occurrence of cross-linking and other useful effects. Innovation in this field has by no means ended; important new products made possible through radiation technology continue to enter the marketplace, and exciting new innovations in the application of radiation to macromolecular materials are under exploration at research institutions around the world [132]. Ionizing radiation may modify physical, chemical and biological properties of materials [133]. Some of the surface characteristics being successfully manipulated by radiation grafting include: chemical resistance, wetability, biocompatibility, dyeability of fabrics, and antistatic properties [132].

Ion implantation is an innovative production technique with which the surface properties of inert materials can be changed easily. It shows distinct advantages because it is environmentally friendly. Ion implantation can be used to induce both surface modifications and bulk property enhancements of textile materials, resulting in improvements to textile products ranging from conventional fabrics to advanced composites. Ion implantation was first done by Rutherford in 1906, when he bombarded aluminum foil with helium ions. Ion implantation has been applied to metals, ceramics, plastics, and polymers [134].

Even though ion implantation is relatively complex in terms of the equipment required, it is a relatively simple process. Ion implantation consists of basically two steps: form plasma of the desired material, and either extract the positive ions from the plasma and accelerate them toward the target, or find a means of making the surface to be implanted the negative electrode of a high voltage system [135]. There are three methods commonly used for ion implantation. They differ in the way in which they either form the plasma or make the surface to be implanted the negative electrode. These methods are mass-analyzed ion implantation, direct ion implantation and plasma source ion implantation [134].

In *mass-analyzed ion implantation*, the plasma that is formed in the ion source is not pure; it contains materials that one does not wish to implant. Thus, these contaminants must be separated from the plasma. To perform this separation, the plasma source is placed at a high voltage and the part to be implanted is placed at ground. This produces a situation where the target is at a negative potential with respect to the plasma source. A negative electrode then extracts the ions from the source. The ions are then accelerated by a high voltage source to the target. Between the ion source and the target there is a large magnet, with magnetic field perpendicular to the direction of ion motion. Ions passing through this magnetic field are bent by the magnetic field. The amount of bending depends on the ion material being implanted and the strength of the magnet [135].

Direct ion implantation eliminates the need for the current limiting magnet found in mass-analyzed ion implantation by using an ion source that produces a plasma and ion beam of just the desired material. In direct ion implantation, the plasma is formed in the ion source and the ions extracted at high energies in a wide beam, passing through a valve directly into the end station, where the ion implant parts within the target area. In such a case, the beam current density can be high (10-50 mA), costs are greatly reduced, and relatively high throughput processing is possible [135].

It is difficult in practice to treat uniformly three-dimensional objects such as sockets in artificial hip joints, without using sophisticated manipulating devices. Recently, however, an innovative hybrid technology, *plasma source ion implantation* has been developed, which can, to a large extent, address the problems with conventional ion implantation of components with complex shapes [136]. In plasma source ion implantation (sometimes referred to as plasma ion immersion), the plasma source floods the chamber of the end station with plasma. Ions are extracted from the plasma and directed to the surface of the part being ion implanted by biasing the part to very high negative voltages using a pulsed, negative high voltage power supply [135].

One of the most recent ion implantation techniques is *metal vapor vacuum arc (MEVVA)*, a type of direct ion implantation. In the mid-1980s, Brown et al. at Lawrence Berkeley Laboratory developed a new type of metal ion source, namely the MEVVA ion source. The MEVVA source makes use of the principle of vacuum arc discharge between the cathode and the anode to create dense plasma from which an intense beam of metal ions of the cathode material is extracted. This metal ion source operates in a pulse mode [134].

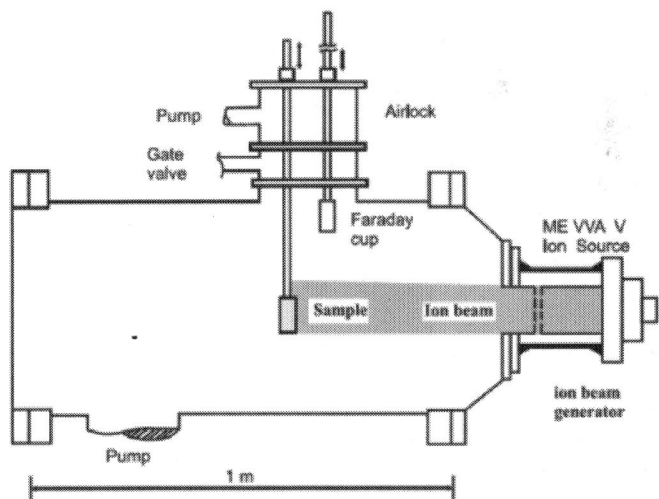

Figure 13. MEVVA ion implanter [134]

Many techniques have been applied to produce the great potential of ion beam modification technology. It desired surface modifications, ranging from conventional flame treatments, "wet" chemical treatments, and electrical treatments (such as corona discharge), to modern plasma treatments and particle beam irradiation (electrons, ions, neutrons and photons) techniques. Among them, particle beam techniques are particularly attractive owing to their flexibility, effectiveness, and environmentally friendly nature compared with conventional techniques. Also, in the domain of particle beam techniques, the ion beam has proven more effective in modifying polymer surfaces than UV-light, c-ray, X-ray and electron beams. This is because energetic ions have a higher cross-section for ionization and larger linear energy transfer (LET, eV nm^{-1}) than these conventional radiation types of comparable energy owing to their deeper range [136].

Numerous papers have appeared in recent years describing surface treatment using ions. Surface modification with ions typically involves fluencies of ~ 10^9 to 10^{14} ions/cm^2 or in some cases, 10^{15} ions/cm^2; higher fluencies may result in destruction of polymer, generally through carbonization. Many different ions have been employed for irradiating polymers, ranging from hydrogen and helium ions up to ions of heavy elements such as gold or uranium [132].

11. Supercritical carbondioxide technology

The enhanced solvent properties provided by supercritical fluids (SCFs) are by no means new. In fact, the various phenomena attributed to SCFs have been known for hundreds of years [137]. However, textile dyeing, using supercritical carbon dioxide as dye solvent, has been developed for the last two decades since it potentially overcomes all the economical and ecological disadvantages derived from the conventional water dyeing process [138]. Supercritical fluids are characterized by a very high solute diffusivity and low viscosity. Consequently, in supercritical fluids all the transport processes are much more rapid. Dyeing in supercritical carbondioxide, for example, entirely avoids the use of water, and consequently there is no possibility of pollution. No auxiliary agents are used and residual dye can be recovered in a reusable form [139].

A supercritical fluid can be defined as a substance above its critical temperature and pressure. Under this condition the fluid has unique properties, in that it does not condense or evaporate to form a liquid or a gas [140]. Supercritical fluids exhibit gas-like viscosities and diffusivities and liquid-like densities [120]. Although a number of substances are useful as supercritical fluids, carbon dioxide has been the most widely used [140], as the critical temperature and pressure are easier to achieve than that of other substances [141]. Supercritical CO_2 is a dyeing medium which is a potential alternative to water as it is inherently nontoxic, inexpensive, and nonflammable, it can be recycled, it has easily accessible critical conditions [142]. The critical point for carbon dioxide occurs at a pressure of 73.8 bar and a temperature of 31.1°C. ScCO$_2$ represent a potentially unique media for either transporting chemical into or out of a polymeric substrate, because of their thermo-physical and transport properties.

The phase diagram of carbon dioxide shown in Fig. 14 represents the interfaces between phases; at the triple point all three phases may coexist. Above the triple point, an increase in temperature drives liquid into the vapor phase, while an increase in pressure drives vapor back to liquid [120]. Above the critical point of carbon dioxide, it retains the free mobility of the gaseous state, but with the increasing pressure its density will increase towards that of a liquid. Solvating power is proportional to density, whilst viscosity remains comparable with that of a normal gas, so the "fluid" has remarkable penetration properties [143].

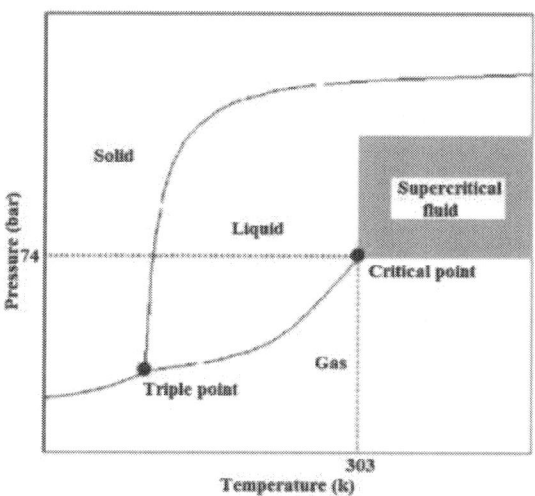

Figure 14. Phase diagram of carbon dioxide [120]

In the dyeing field, Schollmeyer and co-workers, among others, have demonstrated that both synthetic and natural fibres can be dyed with disperse and reactive disperse dyes in supercritical carbondioxide. However, the dyeing of natural fibres with water soluble dyes in supercritical carbondioxide has not yet been successful, since dyes such as reactive, acid and basic dyes have little solubility in this medium due to their high polarity [144]. One approach to this problem was undertaken by Gebert et al. who examined wool and cotton fibers after attempting to open the fiber surface with a swelling auxiliary so that dye molecules could be readily trapped in the fiber [145].

In natural textiles, the dye molecules can be fixed by either physical (e.g. Van der Waals) or chemical (e.g. covalent) bonds. Since the dyes used in a ScCO2-dyeing process are non-polar and natural fibres are polar, the affinity between dyes and textiles is low so physical bonds are weak. Therefore, a dyeing process must be developed for dyeing natural textiles in ScCO2 with reactive dyes that create covalent dye-textile bonds. So far, several reactive dyes known from conventional dyeing in water have been investigated in ScCO2:

• vinylsulphone dyes have been successfully used for silk and wool,

- 2-bromoacrylamide dyes have been successful in dyeing wool and cotton,

- dichlorotriazine dyes have been tested on silk and cotton but showed insufficient fixation [146].

Guzel and Akgerman (2000) dyed the wool fibers with three mordant dyes dissolved in supercritical carbon dioxide. Wool fibers were mordanted with five different metal ions (Cr(III), Al(III), Fe(II), Cu(II) and Sn(II)) using conventional techniques and dyed at 333-353°K temperature and at 150-230 atm pressure. According to the experimental results it was found that dyed materials had excellent wash fastness properties [147].

For water-soluble dyes, attempts were made to dye natural fibers using reverse micelle technique (Fig. 15) in which ionic dye, solubilized in the water-pool, passes into the fiber together with a small amount of water immediately after contact with it. Satisfactory results were obtained for proteinous fibers [120].

Sawada et al. (2002) has developed a reverse micellar system in supercritical carbon dioxide as a dyeing medium. Water-soluble dyes such as reactive dyes and acid dyes could be sufficiently solubilised in the interior of a specially constituted reverse micelle. Protein fabrics, silk and wool, were satisfactorily dyed even in deep shades with conventional acid dyes without any special pretreatment. Compared to previously proposed supercritical dyeing methods, dyeing of fabrics with this system could be performed at low temperatures and pressures in a short time [148].

Jun et al. (2004) investigated the dyeing of wool fabrics with conventional acid dyes in a supercritical CO_2 using a reverse micellar system. A reverse micelle composed of perfluoro 2,5,8,11-tetramethyl-3,6,9,12-tetraoxapentadecanoic acid ammonium salt/CO_2/water had a high potential to solubilize conventional acid dyes and to dye wool fabrics in this system. It was found that dyeability of the acid dye on wool in this system take no influence of the density of CO_2. On the other hand, variation of dyeing temperature resulted in the remarkable differences of the dyeability of the acid dye on wool even though the solubility of dye in the system was not varied by the variation of temperature [149]

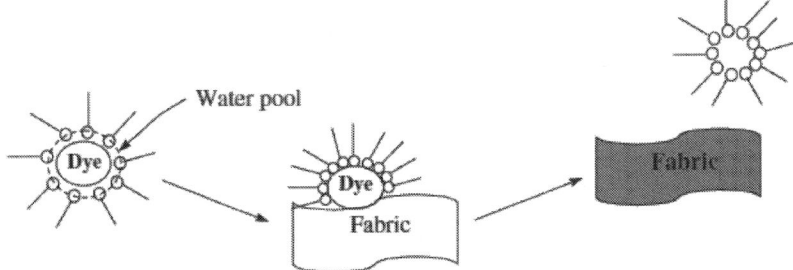

Figure 15. Dyeing of natural fibers using reverse micelle [120]

Jun et al. (2005) investigated the phase behavior of a cationic perfluoropolyether surfactant in supercritical carbon dioxide and its ability to solubilise ionic dyes. This cationic surfactant was found to dissolve satisfactorily in supercritical carbon dioxide and was able to form micellar aggregates incorporating a small amount of water in their interior. Conventional anionic dyes such as acid and reactive dyes were solubilised satisfactorily in the cationic surfactant/water/supercritical carbon dioxide system. The surfactant was more effective in this respect than its anionic analogue [144].

Schmidt et al. (2007) dyed various fibers, among which wool and silk also exist, with C.I. Disperse Yellow 23 modified with 2-bromoacrylic acid and 1,3,5-trichloro-2,4,6-triazine as reactive groups in supercritical carbon dioxide. It was found that on wool and silk, the color depth was higher than on cotton and the wash, rub and light fastness of all dyeings was between 4 and 5 [150].

12. Conclusions

Textile dyeing is the most remarkable process among the wet treatments in textile industry in terms of energy and water consumption and effluent load. In last two decades increased laws related to the environment and competitive market conditions required some new processes to be found in textile dyeing field. This situation increased the interest of the usage of new technologies such as ultrasound, ultraviolet, ozone, plasma, gamma irradiation, laser, microwave, e-beam irradiation, ion implantation, and supercritical carbondioxide in textile industry. These new technologies provide not only decrease in time, energy, and chemical consumption, but also decrease in effluent load. So that all of these new technologies considered to be very interesting future oriented processes because of being environmentally friendly. Although it was proven with many researches that most of these technologies are successful at laboratory scale, there is still need to integrate them into industrial applications. There is no doubt that in future these new technologies will find wide range of applications when their disadvantages (to be expensive, not possible to be used for all fiber types and exc.) will be eliminated.

Author details

Rıza Atav

Namık Kemal University, Textile Engineering Department, Turkey

References

[1] Lewis, D.M., 1992, Wool Dyeing, Society of Dyers and Colorists

[2] Jocic, D., Vilchez, S., Topalovic, T., Molina, R., Navarro, A., Jovancic, P., Julia, M.R. and Erra, P., 2005, Effect of Low-Temperature Plasma and Chitosan Treatment on Wool Dyeing with Acid Red 27, Journal of Applied Polymer Science, 97 (6), 2204–2214p.

[3] Molina, R., Erra, P., Julia, L., and Bertran, E., 2003, Free Radical Formation in Wool Fibers Treated by Low Temperature Plasma, Textile Research Journal, 73, 955-959p.

[4] Onar, N. and Sarışık M., January-March 2005, Use of Enzymes and Chitosan Biopolymer in Wool Dyeing, Fibers & Textiles in Eastern Europe, 13 (1), 54-59p.

[5] Cardamone, J. M. and Damert, W.C., 2006, Low-Temperature Dyeing of Wool Processed for Shrinkage Control, Textile Research Journal, 76 (1), 78–85p.

[6] Kongkachuichay, P., Shitangkoon, A. and Chinwongamorn, N., 2002, Studies on Dyeing of Silk Yarn with Lac Dye: Effects of Mordants and Dyeing Conditions, Science Asia, 28, 161-166p.

[7] Luo, J. and Cao, J., 1997, Low-Temperature Dyeing of Real Silk Fabrics with Liquid Sulphur Dyes, JSDC, 113, 67-69p.

[8] Leksophee, T., Supansomboon, S. and Sombatsompop, N., 2004, Effects of Crosslinking Agents, Dyeing Temperature, and pH on Mechanical Performance and Whiteness of Silk Fabric, Journal of Applied Polymer Science, 91, 1000-1007p.

[9] Tsukada, M., Katoh, H., Wilson, D., Shin, B., Arai, T., Murakami, R. and Freddi, G., 2002, Production of Antimicrobially Active Silk Proteins by Use of Metal-Containing Dyestuffs, Journal of Applied Polymer Science, 86, 1181-1188p.

[10] Shao, J., Liua, J. and Carr, C.M., 2001, Investigation into The Synergistic Effect Between UV/Ozone Exposure and Peroxide Pad-Batch Bleaching on The Printability of Wool, Coloration Technology, 117 (5), 270-275p.

[11] Kamel, M.M., El-Shistawy, R.M., Hana, H.L. and Ahmed, N.S.E., 2003, Ultrasonic-Assisted Dyeing: I. Nylon Dyeability with Reactive Dyes, Polymer International, 52 (3), 373-380p.

[12] Akalın, M., Merdan, N., Kocak, D. and Usta, I, 2004, Effects of Ultrasonic Energy on The Wash Fastness of Reactive Dyes, Ultrasonics, 42, 161-164p.

[13] Vajnhandl, S. and Le Marechal, A.M., 2005, Ultrasound in Textile Dyeing and The Decoloration/Mineralization of Textile Dyes, Dyes and Pigments, 65 (2), 89-101p.

[14] Boz, S., 2008, Ultrasonik Enerjinin Konfeksiyon Sanayiinde Kullanımının İncelenmesi, Master Thesis Ege University Textile Engineering Department

[15] Moholkar, V.S., Nierstrasz, V.A. and Warmoeskerken, M.M.C.G., 2003, Intensification of Mass Transfer in Wet Textile Processes by Power Ultrasound, AUTEX Research Journal, 3 (3), 129-138p.

[16] Jung, C., Budesa, B., Fässler, F., Uehlinger, R., Müller, T. and Wild, M., Effect of Cavitation in Ultrasound Assisted Cleaning of Medical Devices, http://www.kks-ultraschall.ch/assets/files/news/effect-of-caviation-ultrasound.pdf

[17] Kamel, M.M., El-Shishtawy, R.M., Yussef, B.M. and Mashaly, H., 2005, Ultrasonic assisted dyeing III. Dyeing of wool with lac as a natural dye, Dyes and Pigments, 65 (2), 103-110p.

[18] Moholkar, V.S., Rekveld, S. and Warmoeskerken, M.M.C.G., 2000, Modeling of The Acoustic Pressure Fields and The Distribution of The Cavitation Phenomena in A Dual Frequency Sonic Processor, Ultrasonics, 38, 666-670p.

[19] Santos, H.M., Lodeiro, C., Luis, J.L. and Martinez, C., The Power of Ultrasound, http://www.wiley-vch.de/books/sample/3527319344_c01.pdf

[20] Vouters, M., Rumeau, P., Tierce, P. and Costes, S., 2004, Ultrasounds: An Industrial Solution to Optimize Costs, Environmental Requests and Quality for Textile Finishing, Ultrasonics Sonochemistry, 11, 33-38p.

[21] http://www.ntcresearch.org/pdf-rpts/anrp98/c95-g13.pdf, 2008

[22] Ferrero, F. and Periolatto, M., 2012, Ultrasound for Low Temperature Dyeing of Wool with Acid Dye, Ultrasonics Sonochemistry, 19, 601-606

[23] hukla, S.R. and Mathur, M.R., 1995, Low-Temperature Ultrasonic Dyeing of Silk, Journal of the Society of Dyers and Colorists, 111, 342-5p.

[24] http://www.fiber2fashion.com, 2008

[25] Battu, A., Giansetti, M., Rovero, G. and Sicardi, S., 2010, Intensification of Wet Textile Processing By Ultrasound Application, Proceedings of the 22nd IFATCC International Congress, Stresa, Italy

[26] Yukseloğlu, S.M. and Bolat, N., 2010, The Use of Conventional and Ultrasonic Energy in Dyeing of 100% Wool Woven Fabrics, Tekstil ve Konfeksiyon, 2, 162-167p.

[27] McNeil, S.J. and McCall, R.A., 2011, Ultrasound for Wool Dyeing and Finishing, Ultrasonics Sonochemistry, 18, 401-406

[28] Atav, R., Yurdakul, A., 8-10 June-2011, The Use of Ultrasound in Dyeing of Mohair Fibers, 11th World Textile Conference AUTEX, Moulhouse-France, Book of Abstracts, 277-280p.

[29] Srikanth, B., The Basic Benefits of Ultraviolet Technology, http://aquvair.com/Basic%20Benefits%20of%20UV%20-WCP%20Reprint.pdf

[30] http://en.wikipedia.org/wiki/Electromagnetic_spectrum, 2012

[31] http://skincareclub.wordpress.com/2011/02/25/uva-uvb-uvc-rays/, 2012

[32] Bhatti, I.A., Adeel, S. and Abbas, M., Effect of Radiation on Textile Dyeing, http://cdn.intechweb.org/pdfs/25003.pdf

[33]   http://www.epa.gov/sunwise/doc/uvradiation.html, 2012

[34]   http://ec.europa.eu/health/opinions2/glossary/tuv/uv-radiation.htm, 2012

[35]   http://www.google.com.tr/imgres?q=ultraviolet+spec-
       trum&hl=tr&biw=1249&bih=599&tbm=isch&tbnid=lxj5wLLo6i43HM:&imgre-
       furl=http://www.uvlp.ca/&docid=OAkRjqeFw_ZM9M&imgurl=http://www.uvlp.ca/
       images/uvspectrum.jpg&w=450&h=162&ei=3vr7T9XoK8jatAaww8m_Aw&zoom=1,
       2012

[36]   Treigiene, R. and Musnickas, J. R., 2008, Influence of UV Exposure on Properties of
       Wool Fiber Pretreated with Surfactants Solutions, Materials Science, 14 (1), 75-78p.

[37]   Millington, K., 1998, Using Ultraviolet Radiation to Reduce Pilling of Knitted Wool
       and Cotton, Textile Research Journal, 68 (6), 413-421p.

[38]   Millington, K. R., CSIRO Textile and Fiber Technology, UV Technology Applications
       in The Textile Industry, http://www.fiber2fashion.com/industry-article/pdffiles/uv-
       technology-applications-in-the-textile-industry.pdf?PDFPTO-
       KEN=73534d84ba5e495e4164c5a831509f62275d95dc%7C1225281847

[39]   El-Sayed, H. and El-Khatib, E., 2005, Modification of wool fabric using ecologically
       acceptable UV-assisted treatments, Journal of Chemical Technology and Biotechnolo-
       gy, 80 (10), 1111–1117p.

[40]   Millington, K. R., 1998, The Use of Ultraviolet Radiation in an Adsorbable Organo-
       halogen-Free Print Preparation for Wool and in Wool Dyeing: the Siroflash Process
       Journal of the Society of Dyers and Colorists, 114 (10), 286-292p.

[41]   Millington, K. R., 1998, U.S.A. Patent, Application of UV surface treated in textile in-
       dustry, Red Tech 98 North Americas: UNEB Conference Proceeding Chicago ILLI-
       NOIS America

[42]   Xin, J.H., Zhu, R., Hua, J. and Shen, J., 2002, Surface Modification and Low Tempera-
       ture Dyeing Properties of Wool Treated By UV Radiation, Coloration Technology,
       118 (4), 169-173p.

[43]   http://www.sfu.ca/~vwchu/projects/chemistryozone.pdf, 2012

[44]   Polat, D., 2010, Catalytic Ozonation of Industrial Textile Wastewaters in A Three
       Phase Fluidized Bed Reactor, Master Thesis, The Graduate School of Natural Scien-
       ces of Middle East Technical University

[45]   http://water.epa.gov/lawsregs/rulesregs/sdwa/mdbp/upload/
       2001_01_12_mdbp_alter_chapt_3.pdf, 2012

[46]   Parmenter, K., 2004, Global Ozone Handbook Agriculture and Food Industries http://
       www.chimia.it/pdf/Ozone%20Handbook%20-%20Revised%2003-11-05%20Final.pdf

[47]   Balousek, P.J., 1979, The Effects of Ozone Upon A Lignin-Related Model Compound
       Containing A β-Aryl Ether Linkage, PhD Thesis, Lawrence University

[48] http://en.wikipedia.org/wiki/File:Ozone-1,3-dipole.png, 2012

[49] http://astrobioloblog.wordpress.com/2011/08/18/life-in-our-solar-system-%E2%80%93-earth/, 2012

[50] Iglesias, S.C., 2002, Degradation and Biodegradability Enhancement of Nitrobenzene and 2,4-Dichlorophenol By Means of Advanced Oxidation Processes Based on Ozone, Universitat de Barcelona, PhD thesis

[51] Prabaharan, M. and Rao, J.V., 2001, Study on Ozone Bleaching of Cotton Fabric-Process Optimisation, Dyeing and Finishing Properties, Coloration Technolgy, 117 (2), 98-103p.

[52] Song, S., Yao, J., He, Z., Qiu, J. and Chen, J., 2008, Effect of Operational Parameters on The Decolorization of C.I. Reactive Blue 19 in Aqueous Solution By Ozone-Enhanced Electrocoagulation, Journal of Hazardous Materials, 152, 204-210p.

[53] El-Zaher, N.A. and Micheal, M.N., 2002, Time Optimization of Ultraviolet-Ozone Pretreatment for Improving Wool Fabrics Properties, Journal of Applied Polymer Science, 85(7), 1469-1476p.

[54] Atav, R. and Yurdakul, A., 2011, Effect of Ozonation Process on the Dyeability of Mohair Fibers, Coloration Technology, 127(3), 159-166

[55] Wu, C.H. and Ng, H.Y., 2008, Degradation of C.I. Reactive Red 2 (RR2) Using Ozone-Based Systems: Comparisons of Decolorization Efficiency and Power Consumption, Journal of Hazardous Materials, 152, 120-127p.

[56] Eren, H.A. and Aniş, P., 2006, Decolorisation of Textile Effluents by Ozonation, Uludağ University Journal of the Faculty of Engineering and Architecture, 11(1), 83-91p.

[57] Perincek, S., Bahtiyari, M.I., Körlü, A.E. and Duran, K., 2008, Ozone Treatment of Angora Rabbit Fiber, Journal of Cleaner Production, 16, 1900-1906p.

[58] Sargunamani, D. and Selvakumar, N., 2007, Effects of Ozone Treatment on The Properties of Raw And Degummed Tassar Silk Fabrics, Journal of Applied Polymer Science, 104 (1), 147–155p.

[59] Dereli Perincek, S., 2006, An investigation on The Applicability of Ultrasound, Ultraviolet, Ozone and Combination of These Technologies as A Pretreatment Process, Master Thesis, Ege Üniversitesi Fen Bilimleri Enstitüsü

[60] Micheal, M.N. and El-Zaher, N.A., 2003, Efficiency of Ultraviolet/Ozone Treatments in The Improvement of The Dyeability and Light Fastness Of Wool, Journal of Applied Polymer Science, 90 (13), 3668-3675p.

[61] Atav, R. and Yurdakul, A., 2010, Low Temperature Dyeing of Ozonated Angora Fibers, 10th World Textile Conference AUTEX, Vilnius-Litvania, Book of Abstracts, 16p.

[62] Sparavigna, A., Plasma Treatment Advantages for Textiles, http://arxiv.org/ftp/arxiv/papers/0801/0801.3727.pdf

[63] Moore, R., Plasma Surface Functionalization of Textiles, http://www.acteco.org/Acteco/public/results/MOORE_IoN_March08.pdf

[64] http://www.britannica.com/EBchecked/media/148660/States-of-matter, 2012

[65] http://www.fiber2fashion.com/industry-article/17/1636/plasma-treatment-technology-for-textile-industry1.asp, 2012

[66] http://www-alt.igb.fraunhofer.de/www/gf/grenzflmem/schichten/en/TechTextile.en.html, 2012

[67] Shyam Sundar, P., Prabhu, K.H. and Karthikeyan, N.,2007, Fourth State Treatment for Textiles, The Indian Textile Journal, http://www.indiantextilejournal.com/articles/FAdetails.asp?id=290

[68] Desai, A.A., 2008, Plasma Technology: A Review, The Indian Textile Journal http://www.indiantextilejournal.com/articles/FAdetails.asp?id=775

[69] Buyle, G., 2009, Nanoscale Finishing of Textiles Via Plasma Treatment, Materials Technology,   24(1),   46-51p.,   http://www.centexbel.be/files/publication-pdf/MTE_Buyle.pdf

[70] Karahan, H.A., Özdoğan, E., Demir, A., Ayhan, H. and Seventekin, N., 2009, Effects of Atmospheric Pressure Plasma Treatments on Certain Properties of Cotton Fabrics, Fibers & Textiles in Eastern Europe, 17 (2), 19-22p.

[71] Karahan, A., Yaman, N., Demir, A., Ozdoğan, E., Oktem, T. and Seventekin, S., 2006, Tekstilde Plazma Teknolojisinin Kullanım Olanakları: Plazma Nedir?, Tekstil ve Konfeksiyon, 16(1),302-309p.

[72] Jocic, D., Vilchez, S., Topalovic, T., Molina, R., Navarro, A., Jovancic, P., Julia, M.R. and Erra, P., 2005, Effect of Low-Temperature Plasma and Chitosan Treatment on Wool Dyeing with Acid Red 27, Journal of Applied Polymer Science, 97(6), 2204-2214p.

[73] http://www.tno.nl/downloads/def_maritiem_plasmaapplicaties_DV2_05d046.pdf, 2012

[74] Ahmed, A., 2007, A Review of Plasma Treatment for Application on Textile Substrate, BSc. Thesis, The Faculty of Textile Science

[75] Erra, P., Molina, R., Jocic, D., Julia, M.R., Cuesta, A., and Tascon, J.M.D., 1999, Shrinkage Properties of Wool Treated with Low Temperature Plasma and Chitosan Biopolymer, Textile Research Journal, 69(11), 811-815p.

[76] Höcker, H., 2002, Plasma Treatment of Textile Fibers, Pure Appl. Chem., 74(3), 423-427p.

[77] Wakida, T., Tokino, S., Niu, S., Lee, M., Uchiyama, H. and Kaneko, M., 1993, Dyeing Properties of Wool Treated with Low-Temperature Plasma Under Atmospheric Pressure, Textile Research Journal, 63(8), 438-442p.

[78] Yoon, N.M., Lim, Y.J., Tahara, M. and Takagishi, T., 1996, Mechanical and Dyeing Properties of Wool and Cotton Fabrics Treated with Low Temperature Plasma and Enzymes, Textile Research Journal, 66(5), 329-336p.

[79] Jing, Z., 1996, Silk Fabrics Modification Through Plasma Graft Copolymerization, Journal of Textile Research, 4

[80] Wakida, T., Lee, M., Sato, Y., Ogasawara, S., Ge, Y. and Niu, S., 1996, Dyeing Properties of Oxygen Low-Temperature Plasma-Treated Wool and Nylon 6 Fibers with Acid And Basic Dyes, Journal of the Society of Dyers and Colorists, 112(9), 233-236p.

[81] Wakida, T., Cho, S., Choi, S., Tokino, S. and Lee, M., 1998, Effect of Low Temperature Plasma Treatment on Color of Wool and Nylon 6 Fabrics Dyed with Natural Dyes, Textile Research Journal, 68(11), 848-853p.

[82] Kan, C.W., Chan, K., Yuen, C.W.M. and Miao, M.H., 1998, Surface Properties of Low-Temperature Plasma Treated Wool Fabrics, Journal of Materials Processing Technology, 83(1-3), 180-184p.

[83] Kan, C.W., Chan, K., Yuen, C.W.M. and Miao, M.H., 1998, The Effect of Low-Temperature Plasma on The Chrome Dyeing of Wool Fiber, Journal of Materials Processing Technology, 82(1-3), 122-126p.

[84] Kan, C.W., Chan, K., Yuen, C.W.M. and Miao, M.H., 1999, Low Temperature Plasma on Wool Substrates: The Effect of The Nature of The Gas, Textile Research Journal, 69(6), 407-416p.

[85] Kan, C.W., Chan, K. and Yuen, C.W.M., 2004, Surface Characterization of Low Temperature Plasma Treated Wool Fiber-The Effect of The Nature of Gas-, Fibers and Polymers, 5(1), 52-58p.

[86] Iriyama, Y., Mochizuki, T., Watanabe, M. and Utada, M., 2002, Plasma Treatment of Silk Fabrics for Better Dyeability, Journal of Photopolymer Science and Technology, 15(2), 299-306p.

[87] Sun, D. and Stylios, G.K., 2004, Effect of Low Temperature Plasma Treatment on The Scouring and Dyeing of Natural Fabrics, Textile Research Journal, 74(99), 751-756p.

[88] Biniaś, D., Włochowicz, A. and Binias, W., 2004, Selected Properties of Wool Treated By Low-Temperature Plasma, Fibers&Textiles in Eastern Europe, 12(2), 58-62p.

[89] Sun, D. and Stylios, G.K., 2005, Investigating The Plasma Modification of Natural Fiber Fabrics-The Effect on Fabric Surface and Mechanical Properties, Textile Research Journal, 75(9), 639-644p.

[90] Masukuni, M. and Norihiro, I., 2006, Dyeing Properties of Argon-Plasma Treated Wool, Sen'I Gakkaishi, 62(9), 205-211p.

[91] Kan, C. and Yuen, C.M., 2006, Dyeing Behavior of Low Temperature Plasma Treated Wool, Plasma Processes and Polymers, 3 (8), 627-635p.

[92] El-Zawahry, M.M., Ibrahim, N.A. and Eid, M.A., 2006, The Impact of Nitrogen Plasma Treatment upon the Physical-Chemical and Dyeing Properties of Wool Fabric, Polymer-Plastics Technology and Engineering, 45(10), 1123-1132p.

[93] Demir, A., Karahan, A., Ozdogan, E., Oktem, T. and Seventekin, N., 2008, The Synergetic Effects of Alternative Methods in Wool Finishing, Fibers&Textiles in Eastern Europe, 16(2), 89-94p.

[94] Chvalinova, R. and Wiener, 2008, Sorption Properties of Wool Fibers After Plasma Treatment, II Central European Symposium on Plasma Chemistry, Chem. Listy 102, 1473-1477p.

[95] Naebe, M., Cookson, P. G., Rippon, J., Brady, R. P., Wang, X., Brack, N. and van Riessen, G., 2010, Effects of Plasma Treatment of Wool on the Uptake of Sulfonated Dyes with Different Hydrophobic Properties, Textile Research Journal, 80(4), 312-324p.

[96] Demir A., 2010, Atmospheric Plasma Advantages for Mohair Fibers in Textile Applications, Fibers and Polymers, 11(4), 580-585p.

[97] Atav, R. and Yurdakul, A., 2011, Low Temperature Dyeing of Plasma Treated Luxury Fibers. Part I: Results for Mohair (Angora Goat), Fibers & Textiles in Eastern Europe, 19(2), 84-89p.

[98] http://en.wikipedia.org/wiki/Gamma_ray, 2012

[99] http://web.princeton.edu/sites/ehs/osradtraining/radiationproperties/radiationproperties.htm, 2012

[100] Zohdy, M.H., El-Naggar, A.M. and Marie, M.M., 1999, Effect of Copper Treatment on The Dyeability of Gamma Irradiated Acrylic Fabrics with Different Dyestuffs, Materials Chemistry and Physics, 61, 237-243p.

[101] Aytac, A., Sen, M., Deniz, V. and Guven, O., 2007, Effect of Gamma Irradiation on The Properties of Tyre Cords, Nuclear Instruments and Methods in Physics Research, 265, 271-275p.

[102] Zohdy, M.H., El-Naggar, A.M. and Abdallah, W.A., 1997, Silk Screen Printing of Some Reactive Dyes on Gamma Irradiated Wool Fabrics, Polymer Degradation and Stability, 55, 185-189p.

[103] Horio, M., Ogami, K., Kondo, T. and Sekimoto, K., 1962, Effect of Gamma Irradiation upon Wool Fibers, http://repository.kulib.kyoto-u.ac.jp/dspace/bitstream/2433/75949/1/chd041_1_001.pdf

[104] Beevers, R.B. and McLaren, K.G., 1974, The Effect of Low Doses of Co60 Gamma Radiation on Some Physical Properties and The Structure of Wool Fibers, Textile Research Journal, 44(12), 986-994p.

[105] Millington, K.R., 2000, Comparison of The Effects of Gamma and Ultraviolet Radiation on Wool Keratin, Journal of The Society of Dyers and Colourists, 116, 266-272p.

[106] Kan, C.W., 2008, Impact on Textile Properties of Polyester with Laser, Optics & Laser Technology, 40, 113-119p.

[107] http://en.wikipedia.org/wiki/Laser, 2012

[108] ttp://resources.alibaba.com/topic/800007870/Laser_processing_technology_in_Textile_and_Apparel.htm, 2012

[109] Montazer, M., Taheri, S.J. and Harifi, T., 2011, Effect of Laser CO2 Irradiation on Various Properties of Polyester Fabric: Focus on Dyeing, Journal of Applied Polymer Science, 124(1), 342-348p.

[110] Esteves, F. and Alonso, H., 2007, Effect of CO2 Laser Radiation on Surface and Dyeing Properties of Synthetic Fibers, Research Journal of Textile and Apparel, 11(3), 42-47p.

[111] http://en.wikipedia.org/wiki/Excimer_laser, 2012

[112] Knittel, D. and Schollmeyer, E., 1998, Surface Structuring of Synthetic Fibers by UV Laser Irradiation. Part III. Surface Functionality Changes Resulting from Excimer-Laser Irradiation, Polymer International, 45, 103-109p.

[113] Nourbakhsh, S. and Ebrahimi, I., 2012, Different Surface Modification of Poly (Ethylene Terephthalate) and Polyamide 66 Fibers by Atmospheric Air Plasma Discharge and Laser Treatment: Surface Morphology and Soil Release Behavior, Journal of Textile Science&Engineering, 2(2), http://dx.doi.org/10.4172/2165-8064.1000109

[114] Bahtiyari, M.I., 2011, Laser Modification of Polyamide Fabrics, Optics & Laser Technology, 43(1), 114-118p.

[115] Yip, J., Chan, K., Sin, K. and Lau, K., 2002, UV Excimer Laser Modification on Polyamide Materials: Effect on The Dyeing Properties, Materials Research Innovations, 6(2), 73-78p.

[116] Buyukakıncı, B.Y., 2012, Usage of Microwave Energy in Turkish Textile Production Sector, Energy Procedia, 14, 424-431p.

[117] http://en.wikipedia.org/wiki/Microwave, 2012

[118] http://www.gallawa.com/microtech/mwave.html, 2012

[119] http://www.microwavetec.com/theor_basics.php, 2012

[120] Ahmed, N.S.E and El-Shishtawy, R.M., 2010, The Use of New Technologies in Coloration of Textile Fibers, Journal of Material Science, 45(5), 1143-1153p.

[121] http://www.rsc.org/images/evaluserve_tcm18-16758.pdf, 2012

[122] Collins, J.M. and Leadbeater, N.E., 2007, Microwave Energy: A Versatile Tool for The Biosciences, Organic & Bimolecular Chemistry, 5, 1141-1150p.

[123] Haggag, K., Hanna, H.L., Youssef, B.M. and El-Shimy, N.S., March 1995, Dyeing Polyester With Microwave Heating Using Disperse Dyestuffs, American Dyestuff Reporter, 22-35p.

[124] Delaney, M.J. and Seltzer, I., 1972, Microwave Heating for Fixation of Pad-dyeings on Wool, Journal of the Society of Dyers and Colorists, 88(2), 55-59p.

[125] Zhao, X. and He, J.X., 2011, Improvement in Dyeability of Wool Fabric By Microwave Treatment, Indian Journal of Fiber & Textile Research, 36(1), 58-62p.

[126] Leeder, J.D. , McGregor, B.A. and Steadman, R.G., 1998, Properties and Performance of Goat Fiber, RIRDC Publication, No 98/22, RIRDC, Proje No: ULA-8A

[127] Machi, S., 1996, New Trends of Radiation Processing Applications, Radiation Physics and Chemistry, 47(3), 333-336p.

[128] Li, J., Huang, Y., Xu, Z. and Wang, Z., 2005, High-Energy Radiation Technique Treat on The Surface Of Carbon Fiber, Materials Chemistry and Physics, 94, 315-321p.

[129] Alberti, A., Bertini, S., Gastaldi, G., Iannaccone, N., Macciantelli, D., Torri, G. and Vismara, E., 2005, Electron Beam Irradiated Textile Cellulose Fibers. ESR Studies and Derivatisation with Glycidyl Methacrylate (GMA), European Polymer Journal, 41, 1787-1797p.

[130] Fatarella, E., Ciabatti, I. and Cortez, J., 2010, Plasma and Electron-Beam Processes as Pretreatments for Enzymatic Processes, Enzyme and Microbial Technology, 46, 100-106p.

[131] Vahdat, A., Bahrami, H., Ansari, N. and Ziaie, F., 2007, Radiation Grafting of Styrene onto Polypropylene Fibers by A 10mev Electron Beam, Radiation Physics and Chemistry, 76, 787-793p.

[132] Clough, R.L., 2001, High-Energy Radiation and Polymers: A Review of Commercial Process and Emerging Applications, Nuclear Instruments and Methods in Physics Research B, 185, 8-33p.

[133] Chmielewski, A.G. and Haji-Saeid, M., 2004, Radiation Technologies: Past, Present and Future, Radiation Physics and Chemistry, 71, 16-20p.

[134] Oktem, T., Ozdogan, E., Namligoz, S.E., Oztarhan, A., Tek, Z., Tarakcıoglu, I. and Karaaslan, A., 2006, Investigating the Applicability of Metal Ion Implantation Technique (MEVVA) to Textile Surfaces, Textile Research Journal, 76(1), 32-40p.

[135] Ion Beam Processing (IBP) Technologies, Sector Study, FINAL REPORT, Prepared for the North American Technology and Industrial Base Organization (NATIBO), Prepared by BDM Federal, June 1996

[136] Dong, H. and Bell, T., 1999, State-of-The-Art Overview: Ion Beam Surface Modification of Polymers Towards Improving Tribological Properties, Surface and Coatings Technology, 111, 29-40p.

[137] Montero, G., Hinks, D. and Hooker, J., 2003, Reducing Problems of Cyclic Trimer De‐
posits in Supercritical Carbon Dioxide Polyester Dyeing Machinery, Journal of Su‐
percritical Fluids, 26, 47-54p.

[138] Fernandez Cid, M.V., van Spronsen, J., van der Kraan, M., Veugelers, W.J.T., Woer‐
lee, G.F. and Witkamp, G.J., 2007, A significant approach to dye cotton in supercriti‐
cal carbon dioxide with fluorotriazine reactive dyes, J. of Supercritical Fluids, 40,
477-484p.

[139] Hou, A. and Dai, J., 2005, Kinetics of Dyeing of polyester with CI Disperse Blue 79 in
Supercritical Carbondioxide, Coloration Technology, 121, 18-20p.

[140] Özcan, A.S., Clifford, A.A., Bartle, K.D. and Lewis, D.M., 1998, Dyeing of Cotton Fi‐
bres with Disperse Dyes in Supercritical Carbon Dioxide, Dyes and Pigments, 36 (2),
103-110p.

[141] New Technologies in Textile Dyeing and Finishing, http://www.industryhk.org/
english/fp/fp_hki/files/HKI_03_09_textile_e.pdf

[142] Özcan, A.S., and Özcan, A., 2005, Adsorption Behavior of A Disperse Dye on Polyest‐
er in Supercritical Carbon Dioxide, Journal of Supercritical Fluids, 35, 133-139p.

[143] Hou, A., Chen, B., Dai, J. and Zhang, K., 2010, Using Supercritical Carbon Dioxide as
Solvent to Replace Water in Polyethylene Terephthalate (PET) Fabric Dyeing Proce‐
dures, Journal of Cleaner Production, 18,1009-1014p.

[144] Jun, J.H., Ueda, M., Sawada, K., Sugimotod, M. and Urakawaa, H., 2005, Supercritical
Carbon Dioxide Containing A Cationic Perfluoropolyether Surfactant for Dyeing
Wool, Coloration Technology, 121, 315-319p.

[145] Özcan, A.S., Clifford, A.A., Bartle, K.D., Broadbent, P.J. and Lewis, D.M., 1998, Dye‐
ing of Modified Cotton Fibres with Disperse Dyes from Supercritical Carbon Diox‐
ide, Journal of The Society of Dyers and Colourists, 114, 169-173p.

[146] Van der Kraan, M., Bayrak, Ö., Fernandez Cid, M.V., Woerlee, G.F., Veugelers, W.J.T.
and Witkamp, G.J., Textile Dyeing in Supercritical Carbon Dioxide, http://
www.isasf.net/fileadmin/files/Docs/Versailles/Papers/PMt2.pdf

[147] Guzel, B. and Akgerman, A., 2000, Mordant Dyeing of Wool By Supercritical Process‐
ing, Journal of Supercritical Fluids, 18, 247-252p.

[148] Sawada, K., Takagi, T., Jun, J.H., Ueda, M. and Lewis, D.M., 2002, Dyeing Natural Fi‐
bres in Supercritical Carbon Dioxide Using A Nonionic Surfactant Reverse Micellar
System, Coloration Technology, 118(5), 233-237p.

[149] Jun, J.H., Sawada K. and Ueda, M., 2004, Application of Perfluoropolyether Reverse
Micelles in Supercritical CO2 to Dyeing Process, Dyes and Pigments, 61, 17-22p.

[150] Van der Kraan, M., Fernandez Cid, M.V., Woerlee, G.F., Veugelers, W.J.T. and Wit‐
kamp, G.J., 2007, Dyeing of Natural and Synthetic Textiles in Supercritical Carbon
Dioxide with Disperse Reactive Dyes, The Journal of Supercritical Fluids, 40,
470-476p.

The Dyeing and Environment

Chapter 6

Textile Dyes:
Dyeing Process and Environmental Impact

Farah Maria Drumond Chequer,
Gisele Augusto Rodrigues de Oliveira,
Elisa Raquel Anastácio Ferraz,
Juliano Carvalho Cardoso,
Maria Valnice Boldrin Zanoni and
Danielle Palma de Oliveira

Additional information is available at the end of the chapter

http://dx.doi.org/10.5772/53659

1. Introduction

Dyes may be defined as substances that, when applied to a substrate provide color by a process that alters, at least temporarily, any crystal structure of the colored substances [1,2]. Such substances with considerable coloring capacity are widely employed in the textile, pharmaceutical, food, cosmetics, plastics, photographic and paper industries [3,4]. The dyes can adhere to compatible surfaces by solution, by forming covalent bond or complexes with salts or metals, by physical adsorption or by mechanical retention [1,2]. Dyes are classified according to their application and chemical structure, and are composed of a group of atoms known as chromophores, responsible for the dye color. These chromophore-containing centers are based on diverse functional groups, such as azo, anthraquinone, methine, nitro, aril-methane, carbonyl and others. In addition, electrons withdrawing or donating substituents so as to generate or intensify the color of the chromophores are denominated as auxochromes. The most common auxochromes are amine, carboxyl, sulfonate and hydroxyl [5-7].

It is estimated that over 10,000 different dyes and pigments are used industrially and over 7 x 10^5 tons of synthetic dyes are annually produced worldwide [3,8,9]. Textile materials can be dyed using batch, continuous or semi-continuous processes. The kind of process used depends on many characteristics including type of material as such fiber, yarn, fabric, fabric

construction and garment, as also the generic type of fiber, size of dye lots and quality requirements in the dyed fabric. Among these processes, the batch process is the most common method used to dye textile materials [10].

In the textile industry, up to 200,000 tons of these dyes are lost to effluents every year during the dyeing and finishing operations, due to the inefficiency of the dyeing process [9]. Unfortunately, most of these dyes escape conventional wastewater treatment processes and persist in the environment as a result of their high stability to light, temperature, water, detergents, chemicals, soap and other parameters such as bleach and perspiration [11]. In addition, anti-microbial agents resistant to biological degradation are frequently used in the manufacture of textiles, particularly for natural fibers such as cotton [11,12]. The synthetic origin and complex aromatic structure of these agents make them more recalcitrant to biodegradation [13,14]. However, environmental legislation obliges industries to eliminate color from their dye-containing effluents, before disposal into water bodies [9,12].

The textile industry consumes a substantial amount of water in its manufacturing processes used mainly in the dyeing and finishing operations of the plants. The wastewater from textile plants is classified as the most polluting of all the industrial sectors, considering the volume generated as well as the effluent composition [15-17]. In addition, the increased demand for textile products and the proportional increase in their production, and the use of synthetic dyes have together contributed to dye wastewater becoming one of the substantial sources of severe pollution problems in current times [6,9].

Textile wastewaters are characterized by extreme fluctuations in many parameters such as chemical oxygen demand (COD), biochemical oxygen demand (BOD), pH, color and salinity. The composition of the wastewater will depend on the different organic-based compounds, chemicals and dyes used in the dry and wet-processing steps [6,18]. Recalcitrant organic, colored, toxicant, surfactant and chlorinated compounds and salts are the main pollutants in textile effluents [17].

In addition, the effects caused by other pollutants in textile wastewater, and the presence of very small amounts of dyes (<1 mg/L for some dyes) in the water, which are nevertheless highly visible, seriously affects the aesthetic quality and transparency of water bodies such as lakes, rivers and others, leading to damage to the aquatic environment [19,20].

During the dyeing process it has been estimated that the losses of colorants to the environment can reach 10–50% [13,14,17,21,22]. It is noteworthy that some dyes are highly toxic and mutagenic, and also decrease light penetration and photosynthetic activity, causing oxygen deficiency and limiting downstream beneficial uses such as recreation, drinking water and irrigation [13,14,23]

With respect to the number and production volumes, azo dyes are the largest group of colorants, constituting 60-70% of all organic dyes produced in the world [2,24]. The success of azo dyes is due to the their ease and cost effectiveness for synthesis as compared to natural dyes, and also their great structural diversity, high molar extinction coefficient, and medium-to-high fastness properties in relation to light as well as to wetness [2,25]. They have a wide range of applications in the textile, pharmaceutical and cosmetic industries, and are also used in food, paper, leather and paints [26,27]. However, some azo dyes can show toxic effects, especially carcinogenic and mutagenic events [27,28].

The toxic effects of the azo dyes may result from the direct action of the agent itself or of the aryl amine derivatives generated during reductive biotransformation of the azo bond [22]. The azo dyes entering the body by ingestion can be metabolized to aromatic amines by the azoreductases of intestinal microorganisms. If the dyes are nitro, they can be metabolized by the nitroredutases produced by the same microorganisms [29]. Mammalian liver enzymes and other organizations may also catalyze the reductive cleavage of the azo bond and the nitroreduction of the nitro group. In both cases, if N-hydroxylamines are formed, these compounds are capable of causing DNA damage [29, 30].

One of the most difficult tasks confronted by the wastewater treatment plants of textile industries is the removal of the color of these compounds, mainly because dyes and pigments are designed to resist biodegradation, such that they remain in the environment for a long period of time. For example, the half-life of the hydrolyzed dye Reactive Blue 19 is about 46 years at pH 7 and 25°C [31,32].

Carneiro et al. (2010) designed and optimized an accurate and sensitive analytical method for monitoring the dyes C.I. Disperse Blue 373 (DB373), C.I. Disperse Orange 37 (DO37) and C.I. Disperse Violet 93 (DV93) in environmental samples. This investigation showed that DB373, DO37 and DV93 were present in both untreated river water and drinking water, indicating that the effluent treatment (pre-chlorination, flocculation, coagulation and flotation) generally used by drinking water treatment plants, was not entirely effective in removing these dyes. This study was confirmed by the mutagenic activity detected in these wastewaters [33].

In this context, and considering the importance of colored products in present day societies, it is of relevance to optimize the coloring process with the objective of reducing the environmental impact of the textile industry. For this purpose, liposomes could be used to carry several encapsulated dyes, and hence improve the mechanical properties of textile products, resulting in better wash fastness properties and reducing the process temperature, thus economizing energy [34]. Another way is to use ultrasonic energy, studied with the objectives of improving dye productivity and washing fastness, and reducing both energy costs and water consumption [35].

Considering the fact that the textile dyeing process is recognized as one of the most environmentally unfriendly industrial processes, it is of extreme importance to understand the critical points of the dyeing process so as to find alternative, eco-friendly methods.

2. Dyeing process

The dyeing process is one of the key factors in the successful trading of textile products. In addition to the design and beautiful color, the consumer usually looks for some basic product characteristics, such as good fixation with respect to light, perspiration and washing, both initially and after prolonged use. To ensure these properties, the substances that give color to the fiber must show high affinity, uniform color, resistance to fading, and be economically feasible [36].

Modern dyeing technology consists of several steps selected according to the nature of the fiber and properties of the dyes and pigments for use in fabrics, such as chemical structure, classification, commercial availability, fixing properties compatible with the target material to be dyed, economic considerations and many others [36].

Dyeing methods have not changed much with time. Basically water is used to clean, dye and apply auxiliary chemicals to the fabrics, and also to rinse the treated fibers or fabrics [37]. The dyeing process involves three steps: preparation, dyeing and finishing, as follows:

Preparation is the step in which unwanted impurities are removed from the fabrics before dyeing. This can be carried out by cleaning with aqueous alkaline substances and detergents or by applying enzymes. Many fabrics are bleached with hydrogen peroxide or chlorine-containing compounds in order to remove their natural color, and if the fabric is to be sold white and not dyed, optical brightening agents are added [37].

Dyeing is the aqueous application of color to the textile substrates, mainly using synthetic organic dyes and frequently at elevated temperatures and pressures in some of the steps [37,38]. It is important to point out that there is no dye which dyes all existing fibers and no fiber which can be dyed by all known dyes [39]. During this step, the dyes and chemical aids such as surfactants, acids, alkali/bases, electrolytes, carriers, leveling agents, promoting agents, chelating agents, emulsifying oils, softening agents etc [23,38] are applied to the textile to get a uniform depth of color with the color fastness properties suitable for the end use of the fabric [37].This process includes diffusion of the dye into the liquid phase followed by adsorption onto the outer surface of the fibers, and finally diffusion and adsorption on the inner surface of the fibers [40]. Depending on the expected end use of the fabrics, different fastness properties may be required. For instance, swimsuits must not bleed in water and automotive fabrics should not fade after prolonged exposure to sunlight [37]. Different types of dye and chemical additives are used to obtain these properties, which is carried out during the finishing step. Dyeing can also be accomplished by applying pigments (pigments differ from dyes by not showing chemical or physical affinity for the fibers) together with binders (polymers which fix the pigment to the fibers) [39,41].

Finishing involves treatments with chemical compounds aimed at improving the quality of the fabric. Permanent press treatments, water proofing, softening, antistatic protection, soil resistance, stain release and microbial/fungal protection are all examples of fabric treatments applied in the finishing process [37].

Dyeing can be carried out as a continuous or batch process [37]. The most appropriate process to use depends on several factors, such as type of material (fiber, yarn, fabric, fabric construction, garment), generic type of fiber, size of dye batch and quality requirements for the dyed fabric, but batch processes are more commonly used to dye textile materials [10].

In continuous processing, heat and steam are applied to long rolls of fabric as they pass through a series of concentrated chemical solutions. The fabric retains the greater part of the chemicals while rinsing removes most of the preparation chemicals. Each time a fabric is passed through a solution, an amount of water equivalent to the weight of the fabric must be used [37].

In batch processing, sometimes called exhaust dyeing, since the dye is gradually transferred from the dye bath to the material being dyed over a relatively long period of time [10], the dyeing occurs in the presence of dilute chemicals in a closed equipment such as a kier, kettle, beam, jet or beck [37]. Unlike the continuous process, instead of being passed through various baths in a long series of equipment sections, in the batch process the fabric remains in a single piece of equipment, which is alternately filled with water and then drained, at each step of the process. Each time the fabric is exposed to a separate bath, it uses five to ten times its own weight in water [37].

Some batch dyeing machines only operate at temperatures up to 100ºC. However, the system can be pressurized, allowing for the use of temperatures above 100ºC. Cotton, rayon, nylon, wool and some other fibers dye well at temperatures of 100ºC or below. Polyester and some other synthetic fibers dye more easily at temperatures above 100ºC [10].

Since the degree of dye fixation depends on the nature of the fiber, it is important to consider this topic. The fibers used in the textile industry can be divided into two main groups denominated natural fibers and synthetic fibers [36,37]. Natural fibers are derived from the environment (plants or animals), such as wool, cotton, flax, silk, jute, hemp and sisal, most of which are based on cellulose and proteins. On the other hand, synthetic fibers are organic polymers, mostly derived from petroleum sources, for example, polyester, polyamide, rayon, acetate and acrylic [37,39,42]. The two most important textile fibers are cotton, the largest, and polyester [43,44].

Cotton has been used for over 7000 years, and consists of mainly cellulose, natural waxes and proteins. The large number of hydroxyl groups on the cellulose provides a great water absorption capacity [39].

Several aromatic polyesters have been synthesized and studied. Of these, polyethylene terephthalate (PET) and polybutylene terephthalate (PBT) have been produced commercially for more than 50 years. Amongst other uses, PET has been used worldwide for the production of synthetic fibers due to its good physical properties. PET is manufactured from ethylene glycol (EG) and terephthalic acid (TPA) or dimethyl terephthalate (DMT). The polymerization proceeds in two steps: esterification and condensation reactions [45]. The polymer produced after condensation is solidified by jets of cold water and cut into regular granules which frequently have a cubic form. Then, the polymer melt is spun and the yarns are solidified by a stream of cold air [39].

The dye can be fixed to the fiber by several mechanisms, generally in aqueous solution, and may involve primarily four types of interaction: ionic, Van der Waals and hydrogen interactions, and covalent bonds [5].

Ionic interactions result from interactions between oppositely charged ions present in the dyes and fibers, such as those between the positive center of the amino groups and carboxyl groups in the fiber and ionic charges on the dye molecule, and the ionic attraction between dye cations and anionic groups ($-SO_3^-$ and $-CO_2^-$) present in the acrylic fiber polymer molecules. Typical examples of this type of interaction can be found in the dyeing of wool, silk and polyamide [5,36,41].

Van der Waals interactions come from a close approach between the π orbitals of the dye molecule and the fiber, so that the dye molecules are firmly "anchored" to the fiber by an affinity process without forming an actual bond. Typical examples of this type of interaction are found in the dyeing of wool and polyester with dyes with a high affinity for cellulose [36].

Hydrogen interactions are formed between hydrogen atoms covalently bonded in the dye and free electron pairs of donor atoms in the center of the fiber. This interaction can be found in the dyeing of wool, silk and synthetic fibers such as ethyl cellulose [5,36].

Covalent bonds are formed between dye molecules containing reactive groups (electrophilic groups) and nucleophilic groups on the fiber, for example, the bond between a carbon atom of the reactive dye molecule and an oxygen, nitrogen or sulfur atom of a hydroxy, amino or thiol group present in the textile fiber. This type of bond can be found in the dyeing of cotton fiber [5,36,46].

2.1. Application of liposome-based technology in textile dyeing process

There is increasing interest in the textile industry in the development of eco-friendly textile processing, in which the use of naturally occurring materials such as phospholipids, would become important [47]. Phospholipids are natural surfactants and in the presence of water, they organize themselves so as to reduce unfavorable interactions between their hydrophobic tails and the aqueous solution; their hydrophilic head groups exposed to the aqueous phase forming vesicles. Liposomes or phospholipid vesicles are featured by clearly separate hydrophilic and hydrophobic regions [34,48].

Liposomes were first produced in England in 1961 by Alec D. Bangham, who was studying phospholipids and blood clotting. He found that when phospholipids were added to water, they immediately formed a sphere, because one end of each molecule was water soluble, while the opposite end was water insoluble [49]. From a chemical point of view, the liposome is an amphoteric compound containing both positive and negative charges [34,50].

Liposomes are defined as a structure composed of lipid vesicle bilayers which can encapsulate hydrophobic or hydrophilic compounds in the lipid bilayer or in the aqueous volume, respectively [51]. These structures are usually made up of phosphatidylcholine (PC), which has a hydrophilic part consisting of phosphate and choline groups and a hydrophobic part composed of two hydrocarbon chains of variable length [49, 52,53].

Liposomes are often distinguished according to their number of lamellae and size. Small unilamellar vesicles (SUV), large unilamellar vesicles (LUV) and large multilamellar vesicles (MLV) or multivesicular vesicles (MVV) can be differentiated [49]. The diameters of liposomes vary from a nanometer to a micrometer [34]. Multilamellar liposomes (MLV) usually range from 500 to 10,000 nm. Unilamellar liposomes can be small (SUV) or large (LUV); SUV are usually smaller than 50 nm and LUV are usually larger than 50 nm. Very large liposomes are called giant liposomes (10,000 - 10,00,000 nm). They can be either unilamellar or multilamellar. The liposomes containing encapsulated vesicles are called multi-vesicular and they range from 2,000-40,000 nm. LUVs with an asymmetric distribution of phospholipids in the bilayers are called asymmetric liposomes [54]. The thickness of the membrane (phospholipid bilayer) measures approximately 5 to 6 nm [49].

According to Sivasankar, Katyayani (2011), the preparation of liposomes is based on lipids, and those normally used are:

Natural phospholipids:

- Phosphatidyl choline (PC) - Lecithin

- Phosphatidyl ethanolamine (PE) - Cephalin

- Phosphatidyl serine (PS)

- Phosphatidyl inositol (PI)

- Phosphatidyl glycerol (PG)

Synthetic phospholipids:

For saturated phospholipids:

- Dipalmitoyl phosphatidyl choline (DPPC)

- Distearoyl phosphatidyl choline (DSPC)

- Dipalmitoyl phosphatidyl ethanolamine (DPPE)

- Dipalmitoyl phosphatidyl serine (DPPS)

- Dipalmitoyl phosphatidic acid (DPPA)

- Dipalmitoyl phosphatidyl glycerol (DPPG)

For unsaturated phospholipids:

- Dioleoyl phosphatidyl choline (DOPC)

- Dioleoyl phosphatidyl glycerol (DOPG) [52].

Natural acidic lipids, such as PS, PG, PI, PA (phosphatidic acid) and cardiolipin (CL), are added when anionic liposomes are desired, and cholesterol is often included to stabilize the bilayer. These molecules are derivatives of glycerol with two alkyl groups and one amphoteric group [34].

Phosphatidylcholine is the biological lipid most widely used for producing liposomes. Liposomes based on phosphatidylcholine consist of phosphatidic acid and glycerin, with two alcohol groups esterified by fatty acids and a third group esterified by phosphoric acid, to which the amino alcohol choline is added as a polar group [34,49].

According to Barani, Montazer (2008), normally four different methods can be used for the preparation of liposomes:

1. Dry lipid film;

2. Emulsions;

3. Micelle-forming detergents;

4. Alcohol injection technology [34].

Liposomes have two distinct roles: they can provide an excellent model for biological membranes, and they are being developed as controlled delivery systems for hydrophilic and lipophilic agents [34,53,55]. They are promising candidates for adjuvant and carrier systems for drug delivery, are well-documented, and can be used for the same purpose in textile materials [33].

Encapsulation or liposome technology is applied in numerous fields, such as in pharmaceuticals, cosmetics, foods, detergents, textiles and other applications where it is important to liberate the encapsulated material slowly [34,50]. This new clean technology has already been adopted by some textile industries [52]. In recent years, liposomes have been examined as a way of delivering dyes to textiles in a cost-effective and environmentally sensitive way [56].

Conventional dyeing processes consume a great deal of energy, a significant amount of which is wasted in controlling the process parameters in order to achieve uniform results. With respect to the carrier role of liposomes, they can be used in several textile processes such as textile finishing and dyeing, with several types of dyes and fibers. They are nontoxic, biodegradable, and can encapsulate a wide range of solutes [34]. In addition, the main advantages of liposomes are a clear reduction in dyeing temperature (about 10°C as compared to conventional dyeing), improved quality of the textiles produced, with additional benefits with respect to material weight yield during subsequent spinning, improved smoothness and mechanical properties of the dyed textiles, and a clear reduction in the contamination load of the dye baths [52,57]. Low temperature gives a more natural feel and improved quality, with lower environmental impact [34].

In recent years, liposomes have been used in the textile industry as a carrier for auxiliary materials (leveling, retarding and wetting agents) in dyeing, mainly for wool dyeing, and for finishing processes [34,55]. One of the most common problems with textile auxiliaries is that they fail to form a complex in the solution bath. This problem can be solved by using liposomes with selected positive or negative charges. Liposomes can be prepared according to the type of process, solute material and fiber structure [34]. Liposomes from phospholipids have been widely used as a dye carrier in the dyeing process, and create eco-friendly textile processes. Due to their structural properties, liposomes can encapsulate hydrophilic dyes (reactive, acid and basic dyes) in the aqueous phase, and hydrophobic dyes (disperse dyes) in the phospholipid bilayers [58]. Liposomes containing a dye are generally large, irregular and unilamellar [50].

According to Barani & Montazer (2008), the application of liposomes in textile processing can be useful when the release of the solute material is important, and improves the final properties of the products. A wetting agent is required in the conventional bleaching bath of cotton fabrics, but this step can be eliminated by using liposomes. The presence of liposomes in the peroxide bleaching bath can improve the mechanical properties of fabrics and their brightness. Liposomes contain particles of oxidant present in the bleaching solution that represent an unusual reservoir, and release the bleaching agent gradually into the bleaching bath. Moreover, the encapsulation of catalysts used for the decomposition of hydrogen peroxide radicals can be another factor in retarding the rate of decomposition. In this way, liposomes act as a stabilizing agent in the bleaching bath [34].

The role of auxiliary products is very important in textile dyeing with disperse dyes [53]. These compounds show extremely low solubility in water and dispersing agents are needed to maintain a fine, stable dispersion throughout the whole dyeing process at the different temperatures. Martí et al. (2007) analyzed the usefulness of commercial textile liposomes as dispersing agents, and observed that liposomes could be considered as suitable dispersing auxiliaries for polyester dyeing at high temperatures, considering their capacity to stabilize dye dispersions and achieve a suitable dye exhaustion level, with the added value of their environmentally friendly nature [53]. Liposomes clearly improve the dispersion efficiency as compared to conventional dispersing agents [34].

Additionally, liposomes for textile use show a similar price to that of synthetic surfactants used in the dyeing of polyester with disperse dyes. However, the new technology is more environmentally friendly, and hence the reduction in the environmental problem can lead to economic advantages [53]. In addition, liposome preparations tend not to foam. This is an advantage that distinguishes liposomes from other textile auxiliaries [34].

According to Martí et al. (2010), the dyeing of wool and wool blends with the aid of liposomes has demonstrated better quality, energy saving and a reduction in the environmental impact and also the temperature could be reduced, resulting in less fiber damage. Moreover, dye bath exhaustion was shown to be over 90% at the lower temperature (80°C) used, resulting in significant savings in energy costs [55]. The impact of the dyeing process on the environment was also considerably lower, the COD being reduced by about 1000 units [53].

Therefore, liposome-based technology is an alternative, eco-friendly method, which could reduce the environmental impact, offering technical and economic advantages for the textile industry.

2.2. Effect of ultrasonic energy on the dyeing process

Ultrasound-assisted textile dyeing was first reported by Sokolov and Tumansky in 1941[59]. The basic idea of this technology is that ultrasound can enhance mass transfer by reducing the stagnant cores in the yarns. The improvements observed are generally attributed to cavitation phenomena and to other resulting physical effects such as dye dispersion (breaking up of aggregates with high relative molecular mass), degassing (expulsion of dissolved or entrapped air from the fiber capillaries), strong agitation of the liquid (reduction in thickness of the fiber-liquid boundary layer), and swelling (enhancement of dye diffusion rate inside the fiber) [59,60].

According to Vankar & Shanker (2008), ultrasound allows for process acceleration, obtaining the same or better results than existing techniques, but under less extreme conditions, i.e., lower temperatures and lower concentrations of the chemicals used. Wet textile processes assisted by ultrasound are of great interest to the textile industry for this reason [61], and Khatri et al. (2011) showed that the dyeing of polyester fiber using ultrasonic energy resulted in an increased dye uptake and enhanced dyeing rate [35].

Due to the revolution in environmental protection, the use of ultrasonic energy as a renewable source of energy in textile dyeing has been increased, due to the variety of advantages associated with it. On the other hand, there is a growing demand for natural, eco-friendly dyeing for the health sensitive application to textile garments as an alternative to harmful synthetic dyes, which poses a need for suitable effective dyeing methodologies [62].

Ultrasonic energy can clean or homogenize materials, accelerating both physical and chemical reactions, and these qualities can be used to improve textile processing methods. Environmental concern has been focused on textile processing methods for quite some time, and the use of ultrasonic energy has been widely studied in terms of improving washing fastness. The textile dyeing industry has long been struggling to cope with high energy costs, rapid technological changes and the need for a faster delivery time, and the effective management of ultrasonic energy could reduce energy costs and improve productivity [35]. Ultrasonic waves are vibrations with frequencies above 17 kHz, out of the audible range for humans, requiring a medium with elastic properties for propagation. The formation and collapse of the bubbles formed by ultrasonic waves (known as cavitation) is generally considered to be responsible for most of the physical and chemical effects of ultrasound in solid/liquid or liquid/liquid systems [63]. Cavitation is the formation of gas-filled microbubbles or cavities in a liquid, their growth, and under proper conditions, their implosive collapse [59].

It has been reported that ultrasonic energy can be applied successfully to wet textile processes, for example laundering, desizing, scouring, bleaching, mercerization of cotton fabrics, enzymatic treatment, dyeing and leather processing, together with the decoloration/mineralization of textile dyes in waste water [60].

In addition, ultrasonic irradiation shows promise, and has the potential, for use in environmental remediation, due to the formation of highly concentrated oxidizing species such as hydroxyl radicals (HO•), hydrogen radicals (H•), hydroperoxyl radicals (HO$_2$•) and H$_2$O$_2$, and localized high temperatures and pressures [59]. Therefore, the use of ultrasonic energy could indeed reduce the environmental impact caused by the textile industry.

3. Finishing and waste water

The contamination of natural waters has become one of the biggest problems in modern society, and the economical use of this natural resource in production processes has gained special attention, since in predictions for the coming years, the amount of water required per capita is of concern. This environmental problem is related not only to its waste through misuse, but also to the release of industrial and domestic effluents [64].

Of the industries with high-polluting power, the textile dyeing industry, responsible for dyeing various types of fiber, stands out. Independent of the characteristics of the dyes chosen, the final operation of all dyeing process involves washing in baths to remove excesses of the original or hydrolyzed dyes not fixed to the fiber in the previous steps [36]. In these baths, as previously mentioned, it is estimated that approximately 10-50% of the dyes used in the dyeing process are lost, and end up in the effluent [17,21,22], contaminating the environment with about one million tons of these compounds [65]. The dyes end up in the water bodies due mainly to the use of the activated sludge treatment in the effluent treatment plants, which has been shown to be ineffective in removing the toxicity and coloring of some types of dye [33,60,66,67]. Moreover, the reduction of azo dyes by sodium hydrosulfite and the successive chlorination steps with hypochlorous acid, can form 2-benzotriazoles fenil-benzotriazol (PBTA) derivatives and highly mutagenic aromatic amines, often more mutagenic than the original dye [68]. In an aquatic environment, this dye reduction can occur in two phases: 1) The application of reducing agents to the newly-dyed fibers to remove the excess unbound dye, which could lead to "bleeding" of the fabrics during washing, and 2) The use of reducing agents in the bleaching process, in order to make the effluent colorless and conform with the legislation. This reduced colorless effluent containing dyes is sent to the municipal sewage treatment plant, where they chlorinate the effluents before releasing them into water bodies where they may generate PBTAs. Several different PBTAs are already described in the literature, and their chemical structures vary depending on the dyes that originated them [63,69].

So the release of improperly treated textile effluents into the environment can become an important source of problems for human and environmental health. The major source of dye loss corresponds to the incomplete fixation of the dyes during the textile fiber dyeing step [36].

In addition to the problem caused by the loss of dye during the dyeing process, within the context of environmental pollution, the textile industry is also focused due to the large volumes of water used by its industrial park, consequently generating large volumes of effluent [64]. It has been calculated that approximately 200 liters of water are needed for each kilogram of cotton produced [70]. These effluents are complex mixtures of many pollutants, ranging from original colors lost during the dyeing process, to associated pesticides and heavy metals [71], and when not properly treated, can cause serious contamination of the water sources [64]. So the materials that end up in the water bodies are effluents containing a high organic load and biochemical oxygen demand, low dissolved oxygen concentrations, strong color and low biodegradability. In addition to visual pollution, the pollution of water bodies with these compounds causes changes in the biological cycles of the aquatic biota, particularly affecting the photosynthesis and oxygenation processes of the water body, for example by hindering the passage of sunlight through the water [72].

Moreover, studies have shown that some classes of dye, especially azo dyes and their by-products, may be carcinogenic and / or mutagenic [27,33,67,73-77], endangering human health, since the wastewater treatment systems and water treatment plants (WTP) are inef-fective in removing the color and the mutagenic properties of some dyes [78,79]. The diffi-culty in removing them from the environment can be attributed to the high stability of these compounds, since they are designed to resist biodegradation to meet the demands of the consumer market with respect to durability of the colors in the fibers, consequently imply-ing that they also remain in the environment for a long time [32].

With respect to the legislation, there is no consensus amongst the different countries con-cerning effluent discharge, and there is no official document listing the different effluent limit values applied in different countries. Many federal countries, such as the United States of America, Canada and Australia have national environmental legislation, which, as in Eu-rope, establishes the limits that must be complied with. Some countries, such as Thailand, have copied the American system, whereas others, such as Turkey or Morocco, have copied the European model. In some countries, for example India, Pakistan and Malaysia, the emis-sion limits are recommended, but are not mandatory [80]. With respect to the color, in some countries such as France, Austria and Italy, there are limits for the color of the effluent, but since they use different units, a comparison is impossible. The oldest unit is the Hazen, in use since the beginning of the 20th century, but in France, the current unit is (mg $L^{-1}Pt$–Co). The coloration values are determined by a comparative analysis with model solutions pre-pared according to defined procedures [80].

Based on all the problems cited above regarding the discharge of effluents into the environ-ment, it is obvious there is a need to find alternative treatments that are effective in remov-ing dyes from effluents.

4. Waste water treatment

These days, environmental pollution can undoubtedly be regarded as one of the main prob-lems in developed and developing countries. This is due, not just to one, but to a number of factors, such as the misuse of natural resources, inefficient legislation and a lack of environ-mental awareness. Fortunately, in recent years there has been a trend for change and a series of scientific studies are being used as an important tool in the development of new treat-ment technologies and even in the implementation of processes and environmentally friend-ly actions [64,81-84].

Every industrial process is characterized by the use of inputs (raw materials, water, energy, etc.) that undergo transformation giving rise to products, byproducts and waste. The wastes produced at all stages of the various types of human activity, both in terms of composition and volume, vary according to the consumption practices and production methods. The main concerns are focused on the impact these can have on human health and the environ-ment. Hazardous waste, produced mainly by industry, is particularly worrying, because when incorrectly managed, it becomes a serious threat to the environment and therefore to human health. Thus the study of new alternatives for the treatment of different types of industrial effluent continues to be a challenge to combat anthropogenic contamination.

Amongst that of several other industries, the textile sector waste has received considerable attention in recent years, since it can generate large volumes of effluents that, if not correctly treated before being disposed into water resources, can be a problem, as previously mentioned. Effluents from the textile industry are extremely complex, since they contain a large variety of dyes, additives and derivatives that change seasonally, increasing the challenge to find effective, feasible treatments. Currently, the processes developed and available for these industries are based on methods that were designed for other waste, and have limitations when applied to textile effluents. As a consequence, these industries produce colored wastewater with a high organic load, which can contribute enormously to the environmental pollution of surface water and treatment plants if not properly treated before disposal into the water resources [85]. The ingestion of water contaminated with textile dyes can cause serious damage to the health of humans and of other living organisms, due to the toxicity, highlighting mutagenicity of its components [86,87]. Therefore treatments that are more efficient and economical than those currently available are required.

There are several techniques for the treatment of effluents, such as incineration, biological treatment, absorption onto solid matrices, etc. However, these techniques have their drawbacks, such as the formation of dioxins and furans, caused by incomplete combustion during incineration; long periods for biological treatment to have an effect, as also the adsorptive process, that is based on the phase transfer of contaminants without actually destroying them [88,89]. The problem is further aggravated in the textile industry effluents, due to the complexity of their make-up. Thus it can be seen that processes are being used that are not entirely appropriate for the treatment of textile effluents, thereby creating a major challenge for the industry and laundries that need to adapt to current regulations for the control of the color of effluents with a high organic load.

The use of filtration membranes and/or separation [90] and biological methods [91], in addition to incineration processes involving adsorption onto solid matrices, has also being adopted by the textile industry and is receiving considerable attention. However, all these processes only involve phase transfer, generating large amounts of sludge deposited at the end of the tanks and low efficiency in color removal and reduction of the organic load. According to this scenario, many studies have been carried out with the aim of developing new technologies capable of minimizing the volume and toxicity of industrial effluents. Unfortunately, the applicability of these types of system is subject to the development of modified procedures and the establishment of effluent recycling systems, activities that imply evolutionary technologies and which are not yet universally available. Thus the study of new alternatives for the treatment of many industrial effluents currently produced is still one of the main weapons to combat the phenomenon of anthropogenic contamination.

Due to their considerable danger, several authors have attempted to find new forms of treatment to reduce the serious environmental and toxicological risks caused by various organic compounds. Amongst the many reported cases are those based on the use of specific microorganisms, and degradation using advanced oxidation processes (AOP) such as Fenton, photo-Fenton and heterogeneous photocatalysis, which are highlighted below.

The use of microorganisms cultivated specifically for the degradation of polluents to increase the yield of degradation, has been reported by some authors. For example, Flores et al. (1997) examined the behavior of 25 N-substituted aromatic compounds such as organic compounds,azo dyes and nitro, using the methanogenic bacteria acetoclastic, and found that under anaerobic conditions it was easy to mineralize various of the compounds evaluated with a good yield, especially the nitroaromatic and azo dyes [91].

Bornick et al. (2001) evaluated the use of aerobic microorganisms to degrade aromatic amines present in sediments of the river Elbe in Germany. The results obtained showed that it was possible to predict the qualitative degradation of the aromatic amines using degradation constants [92].

Using a structural design, Wang et al.(2007) isolated a bacterium capable of promoting the degradation of the compounds pentyl amine and aniline present in water oil extraction in China. Under conditions of neutral pH and complete aeration of 6 mg O_2 / l at a temperature of 30 ° C, they obtained degradation yields of 82% and 78%, respectively, for pentyl amine and aniline [93].

However, in general, although the use of microorganisms in the treatment of industrial and laboratory wastes containing aromatic amines deserves attention, mainly due to the low investment and maintenance costs, the results are far from ideal, due to the low biodegradation yields, long treatment times, and the generation of sludge deposited at the bottom of the treatment ponds [94].

Of the studies carried out using a source of hydroxyl radicals with oxidizing agents, the photo Fenton system developed by Fukushima et al., 2000 to promote the degradation of aniline stands out. This method has shown promise for the mineralization of aromatic amines, obtaining a reduction of approximately 85%. However, high performance liquid chromatography (HPLC) identified a number of intermediate species formed during the degradation of the aniline, such as p-aminophenol, p-hydroquinone, maleic and fumaric acids and NH_4^+ [95].

Studies involving heterogeneous photocatalysis also deserve attention: for example Pramauro et al., (1995) promoted the degradation of various aniline derivatives using TiO_2 particles suspended in a solution. Under optimal conditions, the method developed showed rapid mineralization of the aromatic amines examined in less than 1 hour of analysis. By-products generated during the degradation of these compounds were identified at the start of the reaction, but none were identified at the end of the reaction. The use of solar radiation was also evaluated, but was shown to be less efficient than artificial radiation [96].

Augugliaro et al.(2000), confirmed that heterogeneous photocatalysis using TiO_2 as a semiconductor may be a suitable method for the complete photodegradation of aniline, 4-ethylaniline and 4-chloroaniline in an aqueous medium. The kinetic parameters for the Langmuir-Hinshelwood model were used to describe the importance of the adsorption results, which proved to be independent of the pH of the solution and of the type of substituent on the aromatic ring of the amine [97].

Canle et al. (2005) studied the behavior of the adsorption of aniline and dimethyl aniline onto three species of TiO_2 (P25, anatase and rutile) and established that the greatest adsorption occurred onto TiO_2 P25. The effect of pH on the degradation provided by these compounds was also evaluated, and lower mineralization percentages were observed in acidic media. This type of behavior can be attributed to the positive charges on both the aniline and the semiconductor, providing electrostatic repulsion between the two species. Thus, an alkaline medium has been recommended as the most appropriate one to promote the mineralization of aromatic amines using a TiO_2 P25 type semiconductor [98].

Chu et al. (2007) observed the effects of pH variation and the addition of hydrogen peroxide on the degradation of 2-chloroaniline using TiO_2 as a semiconductor, with and without the application of UV radiation. The results showed that the addition of low concentrations of H_2O_2 to the UV/TiO_2system provided a significant increase in degradation of the aromatic amine. The addition of an excess of H_2O_2 promoted no increase in degradation, as expected; to the contrary, a reduction in the reaction rate was observed. The variation in pH was evaluated in both systems, and the condition leading to the highest percentages of mineralization was obtained in an alkaline medium using H_2O_2/UV/TiO_2 [99].

Low et al. (1991) monitored the inorganic products resulting from the degradation of several organic nitrogenated, sulfured and halogenated compounds. Degradation was carried out using TiO_2 as the semiconductor and artificially illuminated UV radiation. Ammonium ions were found to be present in higher concentrations than nitrate ions, which can be explained by the fact that compounds having a nitrogen element in their structure pass through a complex degradation step where the generation of ammonium ions is more favorable than the generation of nitrate ions. In turn, compounds with nitro groups in their structures, showed higher concentrations of nitrate ions. Under ideal conditions, all the elements were converted into their respective inorganic forms. Organic carbon was converted to CO_2, the halogens to their corresponding halide, sulfur compounds to sulfate, the phosphate to phosphorus and nitrogen to ammonium and nitrate [100]. However, all these studies used a photocatalytic titanium suspension, requiring a subsequent step to remove the semiconductor.

With a view to these problems, a new technique that has been studied recently, with significant success, is the oxidation of organic matter via the generation of hydroxyl radicals (OH•). This kind of effluent treatment has been highlighted for its destructive character with respect to the organic matter. The advanced oxidation processes (AOP) are processes based on the generation of highly oxidizing species, the hydroxyl radical (OH•), which can oxidize the organic contaminants present in water, air or soil. These radicals have a high oxidation potential (E * = +2.72 V vs. the normal hydrogen electrode, NHE), which results in high reactivity with organic pollutants. This may initiate different types of reaction with different functional groups in organic compounds, forming unstable organic radicals which are then easily oxidized to CO_2, H_2O and inorganic acids, derived from this heteroatom. Many techniques have being developed using this principle as the method of treatment, and have shown potential success in organic residue mineralization processes [92,96, 101-103].

The association of the electrochemical properties with the photocatalytic properties of a semiconductor, have allowed for the development of a promising system, representing a very efficient technique as an AOP method that has received much attention in recent years, assigned as a photoelectrochemical process. The process gained notoriety for the possibility of forming hydroxyl radicals via the oxidation of water. The technique is based on the action of ultraviolet light (hυ) on a semiconductor capable of generating charges and e^-/ H^+, whose separation is facilitated by applying a positive potential (E_{APP}), greater than the potential of a flat band photocatalytic material. The generation of a potential gradient in the photoactivated semiconductor directs the electrons to an auxiliary electrode (cathode), delaying recombination between the holes (h^+) generated in the valence band (VB), and making them available for oxidation processes and the generation of hydroxyl radicals of interest [104].

The results of this process were very promising because of the relatively short treatment time but with great efficiency, both in the removal of color and in the reduction of the organic load. However, the limitations of this technique are related mainly to the choice of the ideal catalyst for promoting the generation of these oxidizing species. Catalysts that promote the generation of radicals absorbing radiation in the visible spectral region are the most desirable for this type of reaction, due to the large percentage emitted in the solar spectrum (approximately 45%) [104].

Thus, the development of an ideal process that promotes color removal and a reduction in the organic load of wastewater from the textile industry with great efficiency is a major challenge in all fields of science, since the synthesis of the best catalyst to take advantage of solar radiation, thus reducing the operating costs, and at the same time solve the problems involved in the hydrodynamics of the reactors, is of importance in the development of the treatment.

The expectations for developing an effective method for the treatment of these wastes are quite promising, but require continuous optimization and knowledge of new aspects. These include better fixation of the dyes to the fibers, process with less water consumption, less hazardous dyes with respect to human health and methods capable of identifying these compounds with more efficacy and rapidity and assays to identify any potential carcinogenic and / or mutagenic properties in the dyes and their derivatives; genetic improvements to produce more efficient culture mediums and resistant biological treatments, leading to a reduction in the generation of sludge; the synthesis of materials that catalyze reactions in the visible spectral regions, leading to a more economic photoeletrochemical method, and also new engineering advances for the construction of more effective reactors, which can take advantage of all these developments in an integrated system, extending the performance of a process more appropriate for the treatment of such a complex effluent.

5. Optimization of the dyeing processes to reduce the environmental impact of the textile industry

The search and development of new methods to promote the treatment of effluents from the textile industry with a maximum of efficiency of the process of decolorization and / or removal of these compounds present in the medium can trigger further damage human health and the environment is fundamental importance. The understanding of the composition of waste generated is extremely significant to develop these methods of treatment due to the high complexity by virtue of huge number of compounds which are added at different stages of the dyeing fabrics.

Environmental problems with used dye baths are related to the wide variety of different components added to the dye bath, often in relatively high concentrations. In the future, many of textile factories will face the requirement of reusing a significant part of all incoming freshwater because traditionally used methods are insufficient for obtaining the required water quality.

However, due to dwindling supply and increasing demand of water in the textile industries, a better alternative is to attempt to further elevate the water quality of wastewater effluent from a secondary wastewater treatment plant to a higher standard for reuse. Thus far very little attention has been paid to this aspect [105].

Therefore, the investment in the search for methodologies to more effective treatment of these effluents can be much smaller than that spent in tertiary treatment to remove these products in low level of concentrations and in the presence of much other interference. This requires action that the cost / benefit are reviewed and the development of new techniques for wastewater treatment capable of effective removal of these dyes is intensified and made economically viable [105,106].

An alternative to minimize the problems related to the treatment of textile effluents would be the development of more effective dye that can be fixed fiber with higher efficiency decreasing losses on tailings waters and reducing the amount of dye required in the dyeing process, reducing certainly improve the cost and quality of the effluent.

6. Conclusion

It was concluded that the synthetic textile dyes represent a large group of organic compounds that could have undesirable effects on the environment, and in addition, some of them can pose risks to humans. The increasing complexity and difficulty in treating textile wastes has led to a constant search for new methods that are effective and economically viable. However, up to the present moment, no efficient method capable of removing both the color and the toxic properties of the dyes released into the environment has been found.

Acknowledgments

This work was supported by the Faculty of Pharmaceutical Sciences at Ribeirão Preto - University of São Paulo, Brazil, and by FAPESP, CAPES and CNPq.

Author details

Farah Maria Drumond Chequer[1], Gisele Augusto Rodrigues de Oliveira[1],
Elisa Raquel Anastácio Ferraz[1], Juliano Carvalho Cardoso[2],
Maria Valnice Boldrin Zanoni[2] and Danielle Palma de Oliveira[1]

1 USP, Department of Clinical, Toxicological and Bromatological Analyses, Faculty of Pharmaceutical Sciences at Ribeirão Preto, University of São Paulo, Ribeirão Preto, SP, Brazil

2 UNESP, Department of Analytical Chemistry, Institute of Chemistry at Araraquara, State of São Paulo University Júlio de Mesquita Filho, Araraquara, SP, Brazil

References

[1] Kirk-Othmer. Encyclopedia of Chemical Technology, v. 7, 5th Edition. Wiley-Interscience; 2004.

[2] Bafana A, Devi SS, Chakrabarti T. Azo dyes: past, present and the future. Environmental Reviews 2011; 19 350–370.

[3] Zollinger, H. Synthesis, Properties of Organic Dyes and Pigments. In: Color Chemistry. New York, USA: VCH Publishers; 1987. p. 92-102.

[4] Carneiro PA, Nogueira, RFP, Zanoni, MVB. Homogeneous photodegradation of C.I. Reactive Blue 4 using a photo-Fenton process under artificial and solar irradiation. Dyes and Pigments 2007; 74 127-132.

[5] Christie R. Colour Chemistry. Cambridge, United Kingdom: The Royal Society of Chemistry; 2001.

[6] Dos Santos AB, Cervantes FJ, van Lier JB. Review paper on current technologies for decolourisation of textile wastewaters: Perspectives for anaerobic biotechnology. Bioresource Technology 2007; 98 (12) 2369-2385.

[7] Arun Prasad AS, Bhaskara Rao KV. Physico chemical characterization of textile effluent and screening for dye decolorizing bactéria. Global Journal of Biotechnology and Biochemistry 2010; 5(2) 80-86.

[8] Robinson T, McMullan G, Marchant R, Nigam P. Remediation of dyes in textile effluent: a critical review on current treatment technologies with a proposed alternative. Bioresource Technology 2001; 77 (12) 247-255.

[9] Ogugbue CJ, Sawidis T. Bioremediation and Detoxification of Synthetic Wastewater Containing Triarylmethane Dyes by Aeromonas hydrophila Isolated from Industrial Effluent. Biotechnology Research International 2011; DOI 10.4061/2011/967925.

[10] Perkins WS. A Review of Textile Dyeing Processes. American Association of Textile Chemists and Colorists 1991; 23 (8): 23–27.http://www.revistavirtualpro.com/files/TIE08_200704.pdf (accessed 12 May 2012)

[11] Couto SR. Dye removal by immobilised fungi. Biotechnology Advances 2009; 27(3) 227-235.

[12] O'Neill C, Hawkes FR, Hawkes DL, Lourenço ND, Pinheiro HM, Delée W. Colour in textile effluents – sources, measurement, discharge consents and simulation: a review. Journal of Chemical Technology and Biotechnology 1999; 74 (11) 1009-1018.

[13] Forgacs E, Cserháti T, Oros G. Removal of synthetic dyes from wastewaters: a review. Environment International 2004; 30 (7) 953- 971

[14] Przystaś W, Zabłocka-Godlewska E, Grabińska-Sota E. Biological Removal of Azo and Triphenylmethane Dyes and Toxicity of Process By-Products. Water Air Soil Pollut 2012; 223 (4) 1581-1592.

[15] Sen S, Demirer GN. Anaerobic treatment of real textile wastewater with a fluidized bed reactor. Water Research 2003; 37 (8) 1868-1878.

[16] Dos Santos, A B. Reductive Decolourisation of Dyes by Thermophilic Anaerobic Granular Sludge. PhD-Thesis. Wageningen University, The Netherlands, 2005.

[17] Ben Mansour H, Houas I, Montassar F, Ghedira K, Barillier D, Mosrati R, Chekir-Ghedira L. Alteration of in vitro and acute in vivo toxicity of textile dyeing wastewater after chemical and biological remediation. Environmental science and pollution research international 2012; DOI 10.1007/s11356-012-0802-7.

[18] Talarposhti AM, Donnelly T, Anderson GK. Colour removal from a simulated dye wastewater using a two-phase anaerobic packed bed reactor. Water Research 2001; 35 (2) 425-432.

[19] Ibrahim MB, Poonam N, Datel S, Roger M. Microbial decolorization of textile dye-containing effluents: a review, Bioresource Technology 1996; 58(3) 217-227.

[20] Wijetunga S, Li XF, Jian C. Effect of organic load on decolourization of textile wastewater containing acid dyes in upflow anaerobic sludge blanket reactor. Journal of Hazardous Materials 2010; 177 (1-3) 792-798.

[21] Vaidya AA, Datye KV. Environmental pollution during chemical processing of synthetic fibers. Colourage 1982; 14 3-10.

[22] Rajaguru P, Fairbairn LJ, Ashby J, Willington MA, Turner S, Woolford LA, Chinnasamy N, Rafferty JA. Genotoxicity studies on the azo dye Direct Red 2 using the in vivo mouse bone marrow micronucleus test. Mutation Research 1999; 444(1) 175-180.

[23] Hubbe MA, Beck KR, O'Neal WG, Sharma YC. Cellulosic substrates for removal of pollutants from aqueous systems: a review. 2. Dyes. Dye biosorption: Review. BioResources 2012; 7 (2), 2592-2687.

[24] Carliell CM, Barclay SJ, Shaw C, Wheatley AD, Buckley C A. The effect of salts used in textile dyeing on microbial decolourisation of a reactive azo dye. Environmental Technology 1998; 19 (11) 1133-1137.

[25] Seesuriyachan P, Takenaka S, Kuntiya A, Klayraung S, Murakami S, Aoki K. Metabolism of azo dyes by Lactobacillus casei TISTR 1500 and effects of various factors on decolorization. Water Research 2007; 41(5) 985-992.

[26] Ben Mansour H, Corroler D, Barillier D, Ghedira K, Chekir L, Mosrati R. Evaluation of genotoxicity and pro-oxidant effect of the azo dyes: Acids yellow 17, violet 7 and orange 52, and of their degradation products by Pseudomonas putida mt-2. Food and Chemical Toxicology 2007; 45 (9) 1670-1677.

[27] Chung KT, Cerniglia CE. Mutagenicity of azo dyes: Structure-activity relationships. Mutation Research, 1992; 277 (3) 201-220.

[28] Pinheiro HM, Touraud E, Thomas O. Aromatic amines from azo dye reduction: status review with emphasis on direct UV spectrophotometric detection in textile industry wastewaters. Dyes and Pigments 2004; 61 (2) 121-139.

[29] Umbuzeiro GA, Freeman H, Warren SH, Kummrow F, Claxton LD. Mutagenicity evaluation of the commercial product CI Disperse Blue 291 using different protocols of the Salmonella assay. Food and Chemical Toxicology 2005; 43 (1) 49-56.

[30] Arlt VM, Glatt H, Muckel E, Pabel U, Sorg BL, Schmeiser HH, Phillips DH. Metabolic activation of the environmental contaminant 3-nitrobenzanthrone by human acetyltransferases and sulfotransferase. Carcinogenesis 2002; 23 (11) 1937-1945.

[31] Hao OJ, Kim H, Chiang PC, 2000. Decolorization of wastewater. Critical Reviews in Environmental Science and Technology 2000; 30 (4) 449-505.

[32] Firmino PIM, Silva MER, Cervantes FJ, Santos AB. Colour removal of dyes from synthetic and real textile wastewaters in one- and two-stage anaerobic systems. Bioresource Technology 2010; 101(20) 7773-7779.

[33] Carneiro PA, Umbuzeiro GA, Oliveira DP, Zanoni MVB. Assessment of water contamination caused by a mutagenic textile effluent/dyehouse effluent bearing disperse dyes. Journal of Hazardous Materials 2010; 174 (1-3) 694-699.

[34] Barani H, Montazer M. A Review on Applications of Liposomes in Textile Processing. Journal of Liposome Research 2008; 18 (3) 249-262.

[35] Khatri Z, Memon MH, Khatri A, Tanwari A. Cold Pad-Batch dyeing method for cotton fabric dyeing with reactive dyes using ultrasonic energy. Ultrasonics Sonochemistry 2011 18 (6) 1301-1307.

[36] Guaratini CCI, Zanoni MVB. Textile dyes. Química Nova 2000;23(1): 71-78. http://www.scielo.br/pdf/qn/v23n1/2146.pdf (accessed 01 May 2012).

[37] Moore SB, Ausley LW. Systems thinking and green chemistry in the textile industry: concepts, technologies and benefits. Journal of Cleaner Production 2004;12 585–601. DOI: 10.1016/S0959-6526(03)00058-1

[38] Reddy SS, Kotaiah B, Reddy NSP. Color pollution control in textile dyeing industry effluents using tannery sludge derived activated carbon. Bulletin of the Chemical Society of Ethiopia 2008;22 (3): 369-378. http://www.ajol.info/index.php/bcse/article/viewFile/61211/49389 (accessed 16 May 2012).

[39] Alcantara MR, Daltin D. A química do processamento têxtil. Química Nova 1996;19(3):   320-330   http://www.quimicanova.sbq.org.br/qn/qnol/1996/vol19n3/v19_n3_17.pdf (accessed 19 April 2012).

[40] Vassileva V, Valcheva E, Zheleva Z. The kinetic model of reactive dye fixation on cotton fibers. Journal of the University of Chemical Technology and Metallurgy 2008;43 (3): 323-326. http://www.uctm.edu/j2008-3/8_V_Vasileva_323-326.pdf (accessed 10 May 2012).

[41] Zollinger H. Color Chemistry. Syntheses,properties and applications of organic dyes and pigments. Weinheim: VCH Publishers; 2003.

[42] Candlin J. Polymers. Dyestuffs in the Chemical Industry. London: Chapman & Hall; 1994.

[43] Gregory P. Dyestuffs. In: Heaton C.A. (ed.) The Chemical Industry. London: Chapman & Hall; 1994. p143-188.

[44] Gregory P. 2009. Dyes and Dye Intermediates. In: Kirk-Othmer (ed.) Encyclopedia of Chemical Technology. New Jersey: John Wiley & Sons; 2009. p1–66. DOI: 10.1002/0471238961.0425051907180507.a01.pub2

[45] Pang K, Kotek R, Tonelli A. Review of conventional and novel polymerization processes for polyesters. Progress in Polymer Science 2006;31(11) 1009–1037. DOI: 10.1016/j.progpolymsci.2006.08.008

[46] Hunger K., editor. Industrial Dyes: Chemistry, Properties, Applications. Weinheim: Wiley-VCH Verlag GmbH & Co. KGaA; 2004. DOI: 10.1002/3527602011

[47] Baptista ALF, Coutinho PJG,: Real Oliveira MECD, Rocha Gomes JIN. Effect of surfactants in soybean Lecithin liposomes studied by energy transfer between NBD-PE and N-Rh-PE. Journal of Liposome Research 2000; 10 (4) 419-429.

[48] Lesoin L, Crampon C, Boutin O, Badens E. Preparation of liposomes using the supercritical anti-solvent (SAS) process and comparison with a conventional method. The Journal of Supercritical Fluids 2011; 57 162-174

[49] Sivasankar, M, Katyayani, T. Liposomes – The future of formulations. International Journal of Research in Pharmacy and Chemistry 2011; 1(2) 259-267.

[50] El-Zawahry MM, El-Shami S, El-Mallah S.H. Optimizing a wool dyeing process with reactive dye by liposome microencapsulation. Dyes and Pigments 2007; 74 684-691.

[51] Martí M, de la Maza A, Parra JL, Coderch L. Liposome as dispersing agent into disperse dye formulation. Textile Research Journal 2011; 81(4) 379-387.

[52] Montazer M., Validi M., Toliyat T. Influence of Temperature on Stability of Multilamellar Liposomes in Wool Dyeing. Journal of Liposome Research, 2006; 16:81-89.

[53] Martí M, Coderch L, de la Maza A, Parra JL. Liposomes of phosphatidylcholine: a biological natural surfactant as a dispersing agent. Color Technology 2007; 123 237-241.

[54] Riaz M. Liposomes preparation methods. Pakistan Journal of Pharmaceutical Sciences, 1996; 19(1) 65-77.

[55] Martí M, de la Maza A, Parra JL, Coderch L. Liposome as dispersing agent into disperse dye formulation. Textile Research Journal 2010; 81(4) 379–387.

[56] Nelson, G. Application of microencapsulation in textile. International Journal of Pharmaceutics 2002; 242 55-62.

[57] Martí M, de la Maza A, Parra JL, Coderch L. Dyeing wool at low temperatures: new method using liposomes. Textile Research Journal 2001; 71(8) 678-682.

[58] El-Zawahry MM, El-Mallahb MH, El-Shamib S. An innovative study on dyeing silk fabrics by modified phospholipid liposomes. Coloration Technology 2009; 125 164-171.

[59] Vajnhandl S, Le Marechal AM. Ultrasound in textile dyeing and the decolouration/ mineralization of textile dyes. Dyes and Pigments 2005; 65 89-101.

[60] Ferrero F., Periolatto M. Ultrasound for low temperature dyeing of wool with acid dye. Ultrasonics Sonochemistry 2012; 19 601-606.

[61] Vankar PS, Shanker R. Ecofriendly ultrasonic natural dyeing of cotton fabric with enzyme pretreatments. Desalination 2008; 230 62-69.

[62] Mansour HF, Heffernan S. Environmental aspects on dyeing silk fabric with sticta coronate lichen using ultrasonic energy and mild mordants. Clean Technologies Environmental Policy 2011; 13 207-213.

[63] Abou-Okeil A, El-Shafie A, El Zawahry MM. Ecofriendly laccase–hydrogen peroxide/ultrasound-assisted bleaching of linen fabrics and its influence on dyeing efficiency. Ultrasonics Sonochemistry 2010; 17 383-390.

[64] Kunz A, Zamora PP, Moraes SG, Durán N. New tendencies on the textile effluent treatment. Quimica Nova 2002;25(1): 78-82. http://www.scielo.br/scielo.php?script=sci_arttext&pid=S0100-40422002000100014 (accessed 01 May 2012).

[65] Associação Brasileira de Química. ABQ. Processos oxidativos avançados e tratamento de corantes sintéticos. Anais da ABQ. http://www.unb.br/resqui/abq2004-1.pdf (accessed 18 May 2012).

[66] Umbuzeiro GA, Roubicek DA, Rech CM, Sato MIZ, CLAXTON LD. Investigating the sources of the mutagenic activity found in a river using the Salmonella assay and dif-

ferent water extraction procedures. Chemosphere 2004; 54(11) 1589-1597. DOI: 10.1016/j.chemosphere.2003.09.009

[67] Umbuzeiro GA, Freeman HS, Warren SH, Oliveira DP, Terao Y, Watanabe T, Claxton LD. The contribution of azo dyes to the mutagenic activity of the Cristais river. Chemosphere 2005; 60 (1) 55-64. DOI:10.1016/j.chemosphere.2004.11.100

[68] Shiozawa T, Suyama K, Nakano K, Nukaya H, Sawanishi H, Oguri A, Wakabayashi K, Terao Y. Mutagenic activity of 2 phenylbenzotriazole derivatives related to a mutagen, PBTA-1, in river water. Mutation Research 1999; 442 (2) 105-111.

[69] Oliveira DP. Corantes como importante classe de contaminantes ambientais – um estudo de caso. PhD thesis. Universidade de São Paulo; 2005.

[70] Carneiro PA, Osugi ME, Sene JJ, Anderson MA, Zanoni MVB. Evaluation of color removal and degradation of a reactive textile azo dye on nanoporous TiO thin-film electrodes. Electrochimica Acta 2004; 49 (22-23) 3807–3820. DOI: 10.1016/j.electacta. 2003.12.057

[71] McMullan G, Meehan C, Conneely A, Kirby N, Robinson T, Nigam P, Banat IM, Marchant R, Smyth WF. Mini-review: microbial decolorisation and degradation of textile dyes. Applied Microbiology and Biotechnology 2001; 56 (1-2): 81-87. DOI: 10.1007/s002530000587. http://www.springerlink.com/content/04vuqljhajpa32rv/ (accessed 10 April 2012).

[72] Pereira WS, Freire RS. Ferro zero: Uma nova abordagem para o tratamento de águas contaminadas com compostos orgânicos poluentes. Química Nova 2005;28(1): 130-136. DOI: 10.1590/S0100-40422005000100022. http://www.scielo.br/scielo.php? script=sci_arttext&pid=S0100-40422005000100022 (accessed 10 March 2012).

[73] Houk VS. The genotoxicity of industrial wastes and effluents. Mutation Research 1992; 277 (2) 91-138.

[74] Jager I, Hafner C, Schneider K. Mutagenicity of different textile dye products in Salmonella typhimurium and mouse lymphoma cells. Mutation Research 2004;561 (1-2) 35-44.

[75] Chequer FMD, Angeli JPF, Ferraz ERA, Tsuboy MS, Marcarini JC, Mantovani MS, Oliveira DP. The azo dyes Disperse Red 1 and Disperse Orange 1 increase the micronuclei frequencies in human lymphocytes and in HepG2 cells. Mutation Research 2009; 676 (1-2) 83-86.

[76] Chequer FMD, Lizier TM, Felicio R, Zanoni MVB, Debonsi HM, Lopes NP, Marcos R, Oliveira DP. Analyses of the genotoxic and mutagenic potential of the products formed after the biotransformation of the azo dye Disperse Red 1. Toxicology in Vitro 2011; 25 (8) 2054-2063

[77] Ferraz ERA, Umbuzeiro GA, de-Almeida G, Caloto-Oliveira A, Chequer FMD, Zanoni MVB, Dorta DJ, Oliveira DP. Differential Toxity of Disperse Red 1 and Disperse

Red 13 in the Ames Test, HepG2 Cytotoxicity Assay, and Daphnia Acute Toxicity Test. Environmental Toxicology 2011; 26 (5) 489-497

[78] Oliveira GAR, Ferraz ERA, Chequer FMD, Grando MD, Angeli JPF, Tsuboy MS, Marcarini JC, Mantovani MS, Osugi ME, Lizier TM, Zanoni, MVB, Oliveira DP. Chlorination treatment of aqueous samples reduces, but does not eliminate, the mutagenic effect of the azo dyes Disperse Red 1, Disperse Red 13 and Disperse Orange 1. Mutation Research 2010 703 (2) 200-208

[79] Konstantinou IK, Albanis TA. TiO_2-assisted photocatalytic degradation of azo dyes in aqueous solution: kinetic and mechanistic investigations - A review. Applied Catalysis B: Environmental 2004; 49 (1) 1-14. DOI:10.1016/j.apcatb.2003.11.010.

[80] Hessel C, Allegre C, Maisseu M, Charbit F, Moulin P. Guidelines and legislation for dye house effluents. Journal of Environmental Management 2007; 83 (2) 171-180. DOI: 10.1016/j.jenvman.2006.02.012.

[81] Carneiro PA, Osugi ME, Fugivara CS, Boralle N, Furlan M, Zanoni MVB. Evaluation of different electrochemical methods on the oxidation and degradation of Reactive Blue 4 in aqueous solution. Chemosphere. 2005; 59(3) 431-439.

[82] Mohan SV, Bhaskar YV, Karthikenyan J. Biological decolourization of simulated azo dye in aqueous phase by algae Spirogyra species. International Journal of Environment and Pollution. 2004; 21 (3) 211-222.

[83] Vaghela SS, Jethva AD, Mehta BB, Dave SP, Adimurthy S, Ramachandraiah G. Laboratory studies of electrochemical treatment of industry azo dye effluent. Environmental Science and Technology. 2005; 39 (8) 2848-2855.

[84] Cardoso JC, Lizier TM, Zanoni MVB. Highly ordered TiO2 nanotube arrays and photoelectrocatalytic oxidation of aromatic amine. Applied Catalysis, B: Environmental. Applied Catalysis, B: Environmental. 2010; 99 (1-2) 96-102.

[85] Pereira RS. Identificação e caracterização das fontes de poluição em sistemas hídricos. Revista Eletrônica de Recursos Hídricos. IPH-UFRGS 2004; 1(1): 20-36. http://www.abrh.org.br/informacoes/rerh.pdf (accessed 06 June 2012).

[86] Baumgarten, MGZ.; Pozza, SA. Qualidade de Águas. Descrição de Parâmetros Químicos referidos na Legislação Ambiental. Ed: FURG; 2001.

[87] Prado AGS, Torres JD, Faria EA, Dias SCL. Comparative adsorption studies of indigo carmine dye on chitin and chitosan. Journal of Colloid and Interface Science. 2004; 1(1) 43-47.

[88] Babel S, Kurniawan TA. Low-cost adsorbents for heavy metals uptake from contaminated water: a review. Journal of Hazardous Materials. 2003; 97(1-3) 219-243.

[89] Associação Brasileira da Indústria Têxtil e de Confecção. Economia. <http://www.abit.org.br/site/navegacao.asp?id_menu=8&IDIOMA=PT> (accessed 19 August 2011).

[90] Walsh FC. Electrochemical technology for environmental treatment and clean energy conversion. Pure and Applied Chemistry. 2001; 73 (12) 1819-1837.

[91] Flores ER, Perez F, Torre M. Scale-up of Bacillus thuringiensis fermentation based on oxygen transfer. Journal of Fermentation and Bioengineering. 1997; 83 (6) 561-564.

[92] Bornick H. Simulation of biological degradation of aromatic amines in river bed sediments. Water Research. 2001; 35 (3) 619-624.

[93] Wang L, Barrington S, Kim JW. Biodegradation of pentylamine and aniline from petrochemical wastewater. Journal of Environmental Management. 2007; 83 (2) 191-197.

[94] Orshansky F, Narkis N. Characteristics of organics removal by pact simultaneous adsorption and biodegradation. Water Research. 1997; 31(3) 391-398.

[95] Fukushima M, Tatsumi K, Morimoto K. The fate of aniline after a photo-fenton reaction in an aqueous system containing iron(III), humic acid, and hydrogen peroxide. Environmental Science and Technology. 2010; 34 (10) 2006-2013.

[96] Pramauro E. et al. Photocatalytic treatment of laboratory wastes containing aromatic amines. Analyst. 1995; 120 (2) 237-242.

[97] Augugliaro V. et al. Photodegradation kinetics of aniline, 4-ethylaniline, and 4-chloroaniline in aqueous suspension of polycrystalline titanium dioxide. Research on Chemical Intermediates. 2000; 26 (5) 413-426.

[98] Canle LM, Santaballa JA, Vulliet E. On the mechanism of TiO2-photocatalyzed degradation of aniline derivatives. Journal of Photochemistry and Photobiology, A: Chemistry. 2005; 175 (2-3) 192-200.

[99] Chu W, Choy WK, So TY. The effect of solution pH and peroxide in the TiO2-induced photocatalysis of chlorinated aniline. Journal of Hazardous Materials. 2007; 141 (1) 86-91.

[100] Low, G. K. C.; McEvoy, S. R.; Matthews, R. W. Formation of nitrate and ammonium ions in titanium dioxide mediated photocatalytic degradation of organic compounds containing nitrogen atoms. Environmental Science and Technology, v. 25, n. 3, p. 460-467, 1991.

[101] Pinheiro HM, Touraud E, Thomas O. Aromatic amines from azo dye reduction: status review with emphasis on direct UV spectrophotometric detection in textile industry wastewaters. Dyes and Pigments. 2004; 61(2) 121-139.

[102] Fukushima M, Tatsumi K, Morimoto K. The Fate of Aniline after a Photo-Fenton Reaction in an Aqueous System Containing Iron(III), Humic Acid, and Hydrogen Peroxide. Environmental Science and Technology. 2006; 34 (10) 2006-2013.

[103] Chu W, Choy WK, So TY. The effect of solution pH and peroxide in the TiO2-induced photocatalysis of chlorinated aniline. Journal of Hazardous Materials. 2007; 141 (1) 86-91.

[104] Finklea HO. Semiconductor electrodes. New York: Elsevier; 1998.

[105] Lin SH, Chen ML. Purification of textile wastewater effluents by a combined Fenton process and the ion exchange. Desalination 1997; 109 (1) 121-130.

[106] Mozia S. Photocatalytic membrane reactor (PMRs) in water and wastewater treatment. A review. Separation and Purification Technology 2010; 73 (2) 71-91.

Chapter 7

Decolorization of Dyeing Wastewater Using Polymeric Absorbents - An Overview

George Z. Kyzas, Margaritis Kostoglou,
Nikolaos K. Lazaridis and Dimitrios N. Bikiaris

Additional information is available at the end of the chapter

http://dx.doi.org/10.5772/52817

1. Introduction

The exact amount of dyes produced in the world is not known. Exact data on the quantity of dyes discharged in the environment are also not available. It is assumed that a loss of 1-2% in production and 1-10% loss in use (after being garment) are a fair estimate [1]. For reactive dyes, this figure can be about 10-20% due to low fixation. Due to large-scale production and extensive application, synthetic dyes can cause considerable environmental pollution and are serious health-risk factors. Although, the growing impact of environmental protection on industrial development promotes the development of eco-friendly technologies, reduced consumption of freshwater and lower output of wastewater [1], the release of important amounts of synthetic dyes to the environment causes public concern, legislation problems and is a serious challenge to environmental scientists.

Globally, accessing to freshwater is becoming more acute every day. In the dyeing of textile materials, water is used firstly in the form of steam to heat the treatment baths, and secondly to enable the transfer of dyes to the fibers. Cotton, which is the world's most widely used fiber, is also the substrate that requires the most water in its processing [2]. The dyeing and rinsing of 1 kg of cotton with reactive dyes demands from 70 to 150 L water, 0.6 to 0.8 kg NaCl and anywhere from 30 to 60 g dyestuff. More than 80,000 tn of reactive dyes are produced and consumed each year, making it possible to estimate the total pollution caused by their use. After the dyeing is completed, the various treatment baths are drained out, including the first dye bath, which has a very high salt concentration, is heavily coloured and contains a substantial load of organic substances [2]. One solution to this problem consists in mixing together all the different aqueous effluents, then concentrating the pollution and re-using the water either as rinsing water or as processing water, depending on the treatment

selected (either nanofiltration or reverse osmosis for the membrane processes). These treatments apply only to very dilute dye baths [2]. This is generally not the case of the first dye baths recovered which are the most heavily polluted ones. The wastewater produced by a reactive dyeing contains [2,3]: (i) hydrolyzed reactive dyes not fixed on the substrate, representing 20-30% of the reactive dyes applied (on average 2 g/L) (this residual amount is responsible for the coloration of the effluents and cannot be recycled); (ii) dyeing auxiliaries or organic substances, which are non-recyclable and responsible for the high BOD/COD of the effluents; (iii) textile fibres, and (iv) 60-100 g/L electrolyte, essentially NaCl and Na_2CO_3, which is responsible for the very high saline content of the wastewater.

In addition, these effluents exhibit a pH of 10-11 and a high temperature (50-70 °C). The legal regulations respecting the limit values for the release of wastewater are changing and are becoming increasingly severe, including the limits with respect to salinity.

A typical effluent treatment is broadly classified into preliminary, primary, secondary, and tertiary stages [4,5]. The preliminary stage includes equalization and neutralization. The primary stage involves screening, sedimentation, flotation, and flocculation. The secondary stage reduces the organic load and facilitates the physical/chemical separation (biological oxidation). The tertiary stage is focused on decolorization. In the latter, adsorption onto various materials can be broadly used to limit the concentration of colour in effluents [3].

1.1. Dyes

The first synthetic dye, Mauveine was discovered by the Englishman, William Henry Perkin by chance in 1856. Since then the dyestuffs industry has matured [6].

A dye or dyestuff is a coloured compound that can be applied on a substrate. With few exceptions, all synthetic dyes are aromatic organic compounds. A substrate is the material to which a colorant is applied by one of the various processes of dyeing, printing, surface coating, and so on. Generally, the substrate includes textile fibers, polymers, foodstuffs, oils, leather, and many other similar materials [7].

Not all coloured compounds are dyestuffs because a coloured compound may not have suitable application on a substrate. For example, a chemical such as copper sulphate, which is coloured, finds no application on any substrate. If it is applied on a substrate, it will not have retaining power on the substrate and for this reason copper sulphate cannot be termed as a dye. On the other hand, congo red, a typical organic coloured compound. When applied to cotton under suitable conditions can be retained on this natural fibre and due to this finds useful application on this fibre. It is termed as a dyestuff [7].

In the field of chemistry, chromophores and auxochromes are the major component element of dye molecule. Dyes contain an unsaturated group basically responsible for colour and designated it as chromophore ("chroma" means colour and "phore" means bearer) (Table 1). Auxochromes ("Auxo" means augment) are the characteristic groups which intensify colour and/or improve the dye affinity to substrate [7] (Table 1).

Chromophore group	Name	Auxogroup	Name
$-N=N-$	Azo	$-NH_2$	Amino
$-N=N^+-O^-$	Azoxy	$-NHCH_3$	Methyl amino
$-N=N-NH$	Azoamino	$-N(CH_3)_2$	Dimethyl amino
$-N=O, N-OH$	Nitroso	$-SO_3H$	Sulphonic acid
$>C=O$	Carbonyl	$-OH$	Hydroxy
$>C=C<$	Ethenyl	$-COOH$	Carboxylic acid
$>C=S$	Thio	$-Cl$	Chloro
$-NO_2$	Nitro	$-CH_3$	Methyl
$>C=NH, >C=N-$	Azomethine	$-OCH_3$	Methoxy
		$-CN$	Cyano
		$-COCH_3$	Acetyl
		$-CONH_2$	Amido

Table 1. Names of chromophore and auxochrome groups of dyes

To further examine the interactions between dyes and substrates, the classification of dyes is required. It is very important to know the chemistry of the dyes in dyeing effluents, in order to synthesize a suitable adsorbent with the appropriate functional group. Hunger et al [7] mentioned that dyes are classified in two methods. The main classification is related to the chemical structure of dyes and particularly considering the chromophoric structure presented in dye molecules. Another type of classification is based on their usage or applying. The classification of dyes by usage or application is the most important system adopted by the Colour Index (CI). Briefly [8]:

Reactive dyes. These dyes form a covalent chemical bond with fiber is ether or ester linkage under suitable conditions. Majority of reactive dyes contains azo that includes metallized azo, triphendioxazine, phthalocyanine, formazan, and anthraquinone. The molecular structures of these dyes are much simpler than direct dyes. They also produce brighter shades than direct dyes. Reactive dyes are primarily used for dyeing and printing of cotton fibers.

Direct dyes. In the presence of electrolytes, these anionic dyes are water-soluble in aqueous solution. They have high affinity to cellulose fibers. Most of the dyes in this class are polyazo compounds, along with some stilbenes, phthalocyanines, and oxazines. To improve wash fastness, frequently chelations with metal (such as copper and chromium) salts are applied to the dyestuff. Also, their treatment with formaldehyde or a cationic dye-complexing resin.

Disperse dyes. These are substantially water insoluble nonionic dyes applied to hydrophobic fibers from microfine aqueous dispersion. They are used predominantly on polyester, polyamide, polyacrylonitrile, polypropylene fibers to a lesser, it is used to dye nylon, cellulose acetate, and acrylic fibers. Chemical structures of dyes are mainly consisted of azo and anthraquinonoid groups, having low molecular weight and containing groups which aid in forming stable aqueous dispersions.

Vat dyes. These dyes are water insoluble and can apply mainly to cellulose fiber by converting them to their leuco compounds. The latter was carried out by reduction and solubilization with sodium hydrosulphite and sodium hydroxide solution, which is called "vatting process". The main chemical/structural groups of vat dyes are anthraquinone and indigoid.

Sulfur dyes. They are water insoluble and are applied to cotton in the form of sodium salts by the reduction process using sodium sulphide as the reducing agent under alkaline conditions. The low cost and good wash fastness properties of dyeing makes these dyes economic attractive.

Cationic (Basic dyes). These dyes are cationic and water soluble. They are applied on paper, polyacrylonitrile, modified nylons, and modified polyesters. In addition, they are used to apply with silk, wool, and tannin–mordant cotton when brightness shade was more necessary than fastness to light and washing.

Acid dyes. They are water soluble anionic dyes and are applied on nylon, wool, silk, and modified acrylics. Moreover, they are used to dye paper, leather, inkjet printing, food, and cosmetics.

Solvent dyes. They are water insoluble, but solvent soluble, dyes having deficient polar solubility group for example sulfonic acid, carboxylic acid or quaternary ammonium. They are used for colouring plastics, gasoline, oils, and waxes.

Mordant dyes. These dyes have mordant dyeing properties with good quality in the presence of certain groups in the dye molecule. These groups are capable to hold metal residuals by formation of covalent and coordinate bonds involving a chelate compound. The salts of aluminium, chromium, copper, cobalt, nickel, iron, and tin are used as mordant that their metallic salts.

Aside from mentioned above, there are azoic dyes, ingrain dyes, pigment [6,7,9]. Comparative analysis of dye classes are presented in Table 2:

Class	Fiber type	Chemistry	Characteristics
Acid	nylon, wool, silk, paper ink, leather	Azo, anthraquinone, azine, xanthene, nitro, nitroso	- Anionic compounds - Highly water soluble - Poor wet fastness
Azoic	cotton, rayon, cellulose acetate, polyester	azo	- Contain azo group - Metallic compounds
Basic	leather, wool, silk paper, modified nylon, polyacrylonitrile,	cyanine, azo, azine, triarylmethane, xanthen,	- Cationic - Highly water soluble -

Class	Fiber type	Chemistry	Characteristics
	polyester, inks	acridine, oxazine, anthraquinone	
Direct	cotton, paper, rayon, leather, nylon	azo, phthalocyanine, stilbene, nitro, benzodifuranone	- Anionic compounds - Highly water soluble - Poor wet fastness.
Disperse	polyester, polyamide, acetate, acrylic, plastics	azo, anthraquinone, styryl, nitro, benzodifuranone	- Colloidal dispersion - Very low water solubility - Good wet fastness
Mordant	wool, leather, anodized aluminium	azo, anthraquinone	- Anionic compounds - Water soluble - Good wet fastness
Reactive	cotton, wool, silk, nylon	azo, anthraquinone, phthalocyanine, formazan	- Anionic compounds - Highly water soluble - Good wet fastness.

Table 2. Classification of dyes and their properties

1.2. Decolorization techniques

Methods of dye wastewater treatment have been reported [11-14]. Also, fungal and bacterial decolorization methods have been reviewed [15-18]. There are several reported methods for the removal of pollutants from effluents. The technologies can be divided into three categories: biological, chemical and physical [11]. All of them have advantages and drawbacks. Because of the high cost and disposal problems, many of these conventional methods for treating dye wastewater have not been widely applied at large scale in the textile and paper industries. At the present time, there is no single process capable of adequate treatment, mainly due to the complex nature of the effluents [19,20]. In practice, a combination of different processes is often used to achieve the desired water quality in the most economical way. A literature survey shows that research has been and continues to be conducted in the areas of combined adsorption-biological treatments in order to improve the biodegradation of dyestuffs and minimize the sludge production.

Biological treatment is often the most economical alternative when compared with other physical and chemical processes. Biodegradation methods such as fungal decolorization, microbial degradation, adsorption by (living or dead) microbial biomass and bioremediation systems are commonly applied to the treatment of industrial effluents because many microorganisms such as bacteria, yeasts, algae and fungi are able to accumulate and degrade different pollutants [14,16,17]. However, their application is often restricted because of technical constraints. Biological treatment requires a large land area and is constrained by sensitivity toward diurnal variation as well as toxicity of some chemicals, and less flexibility in design and operation [21]. Biological treatment is incapable of obtaining satisfactory color elimination with current conventional biodegradation processes [11]. Moreover, although many organic molecules are degraded, many others are recalcitrant due to their complex chemical structure and synthetic organic origin [22]. In particular, due to their xenobiotic nature, azo dyes are not totally degraded.

Chemical methods include coagulation or flocculation combined with flotation and filtration, precipitation-flocculation with Fe(II)/Ca(OH)$_2$, electroflotation, electrokinetic coagulation, conventional oxidation methods by oxidizing agents (ozone), irradiation or electrochemical processes. These chemical techniques are often expensive, and although the dyes are removed, accumulation of concentrated sludge creates a disposal problem. There is also the possibility that a secondary pollution problem will arise because of excessive chemical use. Recently, other emerging techniques, known as advanced oxidation processes, which are based on the generation of very powerful oxidizing agents such as hydroxyl radicals, have been applied with success for pollutant degradation. Although these methods are efficient for the treatment of waters contaminated with pollutants, they are very costly and commercially unattractive. The high electrical energy demand and the consumption of chemical reagents are common problems.

Different physical methods are also widely used, such as membrane-filtration processes (nanofiltration, reverse osmosis, electrodialysis) and adsorption techniques. The major disadvantage of the membrane processes is that they have a limited lifetime before membrane fouling occurs and the cost of periodic replacement must thus be included in any analysis of their economic viability. In accordance with the very abundant literature data, liquid-phase adsorption is one of the most popular methods for the removal of pollutants from wastewater since proper design of the adsorption process will produce a high-quality treated effluent. This process provides an attractive alternative for the treatment of contaminated waters, especially if the adsorbent is inexpensive and does not require an additional pre-treatment step before its application. Adsorption is a well known equilibrium separation process and an effective method for water decontamination applications [23]. Adsorption has been found to be superior to other techniques for water re-use in terms of initial cost, flexibility and simplicity of design, ease of operation and insensitivity to toxic pollutants. Adsorption also does not result in the formation of harmful substances.

1.3. Adsorption

Adsorption techniques for wastewater treatment have become more popular in recent years owing to their efficiency in the removal of pollutants, which are difficulty treated with biological methods. Adsorption can produce high quality water while also being a process that is economically feasible. Decolourisation is a result of two mechanisms (adsorption and ion exchange) and is influenced by many factors including dye/adsorbent interaction, adsorbent's surface area, particle size, temperature, pH and contact time.

Physical adsorption occurs when weak interparticle bonds exist between the adsorbate and adsorbent. Examples of such bonds are van der Waals, hydrogen and dipole-dipole. In the majority of cases physical adsorption is easily reversible [24]. Chemical adsorption occurs when strong interparticle bonds are present between the adsorbate and adsorbent due to an exchange of electrons. Examples of such bonds are covalent and ionic bonds. Two means that are useful to be developed are "chemisorption" and "physisorption". Chemisorption is a kind of adsorption which involves a chemical reaction between the surface and the absorbate. New chemical bonds are generated at the adsorbent surface. Examples include macroscopic phenomena that can be very obvious, like corrosion, and subtler effects associated with heterogeneous catalysis. The strong interaction between the adsorbate and the substrate surface creates new types of electronic bonds. In contrast with chemisorption is physisorption, which leaves the chemical species of the adsorbate and surface intact. It is conventionally accepted that the energetic threshold separating the binding energy of "physisorption" from that of "chemisorptions" is about 0.5 eV per adsorbed species. Chemisorption is deemed to be irreversible in the majority of cases [24]. Suzuki [25] covers the role of adsorption in water environmental processes and also the development of newer adsorbents to modernise the treatment systems. Most adsorbents are highly porous materials. As the pores are generally very small, the internal surface area is orders of magnitude greater than the external area.

Separation occurs because either the differences in molecular mass, shape or polarity causes some molecules to be held more strongly on the surface than others or the pores are too small to admit large molecules [25]. However, amongst all the adsorbent materials proposed, activated carbon is the most popular for the removal of pollutants from wastewater [26,27]. In particular, the effectiveness of adsorption on commercial activated carbons (CAC) for removal of a wide variety of dyes from wastewaters has made it an ideal alternative to other expensive treatment options [26]. Because of their great capacity to adsorb dyes, CAC are the most effective adsorbents. This capacity is mainly due to their structural characteristics and their porous texture which gives them a large surface area, and their chemical nature which can be easily modified by chemical treatment in order to increase their properties. However, activated carbon presents several disadvantages [27]. It is quite expensive, the higher the quality, the greater the cost, non-selective and ineffective against disperse and vat dyes. The regeneration of saturated carbon is also expensive, not straightforward, and results in loss of the adsorbent. The use of carbons based on relatively expensive starting materials is also unjustified for most pollution control applications [28]. This has led many workers to search for more economic adsorbents.

1.3.1. Polymeric adsorbents (chitosan)

The majority of commercial polymers and ion exchange resins are derived from petroleum-based raw materials using chemical processes that are not always safe or environmental friendly. Today, there is growing interest in developing natural low-cost alternatives to synthetic polymers [29]. Chitin (Figure 1), found in the exoskeleton of crustaceans, the cuticles of insects, and the cells walls of fungi, is the most abundant aminopolysaccharide in nature [30]. This low-cost material is a linear homopolymer composed of b(1-4)-linked N-acetyl glucosamine. It is structurally similar to cellulose, but it is an aminopolymer and has acetamide groups at the C-2 positions in place of the hydroxyl groups. The presence of these groups is highly advantageous, providing distinctive adsorption functions and conducting modification reactions. The raw polymer is only commercially extracted from marine crustaceans primarily because a large amount of waste is available as a by-product of food processing [30]. Chitin is extracted from crustaceans (shrimps, crabs, squids) by acid treatment to dissolve the calcium carbonate followed by alkaline extraction to dissolve the proteins and by a decolorization step to obtain a colourless product [30].

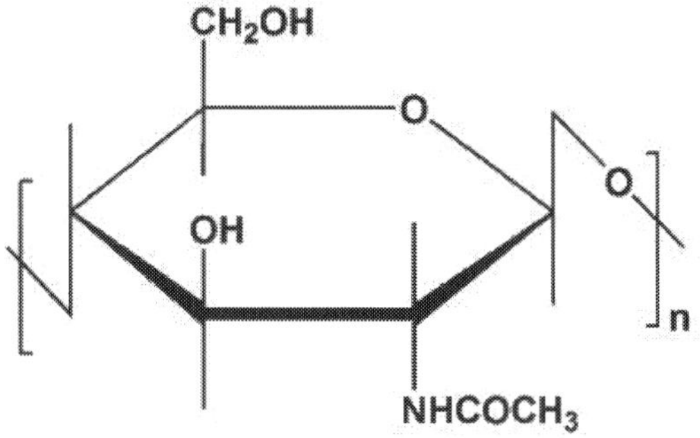

Figure 1. Chemical structures of chitin.

Since the biodegradation of chitin is very slow in crustacean shell waste, accumulation of large quantities of discards from processing of crustaceans has become a major concern in the seafood processing industry. So, there is a need to recycle these by-products. Their use for the treatment of wastewater from other industries could be helpful not only to the environment in solving the solid waste disposal problem, but also to the economy. However, chitin is an extremely insoluble material. Its insolubility is a major problem that confronts the development of processes and uses of chitin [30], and so far, very few large-scale industrial uses have been found. More important than chitin is its derivative, chitosan.

Figure 2. Chemical structures of chitosan.

Partial deacetylation of chitin results in the production of chitosan (Figure 2), which is a polysaccharide composed by polymers of glucosamine and N-acetyl glucosamine (Figure 3).

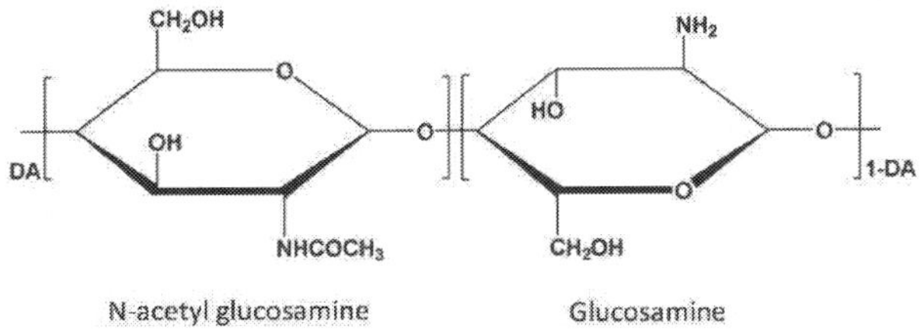

Figure 3. Chemical structures of commercial chitosan composed of N-acetyl glucosamine.

The "chitosan label" generally corresponds to polymers with less than 25% acetyl content. The fully deacetylated product is rarely obtained due to the risks of side reactions and chain depolymerization. Copolymers with various extents of deacetylation and grades are now commercially available. Chitosan and chitin are of commercial interest due to their high percentage of nitrogen compared to synthetically substituted cellulose. Chitosan is soluble in acid solutions and is chemically more versatile than chitin or cellulose. The main reason for this is undoubtedly its appealing intrinsic properties, as documented in a recent review [30], such as biodegradability, biocompatibility, film-forming ability, bioadhesivity, polyfunctionality, hydrophilicity and adsorption properties. Most of the properties of chitosan can be related to its cationic nature [30], which is unique among abundant polysaccharides and natural polymers. These numerous properties lead to the recognition of this polyamine as a promising raw material for adsorption purposes.

The elevated interest in chitin and chitosan is reflected by an increase in the number of articles published and talks given on this topic. Currently, these polymers and their numerous derivatives as described in reviews [29,31] are widely used in pharmacy, medicine, biotechnology, chemistry, cosmetics and toiletries, food technology, and the textile, agricultural, pulp and paper industries and other fields such as oenology, dentistry and photography. The potential industrial use of chitosan is widely recognized. These versatile materials are also widely used in clarification and water purification, water and wastewater treatment as coagulating, flocculating and chelating agents. However, despite a large number of publication on the use of chitosan for pollutant recovery, the research failed to find practical applications on the industrial scale: this aspect will be discussed later.

2. Synthesis of adsorbents

2.1. Grafting reactions

The basic idea of modifications is to make various changes in chitosan structure to enhance its properties (capacity, resistance etc). In particular, several researchers have proposed certain mofidifications in chitosan backbone to improve its adsorption capacity. These modifications are realized with grafting reactions [32-38]. The modifications can improve chitosan's removal performance and selectivity for dyes, alter the physical and mechanical properties of the polymer, control its diffusion properties and decrease the sensitivity of adsorption to environmental conditions. Many scientists suggested chemical grafting specific ligands [39,40]. However the only class for which chitosan [37] has low affinity is basic (cationic) dyes. To overcome this problem, the use of N-benzyl mono- and disulfonate derivatives of chitosan is suggested to enhance its cationic dye hydrophobic adsorbent properties and to improve its selectivity [32,33,41]. These derivatives could be used as hydrophobic adsorbents in acidic media without any cross-linking reactions. To enhance and further develop the high potentials of chitosan, it is necessary to add/introduce chemical substituents at a specific position in a controlled manner [37]. This chemical derivatization promotes new adsorption properties in particular towards basic dyes in acidic medium or reactive/acid dyes in basic medium. Another study deals with the enzymatic grafting of carboxyl groups onto chitosan as a mean to confer the ability to adsorb basic dyes on beads [37]. The presence of new functional groups on the surface of beads results in increased surface polarity and density of adsorption sites and hence improved adsorption selectivity for the target dye. Other studies showed that the ability of chitosan to selectively adsorb dyes could be further improved by chemical derivatization. Novel chitosan-based materials with long aliphatic chains are developed by reacting chitosan with high fatty acids and glycidyl moieties [36]. In this way, these products could be used as effective adsorption materials for both anionic and cationic dyes. Other researchers suggested the use of cyclodextrin-grafted chitosan derivatives as new chitosan derivatives for the removal of dyes [34,35,42]. These materials are characterized by a rate of adsorption and a global efficiency greater than that of the parent chitosan polymer [35].

2.2. Cross-linking reactions

The pure form of chitosan powders (raw) tends to present some disadvantages such as unsatisfactory mechanical properties and poor heat resistance. Another important limitation of the pure form is its solubility in acidic media and therefore it cannot be used as an insoluble adsorbent under these conditions (except after physical and chemical modification). The main technique to overcome these limitations is to transform the raw polymer into a form whose physical characteristics are more attractive. So, cross-linked beads have been developed and proposed. After cross-linking, these materials maintain their properties and original characteristics [43], particularly their high adsorption capacity, although this chemical modification results in a decrease in the density of free amine groups at the surface of the adsorbent in turn lowering polymer reactivity towards metal ions [44,45]. The cross-linking agent is very important. Therefore many researchers studied the chitosan behaviour prepared with different cross-linkers, such as glutaraldehyde (GLA), tripolyphosphate sodium (TPP), epichlorydrine (EPI), ethylene glycol diglycidyl ether (EGDE), etc [46-49]. The change in adsorption capacity was confirmed; the results showed that the chitosan-EPI beads presented a higher adsorption capacity than GLA and EGDE [46,47]. They reported that these materials can be used for the removal of reactive, direct and acid dyes. It was found that 1 g chitosan adsorbed 2498, 2422, 2383 and 1954 mg of various reactive dyes (Reactive Blue 2, Reactive Red 2, Direct Red 81 and Acid Orange 12, respectively) [49]. As a comparison, it is specified that the adsorption capacities of commercially activated carbon for reactive dyes generally vary from 280 to 720 mg/g. Another advantage of EPI is that it does not eliminate the cationic amine function of the polymer, which is the major adsorption site to attract the anionic dyes during adsorption [47]. The cross-linking with GLA (formation of imine functions) or EDGE decreases the availability of amine functions for the complexation of dyes. With a high cross-linking ratio the uptake capacity decreases drastically. Among the conditions of the cross-linking reaction that have a great impact on dye adsorption are the chemical nature of the cross-linker, as mentioned above, but also the extent of the reaction. In general, the adsorption capacity depends on the extent of cross-linking and decreases with an increase in cross-linking density. When chitosan beads were cross-linked with GLA under heterogeneous conditions, it was found that the saturation adsorption capacity of reactive dyes on cross-linked chitosan decreased exponentially from 200 to 50 mg/g as the extent of cross-linking increased from 0 to 1.5 mol GLA/mol of amine. This is because of the restricted diffusion of molecules through the polymer network and reduced polymer chain flexibility. Also the loss of amino-binding sites by reaction with aldehyde is another major factor in this decrease. However, the cross-linking step was necessary to improve mechanical resistance, to enhance the resistance of material against acid, alkali and chemicals, and also to increase the adsorption abilities of chitosan. According to literature [46-49], the adsorption capacity of non cross-linked beads was greater than that of cross-linked beads in the same experimental conditions. The materials, mainly cross-linked using GLA, have been also proposed as effective dye removers by several researchers [34,43]. The reaction of chitosan with GLA leads to the formation of imine groups, in turn leading to a decrease in the number of amine groups, resulting in a lowered adsorption capacity, especially for dyes adsorbed through ion-exchange mechanisms. In heterogeneous conditions, chitosan (solid state) was simply mixed with GLA solution, while in homogeneous conditions chitosan was mixed with GLA solution after being dissolved in acetic acid solution. An optimum aldehyde/amine ratio was determined for dye adsorption that depends on the type of cross-linking (water-soluble or solid-state solution) operation. The initial increase in dye adsorption was attributed to the low levels of cross-linking in the precipitates preventing the formation of closely packed chain arrangements without any great reduction in the swelling capacity. This increase in adsorption was interpreted in terms of the increases in hydrophilicity and accessibility of complexing groups as a result of partial destruction of the crystalline structure of the polymer by cross-linking under homogeneous conditions. At higher levels of cross-linking, the precipitates had lower swelling capacities, and hence lower accessibility because of the more extensive three-dimensional network and also because of its more hydrophobic character with increased GLA content. In general, the adsorption capacity increased greatly at low degrees of substitution but decreased with increasing substitution. This phenomenon is interpreted in terms of increased hydrophilicity caused by the destruction of the crystalline structure at low cross-linking densities, while this can be associated with an accompanying decrease in active sites, accessibility, and swellability of the adsorbent by increasing the level of cross-linking. Furthermore, it is noted that cross-linking can change the crystalline nature of chitosan, as suggested by the XRD diffractograms. After the cross-linking reaction, there was a small increase in the crytallinity of the chitosan beads and also increased accessibility to the small pores of the material.

3. Adsorption conditions

3.1. Equilibrium

3.1.1. pH

The pH of the dye solution plays an important role in the whole adsorption process and particularly on the adsorption capacity; influences the surface charge of the adsorbent, the degree of ionization of the material present in the solution, the dissociation of functional groups on the active sites of the adsorbent, and the chemistry of dye solution. It is important to indicate that while the adsorption on activated carbon was largely independent of the pH, the adsorption of dyes on chitosan was controlled by the acidity of the solution. pH affects the surface charge of the adsorbent. Chitosan is a weak base and is insoluble in water and organic solvents, however, it is soluble in dilute aqueous acidic solution (pH~6.5), which can convert the glucosamine units into a soluble form $R-NH_3$ $^+$. Chitosan is precipitated in alkaline solution or in the presence of polyanions and forms gels at lower pH. Its pKa depends on the deacetylation degree, the ionic strength and the neutralization of amine groups. In practice pKa lies within 6.5-6.7 for fully neutralized amine functions [50]. So, chitosan is polycationic in acidic medium: the free amino groups are protonated and the polymer becomes fully soluble and this facilitates electrostatic interaction between chitosan and the negatively charged anionic dyes. This cationic property, especially in the case of anionic dyes, depending on the charge and functions of the dye under the corresponding experimental conditions will influence the adsorption procedure [51]. In the literature, the ability of the anionic dyes to adsorb onto chitosan beads is often attributed to the surface charge which depends on the pH of the operating batch system, as mentioned. Dye adsorption occurred through electrostactic attraction on protonated amine groups and numerous researchers concluded that the influence of the pH confirmed the essential role of electrostatic interactions between the chitosan and the target dye. For example, chitosan had a positively charged surface below pH=6.5 (point of zero potential) [52], and reducing the pH increased the positivity of the surface, thus making the adsorption process pH sensitive. Decreasing the pH makes more protons available to protonate the amine group of chitosan with the formation of a large number of cationic amines. This results in increased dye adsorption by chitosan due to increased electrostatic interactions. Differences in pH of the solution have also been reported to influence the dye adsorption capacity of chitosan and its mechanism [44,45]. They noted that, at low pH, chitosan's free amino groups are protonated, causing them to attract anionic dyes, demonstrating that pH is one of the most important parameters controlling the adsorption process. Some researchers [32,33,53] also found that the adsorption capacity of cationic dyes on chitosan-grafted materials was strongly affected by the pH of solution and was generally significantly decreased by increasing the pH. pH is also known to affect the structural stability of dye molecules (in particular the dissociation of their ionizable sites), and therefore their colour intensity. As a result, the dye molecule has high positive charge density at a low pH. This indicates that the deprotonation (or protonation) of a dye must be taken into consideration. If the dyes to be removed are either weakly acidic or weakly basic, then the pH of the medium affects their structure and adsorption. Initial pH also influences the solution chemistry of the dyes: hydrolysis, complexation by organic and/or inorganic ligands, redox reactions, and precipitation are strongly influenced by pH, and on the other side strongly influence speciation and the adsorption capacity of the dyes. Some useful structural characteristics must be pointed out, as given below:

The free amine groups in chitosan are much more reactive and effective for chelating pollutants than the acetyl groups in chitin. There is no doubt that amine sites are the main reactive groups for (anionic) dye adsorption, though hydroxyl groups (especially in the C-3 position) may contribute to adsorption. Almost all functional properties of chitosan depend on the chain length, charge density and charge distribution and much of its potential as adsorbent is effected by its cationic nature and solution behaviour. However, at neutral pH, about 50% of total amine groups remain protonated and theoretically available for the adsorption of dyes. The existence of free amine groups may cause direct complexation of dyes co-existing with anionic species, depending on the charge of the dye. As the pH decreases, the protonation of amine groups increases together with the efficiency. The optimum pH is frequently reported in the literature to be around pH 2-4. Below this range, usually a large excess of competitor anions limits adsorption efficiency. This competitor effect is the subject of many studies aiming to develop materials that are less sensitive to the presence of competitor anions and to the pH of the solution, as described in the next two paragraphs.

It is not really the total number of free amine groups that must be taken into account but the number of accessible free amine groups. There are several explanations for this. The availability of amine groups is controlled by two important parameters: (i) the crystallinity of polymer, and (ii) the diffusion properties of dyes. It is known that some of the amine sites on chitosan are included in both the crystalline area and in inter or intramolecular hydrogen bonds. Moreover the residual crystallinity of the polymer may control the accessibility to adsorption sites. The deacetylation degree also controls the fraction of free amine groups available for interactions with dyes. Indeed, the total number of free amine groups is not necessarily accessible for dye uptake. Actually, rather than the fraction or number of free amine groups available for dye uptake, it would be better to consider the number of accessible free amine groups. It is also concluded that the hydrogen bonds linked between monomer units of the same chain (intramolecular bonds) and/or between monomer units of different chains (intermolecular bonds) decrease their reactivity. The weakly porous structure of the polymer and its residual crystallinity are critical parameters for the hydratation and the accessibility to adsorption sites.

3.1.2. Isotherms

Adsorption properties and equilibrium data, commonly known as adsorption isotherms, describe how pollutants interact with adsorbent materials and so, are critical in optimizing the use of adsorbents [31]. In order to optimize the design of an adsorption system to remove dye from solutions, it is important to establish the most appropriate correlation for the equilibrium curve. An accurate mathematical description of equilibrium adsorption capacity is indispensable for reliable prediction of adsorption parameters and quantitative comparison of adsorption behaviour for different adsorbent systems (or for varied experimental conditions) within any given system. Adsorption equilibrium is established when the amount of dye being adsorbed onto the adsorbent is equal to the amount being desorbed. It is possible to depict the equilibrium adsorption isotherms by plotting the concentration of the dye in the solid phase versus that in the liquid phase. The distribution of dye molecule between the liquid phase and the adsorbent is a measure of the position of equilibrium in the adsorption process and can generally be expressed by one or more of a series of isotherm models [54-58]. The shape of an isotherm may be considered with a view to predicting if a sorption system is "favourable" or "unfavourable". The isotherm shape can also provide qualitative information on the nature of the solute–surface interaction. In addition, adsorption isotherms have been developed to evaluate the capacity of chitosan materials for the adsorption of a particular dye molecule. They constitute the first experimental information, which is generally used as a convenient tool to discriminate among different materials and thereby choose the most appropriate one for a particular application in standard conditions. The most popular classification of adsorption isotherms of solutes from aqueous solutions has been proposed by Giles et al. [57,58]. Four characteristic classes are identified, based on the configuration of the initial part of the isotherm (i.e., class S, L, H, C). The Langmuir class (L) is the most widespread in the case of adsorption of dye compounds from water, and it is characterized by an initial region, which is concave to the concentration axis. Type L also suggests that no strong competition exists between the adsorbate and the solvent to occupy the adsorption sites. However, the H class (high affinity) results from extremely strong adsorption at very low concentrations giving rise to an apparent intercept on the ordinate. The H-type isotherms suggest the uptake of pollutants by materials through chemical forces rather than physical attraction. There are several isotherm models available for analyzing experimental data and for describing the equilibrium of adsorption, including Langmuir, Freundlich, Langmuir-Freunldich, BET, Toth, Temkin, Redlich-Peterson, Sips, Frumkin, Harkins-Jura, Halsey, Henderson and Dubinin-Radushkevich isotherms. These equilibrium isotherm equations are used to describe experimental adsorption data. The different equation parameters and the underlying thermodynamic assumptions of these models often provide insight into both the adsorption mechanism, and the surface properties and affinity of the adsorbent. Therefore, it is important to establish the most appropriate correlation of equilibrium curves to optimize the condition for designing adsorption systems. Various researchers have used these isotherms to examine the importance of different factors on dye molecule sorption by chitosan. However, the two most frequently used equations applied in solid/liquid systems for describing sorption isotherms are the Langmuir [55] and the Freundlich [56] models and the most popular isotherm theory is the Langmuir one which is commonly used for the sorption of dyes onto chitosan.

The majority of researchers found significant correlations between dye concentration and the dye-binding capacity of polymeric adsorbents and particularly chitosan [3,41,53]. The amount of the dye adsorbed onto chitosan increased with an increase in the initial concentration of dye solution if the amount of adsorbent was kept unchanged. This is due to the increase in the driving force of the concentration gradient with the higher initial dye concentration. In most cases, at low initial concentration the adsorption of dyes by chitosan is very intense and reaches equilibrium very quickly. This indicates the possibility of the formation of monolayer coverage of the dye molecules at the outer interface of the chitosan. At a fixed adsorbent dose, the amount of dye adsorbed increases with the increase of dye concentration in solution, but the percentage of adsorption decreased. In other words, the residual concentration of dye molecules will be higher for higher initial dye concentrations. In the case of lower concentrations, the ratio of initial number of dye moles to the available adsorption sites is low and subsequently the fractional adsorption becomes independent of initial concentration [3,41,46,47,49,52,53]. At higher concentrations, however, the number of available adsorption sites becomes lower and subsequently the removal of dyes depends on the initial concentration. At the high concentrations, it is not likely that dyes are only adsorbed in a monolayer at the outer interface of chitosan. As a matter of fact, the diffusion of exchanging molecules within chitosan particles may govern the adsorption rate at higher initial concentrations. In another theory, it is claimed that adsorption rate may diminish with an increase in dye concentration [34].This could be ascribed to the accompanying increase in dye aggregation and/or depletion of accessible active sites on the material.

Generally speaking, the adsorption of pollutants increases with the increase in temperature, because high temperatures provide a faster rate of diffusion of adsorbate molecules from the solution to the adsorbent [31,41]. However, it is well known that temperature plays an important role in adsorption onto activated carbon, generally having a negative influence on the amount adsorbed. The adsorption of organic compounds (including dyes) is an exothermic process and the physical bonding between the organic compounds and the active sites of the carbon will weaken with increasing temperature. Also with the increase of temperature, the dye solubility also increases, the interaction forces between the solute and the solvent become stronger than those between solute and adsorbent, consequently the solute is more difficult to adsorb. Both of these features are consistent with the order of Langmuir adsorption capacity. The adsorption of dyes by chitosan is also usually exothermic: an increase in the temperature leads to an increase in the dye adsorption rate, but diminishes total adsorption capacity [3,41,53]. However, these effects are small and normal wastewater temperature variations do not significantly affect the overall decolorization performance [59]. In addition, the adsorption process is not usually operated at high temperature, because this would increase operation costs. The increase in temperature affects not only the solubility of the dye molecule (its solubility increases), but also the adsorptive potential of the material (its adsorption capacity increases). Both effects work in the same direction causing an increase of cost operation in the batch system. In general, this could be confirmed by the thermodynamic parameters. An increase in temperature is also followed by an increase in the diffusivity of the dye molecule, and consequently by an increase in the adsorption rate if diffusion is the rate-limiting step. Temperature could also influence the desorption step and consequently the reversibility of the adsorption equilibrium. So, the temperature (and its variation) is an important factor affecting chitosan adsorption and investigations of this parameter offer interesting results, albeit often contradictory. The usual increase in adsorption may be attributed to the fact that on increasing temperature, a greater number of active sites is generated on the polymer beads because of an enhanced rate of protonation/deprotonation of the functional groups on the beads. The fact that adsorption of dyes on chitosan increases with higher temperature can be surprising. Other authors concluded that an increase in temperature leads to a decrease in the amount of adsorbed dye at equilibrium since adsorption on chitosan is exothermic [60,61].

3.3. Kinetics – Modelling

The conventional approach to quantify the batch adsorption process is the empirical fitting based on the dependence between the global adsorption rate and the driving force for adsorption i.e. the difference between the actual and the equilibrium adsorbate concentration. This dependence can be a linear one but in most cases is taken as quadratic. The parameters extracted from the fitting procedure may be appropriate to represent the experimental data, but they do not have any direct physical meaning (although some researchers relate them to an hypothetical diffusion coefficient). In addition, this type of empirical modelling cannot be extended to describe more complicated process geometries than the simple batch one. A phenomenological approach based on fundamental principles) is needed, in order to develop models with parameters with physical meaning. This type of models can be extended to be used for the design of processes of engineering interest (like fixed bed adsorption) and improve the understanding of the adsorbent structure in line with its interaction with the adsorbate.

The phenomenological approach to the batch adsorption is well-known. Kyzas et al [41] made an attempt for the description of the derivation of the models, of the inherent assumptions and of how several simplified models used in literature are hooked to the general approach. The description of the approach will be made for a general porous solid and then the adjustment to the particular structure of the chitosan derivatives will be presented. The actual difficulty in the modeling of the adsorption process for porous adsorbents is not its physics but its geometry in micro-scale. For extensive discussion on the physics of adsorption process please see other works [62,63]. The physics is well-understood in terms of the occurring processes: (i) diffusion of the adsorbate in the liquid, (ii) adsorption-desorption between the liquid and solid phase, and (iii) surface diffusion of the adsorbate. See a simple schematic on the phenomena occurring in a pore in Figure 4.

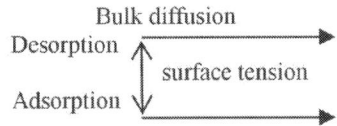

Figure 4. Phenomena occurring in pore scale

On the other hand, the actual shape of the geometry, in which the above phenomena occur, is completely unknown. Recently, several reconstructions or tomographic techniques have been developed for the 3-D digitization of the internal structure of porous media [64]. This digitized structure can be used to solve the equations describing the prevailing phenomena. The above approach is extremely difficult (still inapplicable for the size of pores met in adsorption applications), time consuming and its use is restricted to porous media development applications. A more effective approach is needed for the purpose of adsorption process design. So, the classical approach of "homogenizing" the corresponding partial differential equations and transferring the lack of knowledge of the geometry to the values of the physical parameters of the problem is given below. In the adsorbent particle phase, the adsorbate can be found both as solute in the liquid, filling the pores of the particle (concentration C in kg/m³) and adsorbed on the solid phase (concentration q in kg adsorbate/kg adsorbent). The "homogeneous" equations for the evolution of C and q inside a spherical adsorbent particle of radius R, are the following:

$$\varepsilon_p \frac{\partial C}{\partial t} = \frac{1}{r^2} \frac{\partial}{\partial r} r^2 D_p \frac{\partial C}{\partial r} - \rho_p G(C, q) \tag{1}$$

$$\frac{\partial q}{\partial t} = \frac{1}{r^2} \frac{\partial}{\partial r} r^2 D_s \frac{\partial q}{\partial r} - G(C, q) \tag{2}$$

where t is the time, r is the radial direction, ε_p the porosity of the particle, ϱ_p the density of the particle (kg adsorbent/m³), D_p is the liquid phase diffusivity of the adsorbate and D_s is the corresponding surface diffusivity. The diffusivity D_p for a given pair of adsorbate-fluid depends only on the temperature, whereas the diffusivity D_s depends both on the type of adsorbent and on q also (apart from their dependence on geometry). For particles with non-uniform structure the parameters D_p, D_s, ε_p, ϱ_p are functions of r but this case will not be considered here. The function G(C,q) denotes the rate of the adsorption-desorption process. The boundary conditions for the above set of equations are:

i) mass transfer from the solution to particle:

$$k_m \left(C_b - C \right) = -D_p \left(\frac{\partial C}{\partial r} \right)_{r=R} \quad \text{at } r = R \tag{3}$$

where C_b is the concentration of the adsorbate in the bulk solution, and k_m is the mass transfer coefficient from the bulk solution to the particle.

ii) spherical symmetry:

$$\left(\frac{\partial C}{\partial r} \right) = \left(\frac{\partial q}{\partial r} \right) = 0 \quad \text{at } r = R \tag{4}$$

Having values for the initial concentrations of adsorbate in the particle and for the bulk concentration C_b, the above mathematical problem can be solved for the functions $C(r,t)$ and $q(r,t)$. The above model is the so-called non-equilibrium adsorption model. Models of this structure are used in large scale solute transport applications (e.g. soil remediation), where the phenomena are of different scale from the present application [65]. In the case of adsorption by small particles, the adsorption-desorption process is much faster than the diffusion, leading to an establishment of a local equilibrium (which can be found by setting $G(C,q)=0$ and corresponds to the adsorption isotherm $q=f(C)$). In the present case, the local

$$q = \frac{q_m bC^{1/n}}{1 + bC^{1/n}} \tag{5}$$

where q_m and b corresponds to Q_{max} and K_{LF}, respectively of L-F model.

In the limit of very fast adsorption-desorption kinetics it can be shown by a rigorous derivation that the mathematical problem can be transformed to the following:

$$\frac{\partial q}{\partial t} = \frac{1}{r^2}\frac{\partial}{\partial r}r^2 D(C)\frac{\partial q}{\partial r} \tag{6}$$

$$\left(\frac{\partial q}{\partial r}\right)_{r=0} = 0 \tag{7}$$

$$k_m(C_b - C) = \rho_p D_p\left(\frac{\partial q}{\partial r}\right)_{r=R} \tag{8}$$

An additional assumption considered in the above derivation is that the amount of the adsorbate found in the liquid phase in the pores of the particle is insignificant compared to the respective amount adsorbed on the solid phase (i.e. $\varepsilon_p C \ll \rho_p q$). This assumption always holds, as it can be easily checked by recalling the definition of the adsorption process.

The concentration C in eqs 6, 8 can be found by inverting the relation $q=f(C)$. The diffusivity D is an overall diffusivity, which combines the bulk and surface diffusivity and is given by the relation:

$$D = D_s + \frac{D_p}{\rho_p f'(C)} \tag{9}$$

where the prime denotes the differentiation of a function with respect to its argument. The average concentration of the adsorbed species (dye molecules) can be computed by the relation:

$$q_{ave} = \frac{3}{R^3} \int_0^R qr^2 \, dr \tag{10}$$

In case of batch experiments the concentration of the solute in bulk liquid C_b decreases due to its adsorption; so, the evolution of the C_b must be taken into account by the model. The easier way to do this is to consider a global mass balance of the adsorbate:

$$C_b = C_{b_0} - \frac{m}{V} q_{ave} \tag{11}$$

where C_{bo} is the initial value of C_b; m is the total mass of the adsorbent particles; V is the volume of the tank.

Let us now refer to the several analytical or simple numerical techniques used in literature for the above problem and explain under which conditions they can be employed. In the case of an overall diffusion coefficient D with no concentration dependence (which implies (i) constant D_s, and (ii) zero D_p or linear adsorption isotherm) and a linear adsorption isotherm, the complete problem is linear and it can be solved analytically. The mathematical problem for the case of surface diffusion is equivalent to the homogeneous diffusion or solid diffusion model used by several authors [66]. The complete solution is given by Tien [54], but their simplified versions [67], which ignore external mass transfer and/or the solute concentration reduction through equation 13 (i.e. assuming an infinite liquid volume), are extensively used in the literature with topic of dye adsorption [68]. In case of constant diffusivity, and non-linear adsorption isotherm (which implies no pore diffusion and it is usually used in adsorption from liquid phase studies), several techniques based on the fundamental solutions of the diffusion equation can be used [66,69]. In general, the partial differential equation is transformed to an integro-differential equation for the average concentration q_{ave} (dimensionality reduction).

Another class of approximation methods, which can be used in any case (but with a questionable success for non-linear problems), is the so-called Linear Driving Force model (LDF model) [70]. The concentration profile in the particle is approximated by a simple function (usually polynomial with two or three terms) and using this, the partial differential equation problem can be replaced by an expression for the evolution of q_{ave}. This method is not appropriate for use in batch adsorption studies, but it has been proven to be successful in cases like fixed bed adsorption. In these cases, where the solution of equation 8 in several positions along the case of adsorption bed is required, LDF approximation is very useful. In cases, where no simplifications are possible and high accuracy is required, one has to resort to numerical techniques. A widely used approach is the collocation discretization of the spatial dimension, which is very efficient due to the use of higher order approximation. However, its advantage can be completely lost for the case of very rapid (or even discontinuous) changes of the parameters inside the particle. In order to be able to extend our approach to non-uniform particles, a typical finite difference discretization (permitting arbitrarily dense grid) is performed in the spatial dimension.

The model for adsorption by chitosan derivatives and how it fits to the above general adsorption model will be discussed next. The chitosan particles are homogeneous consisting of swollen polymeric material. The basic (cationic) dyes are diffused in the swollen polymer phase and are incorporated (adsorbed) on the polymer with the interaction of the amino and hydroxyl groups [71]. In the case of reactive (anionic) dyes, the incorporation step is due to strong electrostatic attraction [72]. The kinetics of the whole process have been modeled in two distinct ways: based on diffusion [73] or based on the global chelation (or in general immobilization) reaction [75]. In the present study, the two modeling approaches are unified by assuming a composite mechanism of diffusion in the water phase, adsorption-desorption on the polymeric structure and diffusion on the polymeric structure (transition from one adsorption site to another). The system of equations for the above model is exactly the same, but with different meaning for some symbols. In particular, ε_p is the liquid volume fraction of the particle, D_p is again the liquid phase diffusivity of the adsorbate, and D_s is the diffusivity associated to the transition rate from one adsorption site to another. The function $G(C,q)$ denotes the rate of the reversible adsorption process (chelation reaction or electrostatic attraction). The rest of the model development is exactly the same with that of the general porous solid, assuming that the form of the adsorption-desorption equilibrium isotherm is that of extended Langmuir.

A very useful application to experimental data of the model described above is developed by our research team [41]. The dye adsorbents were grafted chitosan beads with sulfonate groups (abbreviated as Ch-g-Sulf), N-vinylimido groups (abbreviated as Ch-g-VID), and non-grafted beads (abbreviated as Ch). The tests were carried out both for reactive and basic dyes.

Although chitosan has a gel-like structure, the water molecules can be found in relatively large regions, which can be characterized as pores with diameters of 30-50 nm for the pure chitosan. The dye molecules which are diffused through these pores, are adsorbed-desorbed on the pore walls and finally are diffused (transferred) from one adsorption site to the other. All these phenomena are included to the kinetic model described in this study. There are two unknown parameters, namely pore and surface diffusivity (D_p and D_s), which must be found from the kinetic data. The simultaneous determination of both parameters requires experimental data of intraparticle adsorbate concentration, which are not easily available. Thus, the kinetic data can be fitted to the infinite pairs of D_p and D_s by keeping the value of the one parameter fixed and calculating the value of the other one that fits the experimental kinetic data.

At the first stage of model employment, the surface diffusion is ignored (setting $D_s=0$) and the pore diffusivity is calculated and fitted to the experimental kinetic data. In order to make the diffusion coefficients comparative, it is necessary to estimate the diffusivity of each dye in the water (D_w) at the corresponding temperatures. The diffusivity in water are calculated according to the Wilke-Chang correlation (molecular radius can be found by employing codes like the BioMedCAChe 5.02 program by Fujitsu).

The key point of the procedure developed here is to compare the effective pore diffusivity (D_p) of the dye with its diffusivity in the water (D_w). This relation has the well known form [75]:

$$D_p = D_w \frac{\varepsilon}{\tau} \tag{11}$$

where ε is the porosity (water volume fraction), and τ is the tortuosity of the adsorbent particle. Given that the structure of the particle (ε,τ) does not depend on the temperature, and assuming that the pure pore diffusion with no any other type of interactions is responsible for the solute transfer in the adsorbent particle, the temperature's dependence of D_p must follow the temperature's dependence of D_p.

Basic dye. From all the combinations of dye-adsorbent examined up to now, the only one in which D_p follows the temperature dependence of D_w is the combination of basic dye-Ch (adsorption of basic dye onto non-grafted chitosan). The value $D_p/D_w = \tau/\varepsilon$ is found to be temperature independent, suggesting that the actual transport mechanism is the pore diffusion with no specific interactions between adsorbent and solute (τ/ε is found to be about 12). Indeed, Ch has amino and hydroxyl groups in its molecule, but smaller in number compared to that of grafted chitosan. So, the relatively weak interactions between dye and adsorbent (hydrogen bonding, van der Waals forces, pi-pi interactions, chelation) [72,76] do not drastically affect the transportation/diffusion of dye. Taking into account that ε for chitosan derivatives ranges between 0.4 and 0.6, τ is calculated to be about 6; this is a reasonable value located between the generally proposed value ($\tau=3$) and the values that holds for microporous solids as activated carbon ($\tau>10$) [77]. On the contrary, the higher values of the D_p, which were found in the case of grafted chitosan derivatives, indicate that the surface diffusion mechanism takes part. So, the fitting procedure must be repeated using the D_p values calculated (as an approximation assuming similar geometric structures) for the non-grafted chitosan (Ch) and searching for the D_s values that fits these data. Of course, a slight decrease of τ/ε versus temperature, which was found for Ch, suggests that some surface diffusion is still presented, but given the approximate nature of analysis, it is reasonable to ignore it. The surface diffusivity extracted from the experimental data appears to exhibit smaller temperature dependence than pore diffusivity. In surface diffusivity, the increase with temperature is related to the enhancement of the thermal motion of the adsorbed molecules (proportional to the absolute temperature), whereas in pore diffusivity the latter is related to the decrease of the water viscosity. The surface diffusivity also increases with grafting as the density of the adsorption sites increases. It is due to the fact that the larger density of adsorption sites corresponds to the smaller site-to-site distance in the chitosan backbone, and consequently to the higher probability of transition from one site to another.

Reactive dye. The situation (mechanism of diffusion) is different in the case of the reactive dyes. It would be helpful to observe the relation of the structure of this type of dye and the respective structure of the adsorbents used. The dye is composed of sulfonate groups [8], and the adsorbents contain amino groups (the higher basicity of adsorbent, the stronger protonation occurs at acidic conditions) [75]. So, the main adsorption mechanism of reactive dyes onto chitosan is based on the strong electrostatic interactions between dissociated sulfonate groups of the dye and protonated amino groups of the chitosan. Many studies confirm that the above process is favored under acidic conditions, where the total "charge" of chitosan adsorbent is more positive (due to the stronger protonation of amino groups at acidic pH values) [78,79].

Proposing our diffusion concept, it is obvious from the experimental data that: (i) the temperature's dependence of the coefficients D_p suggests that the transport mechanism is not simply a pure diffusion through the pores, and (ii) the small values of D_p versus D_w set questions about the existence of surface diffusivity. The above observations are compatible to the nature of interactions between the reactive dye and the adsorbent, which is described above. The strong electrostatic interaction in the adsorption sites inhibits the surface diffusion. Moreover, the electrostatic forces have a relatively large region of action. Using as reference the non-grafted chitosan (Ch), the existence of charges of opposite sign at the pore walls creates a surface charge gradient in addition to the adsorbate gradient in the adsorbent particle. This charge gradient drags the oppositely charged dye molecules inside the particle and leads to enhanced effective pore diffusivity. This fact could completely explain the increase of diffusivity related with the grafting groups (the more positively charged grafted groups, the stronger attraction of negatively charged dye molecule). As the density of the adsorption sites increases, the charge density in the particle increases, leading to higher effective pore diffusivity values. The temperature's dependence of this electrostatically facilitated diffusion process is weaker than that of the pure diffusion process. However, the opposite phenomenon is occurred in the case of sulfonate-chitosan derivative (Ch-g-Sulf), where the surface charge is of the same sign as that of the dye molecule, inhibiting the diffusion process. The above concept of the adsorption process of reactive dyes on chitosan derivatives suggests the development of models that take into account explicitly the electrostatic interaction between dye and adsorbent, instead of considering them only by the modification, which they create to the effective pore diffusivity D_p. Conclusively, by employing the phenomenological model based on the pair pore-surface diffusion for the transport of the dye in the adsorbent particle information about the actual mechanism of adsorption and on interaction between adsorbent and adsorbate can be extracted.

4. Desorption conditions and Reuse

Chitosan adsorbents present considerable advantages such as their high adsorption capacity, selectivity and also the facility of regeneration. The regeneration of the adsorbent may be crucially important for keeping the process costs down and to open the possibility of recovering the pollutant extracted from the solution. For this purpose, it is desirable to desorb the adsorbed dyes and to regenerate the chitosan derivative for another cycle of application. Generally, the regeneration of saturated chitosan for non-covalent adsorption can be easily achieved by using an acid solution as the desorbing agent. Researchers proposed to desorb the dye from the beads by changing the pH of the solution [52,53] and they showed that the beads could be reused five times without any loss of mechanical or chemical efficacy. The optimum pH of desorption was determined to be the contrast of the optimum of adsorption [53]. Another study has shown the desorption capability of chitosan even at the same pH conditions as those of adsorption [3]. In the same study, 10 cycles were achieved with no significant loss of re-adsorption capacity (~5% loss between 1st and last cycle).

5. Conclusions

The treatment of industrial dyeing effluent that contains the large number of organic dyes by adsorption process, using easily available low-cost adsorbents, is an interesting alternative to the traditionally available aqueous waste processing techniques (chemical coagulation/flocculation, ozonation, oxidation, photodegradation, etc.). Undoubtedly, low-cost adsorbents offer a lot of promising benefits for commercial purposes in the future. The distribution of size, shape, and volume of voids species in the porous materials is directly related to the ability to perform the adsorption application. The comparison of adsorption performance of different adsorbents depend not only on the experimental conditions and analytical methods (column, reactor, and batch techniques) but also the surface morphology of the adsorbent, surface area, particle size and shape, micropore and mesopore volume, etc. Many researchers have made comparison between the adsorption capacities of the adsorbents, but they have nowhere discussed anything about the role of morphology of the adsorbent, even in case of the inorganic material where it plays a major role in the adsorption process. The pH value of the solution is an important factor which must be considered during the designing of the adsorption process. The pH has two kinds of influences on dye: (i) an effect on the solubility, and (ii) speciation of dye in the solution (it depends on dye class). It is well known that surface charge of adsorbent can be modified by changing the pH of the solution. The high adsorption of cationic or acidic dyes at higher pH may be due to the surface of adsorbent becomes negative, which enhances the positively charged dyes through electrostatic force of attraction and vice versa in case of anionic or basic dyes. In case of chitosan-based materials, the literature reveals that maximum removal of dyes from aqueous waste can be achieved in the pH range of 2-4. However, physical and chemical processes such as drying, autoclaving, cross-linking reactions, or contacting with organic or inorganic chemicals proposed for improving the sorption capacity and the selectivity. The production of chitosan also involves a chemical deacetylation process. Chitosan is characterized by its easy dissolution in many dilute mineral acids, with the remarkable exception of sulfuric acid. It is thus necessary to stabilize it chemically for the recovery of dyes in acidic solutions. Several methods have been developed to reinforce chitosan stability. The advantage of chitosan over other polysaccharides is that its polymeric structure allows specific modifications without too many difficulties. The chemical derivatization of the polymer by grafting new functional groups onto the chitosan backbone may be used to increase the adsorption efficiency, to improve adsorption selectivity, and also to decrease the sensitivity of adsorption environmental conditions. It is interesting to note the relationships between physicochemical properties and/or sources of chitosan and the dye-binding properties. Most of the properties and potential of chitosan as adsorbent can be related to its cationic nature, which is unique among abundant polysaccharides and natural polymers, and its high charge density in solution.

The common adsorbent, commercially available activated carbon has good capacity for the removal of pollutants. But its main disadvantages are the high price of treatment and difficult regeneration, which increases the cost of wastewater treatment. Thus, there is a demand for other adsorbents, which are of inexpensive material and do not require any expensive additional pretreatment step. However, there is no direct answer to the question which adsorbent is better: chitosan (raw material, preconditioned chitosan, grafted or cross-linked chitosans) or activated carbons? The best choice depends on the dye and it is impossible to determine a correlation between the chemical structure of the dye and its affinity for either carbon or chitosan. Each product has advantages and drawbacks. In addition, comparisons are difficult because of the scarcity of information and also inconsistencies in data presentation.

Author details

George Z. Kyzas[1,2*], Margaritis Kostoglou[2], Nikolaos K. Lazaridis[2] and Dimitrios N. Bikiaris[2]

*Address all correspondence to: georgekyzas@gmail.com

1 Department of Oenology and Beverage Technology, Technological Educational Institute of Kavala, Greece

2 Department of Chemistry, Division of Chemical Technology, Aristotle University of Thessaloniki, Greece

References

[1] Forgacs, E, Cserhati, T, & Oros, G. (2004). Removal of synthetic dyes from wastewaters: a review,. *Environ. Int.*, 30, 953-971.

[2] Allègre, C, Moulin, P, Maisseu, M, & Chrabit, F. (2006). Treatment and reuse of reactive dyeing effluents. *J. Membrane Sci.*, 269, 15-34.

[3] Kyzas, G. Z, Kostoglou, M, Vassiliou, A. A, & Lazaridis, N. K. (2011). Treatment of real effluents from dyeing reactor: Experimental and modeling approach by adsorption onto chitosan. *Chem. Eng. J.*, 168, 577-585.

[4] Vandenivere, P. C, Bianchi, R, & Vestraete, W. (1998). Treatment and reuse of wastewater from textile wet-processing industry: review of emerging technologies. *J. Chem. Technol. Biot.*, 72, 289-302.

[5] Neill, O, Hawkes, C, Hawkes, F. R, Lourenço, D. L, Pinheiro, N. D, & Delée, H. M. W. (1999). Colour in textile effluents-sources, measurement, discharge consents and sim‐ ulation: a review. J. Chem. Technol. Biot. , 74, 1009-1018.

[6] Hunger, K. (2003). Industrial dyes: Chemistry, properties, applications. Germany: Wiley-VCH;.

[7] Rangnekar, D. W, & Singh, P. P. (1980). An introduction to synthetic dyes. Bombay: Himalaya;.

[8] Zollinger, H. (1987). Color Chemistry. Syntheses, Properties, and Applications of Or‐ ganic Dyes and Pigments. New York: Weinhem- VCH;.

[9] Shenai, V. A. (1977). Technology of textile processing volume II: Chemistry of dyes and principles of dyeing. Bombay: Sevak;.

[10] Pokhrel, D, & Viraraghavan, T. (2004). Treatment of pulp and paper mill wastewater‐ a review. Sci. Total Environ., 333, 37-58.

[11] Robinson, T, Mcmullan, G, Marchant, R, & Nigam, P. (2001). Remediation of dyes in textile effluent: a critical review on current treatment technologies with a proposed alternative. Bioresour. Technol., 77, 247-255.

[12] Slokar, Y. M, Majcen, Le, & Marechal, A. (1998). Methods of decoloration of textile wastewaters. Dyes Pigments, 37, 335-356.

[13] Delee, W, O'Neill, C, Hawkes, F. R., & Pinheiro, H. M. (1998). Anaerobic treatment of textile effluents: a review. J. Chem. Technol. Biot., 73, 323-335.

[14] Banat, I. M, Nigam, P, Singh, D, & Marchant, R. (1996). Microbial decolorization of textile-dye-containing effluents: a review. Bioresour. Technol., 58, 217-227.

[15] Pearce, C. I, Lloyd, J. R, & Guthrie, J. T. (2003). The removal of colour from textile wastewater using whole bacterial cells: a review. Dyes Pigments, 58, 179-196.

[16] Mcmullan, G, Meehan, C, Conneely, A, Kirby, N, Robinson, T, Nigam, P, Banat, I. M, Marchant, R, & Smyth, W. F. (2001). Microbial decolourisation and degradation of textile dyes. Appl. Microbiol. Biotechnol., 56, 81-87.

[17] Fu, Y, & Viraraghavan, T. (2001). Fungal decolorization of dye wastewaters: a re‐ view. Bioresour. Technol., 79, 251-262.

[18] Stolz, A. (2001). Basic and applied aspects in the microbial degradation of azo dyes. Appl. Microbiol. Biotechnol., 56, 69-80.

[19] Pereira, M. F. R, Soares, S. F, Orfao, J. J. M., & Figueiredo, J. L. (2003). Adsorption of dyes on activated carbons: influence of surface chemical groups. Carbon, 41, 811-821.

[20] Marco, A, Esplugas, S, & Saum, G. (1997). How and why to combine chemical and biological processes for wastewater treatment. Water Sci. Technol., 35, 231-327.

[21] Bhattacharyya, K. G, & Sarma, A. (2003). Adsorption characteristics of the dye, Brilliant Green, on Neem leaf powder. *Dyes Pigments*, 57, 211-222.

[22] RaviKumar, M. N. V., Sridhari, TR, Bhavani, KD, & Dutta, PK. (1998). Trends in color removal from textile mill effluents. *Colorage*, 40, 25-34.

[23] Dabrowski, A. (2001). Adsorption, from theory to practice. *Adv. Colloid Int. Sci.*, 93, 135-224.

[24] Ruthven, D. M. (1984). Principles of adsorption and adsorption processes;. Wiley-Int;.

[25] Suzuki, M. (1997). Role of adsorption in water enviroment processes. *Wat. Sci. Tech.*, 35, 1-11.

[26] Ramakrishna, K. R, & Viraraghavan, T. (1997). Dye removal using low cost adsorbents. *Water Sci. Technol.*, 36, 189-196.

[27] Babel, S, & Kurniawan, T. A. (2003). Low-cost adsorbents for heavy metals uptake from contaminated water: a review. *J. Hazardous Mater.*, 97, 219-243.

[28] Streat, M, Patrick, J. W, & Perez, M. J. (1995). Sorption of phenol and para-chlorophenol from water using conventional and novel activated carbons. *Water Res.*, 29, 467-472.

[29] Crini, G. (2006). Non-conventional low-cost adsorbents for dye removal: a review. *Bioresour Technol.*, 60, 67-75.

[30] Rinaudo, M. (2006). Chitin and chitosan: properties and applications. *Prog. Polym. Sci.*, 31, 603-632.

[31] Crini, G, & Badot, P. M. (2008). Application of chitosan, a natural aminopolysaccharide, for dye removal from aqueous solutions by adsorption process using batch studies: a review of recent literature. Prog. Polym. Sci. , 33, 399-447.

[32] Crini, G, Martel, B, & Torri, G. (2008). Adsorption of C.I. Basic Blue 9 on chitosan-based materials. *Int. J. Environ. Pollut.*, 34, 451-465.

[33] Crini, G, Robert, C, Gimbert, F, Martel, B, Adam, O, De Giorgi, F, & Badot, P. M. (2008). The removal of Basic Blue 3 from aqueous solutions by chitosan-based adsorbent: batch studies. *J. Hazard. Mater.*, 153, 96-106.

[34] Gaffar, M. A, Rafie, S. M, & Tahlawy, K. F. (2005). Preparation and utilization of ionic exchange resin via graft copolymerization of b-CD itaconate with chitosan. *Carbohyd. Polym.*, 56, 387-396.

[35] Martel, B, Devassine, M, Crini, G, Weltrowski, M, Bourdonneau, M, & Morcellet, M. (2001). Preparation and adsorption properties of a beta-cyclodextrin-linked chitosan derivative. *J. Polym. Sci. A Polym. Chem.*, 39, 169-176.

[36] Shimizu, Y, Tanigawa, S, Saito, Y, & Nakamura, T. (2005). Synthesis of chemically modified chitosans with a higher fatty acid glycidyl and their adsorption abilities for anionic and cationic dyes. *J. Appl. Polym. Sci.*, 96, 2423-2428.

[37] Chao, A. C, Shyu, S. S, Lin, Y. C, & Mi, F. L. (2004). Enzymatic grafting of carboxyl groups on to chitosan-to confer on chitosan the property of a cationic dye adsorbent. *Bioresour Technol*, 91-157.

[38] Uzun, I, & Guzel, F. (2006). Rate studies on the adsorption of some dyestuffs and p-nitrophenol by chitosan and monocarboxymethylated(MCM)-chitosan from aqueous solution. *Dyes Pigm*, 118, 141-154.

[39] Prabaharan, M, & Mano, J. F. Chitosan derivatives bearing cyclodextrin cavities as novel adsorbent matrices. *Carbohyd. Polym.*, 63, 153-166.

[40] Jayakumar, R, Prabaharan, M, Reis, R. L, & Mano, J. F. (2002). Graft copolymerized chitosan-present and status applications. *Carbohyd. Polym.*, 62, 142-158.

[41] Kyzas, G. Z, Kostoglou, M, & Lazaridis, N. K. (2010). Relating interactions of dye molecules with chitosan to adsorption kinetic data. *Langmuir*, 26, 9617-9626.

[42] El-Tahlawy, K. F, Gaffar, M. A, & Rafie, S. (2006). Novel method for preparation of b-cyclodextrin/grafted chitosan and it's application. Carbohyd. Polym. , 63, 385-392.

[43] Cestari, A. R, Vieira, EFS, dos Santos, AGP, Mota, JA, & de Almeida, VP. (2004). Adsorption on of anionic dyes on chitosan beads. 1. The influence of the chemical structures of dyes and temperature on the adsorption kinetics. *J. Colloid Interf. Sci.*, 280, 380-386.

[44] Gibbs, G, Tobin, J. M, & Guibal, E. (2003). Adsorption of acid green 25 on chitosan: influence of experimental parameters on uptake kinetics and adsorption isotherms. *J. Appl. Polym. Sci.*, 90, 1073-1080.

[45] Gibbs, G, Tobin, J. M, & Guibal, E. (2004). Influence of chitosan preprotonation on reactive black 5 adsorption isotherms and kinetics. *Ind. Eng. Chem. Res.*, 43, 1-11.

[46] Chiou, M. S, & Li, H. Y. (2003). Adsorption behavior of reactive dye in aqueous solution on chemical cross-linked chitosan beads. *Chemosphere*, 50, 1095-1105.

[47] Chiou, M. S, Kuo, W. S, & Li, H. Y. (2003). Removal of reactive dye from wastewater by adsorption using ECH cross-linked chitosan beads as medium. *J. Environ. Sci. Health. A*, 38, 2621-2631.

[48] Chiou, M. S, Ho, P. Y, & Li, H. Y. (2003). Adsorption behavior of dye AAVN and RB4 in acid solutions on chemically crosslinked chitosan beads. *J. Chin. Inst. Chem. Eng.*, 34, 625-634.

[49] Chiou, M. S, Ho, P. Y, & Li, H. Y. (2004). Adsorption of anionic dye in acid solutions using chemically cross-linked chitosan beads. *Dyes Pigm.*, 60, 69-84.

[50] Guibal, E. (2005). Heterogeneous catalysis on chitosan-based materials: a review. *Prog. Polym. Sci.*, 30, 71-109.

[51] Guibal, E, Mccarrick, P, & Tobin, J. M. (2003). Comparison of the adsorption of anionic dyes on activated carbon and chitosan derivatives from dilute solutions. *Sep. Sci. Technol.*, 38, 3049-3073.

[52] Chatterjee, S, Chatterjee, S, Chatterjee, B. P, Das, A. R, & Guha, A. K. (2005). Adsorption of a model anionic dye, eosin Y, from aqueous solution by chitosan hydrobeads. *J. Colloid Interf. Sci.*, 288, 30-35.

[53] Kyzas, G. Z, & Lazaridis, N. K. (2009). Reactive and basic dyes removal by sorption onto chitosan derivatives. *J. Colloid Interf. Sci.*, 331, 32-39.

[54] Tien, C. (1994). Adsorption Calculation and Modeling. Boston: Butterworth-Heinemann;.

[55] Langmuir, I. (1916). The constitution and fundamental properties of solids and liquids. *J. Am. Chem. Soc.*, 38, 2221-2295.

[56] Freundlich, H. M. F. (1906). Uber die adsorption in lo° sungen. *Z. Phys. Chem.*, 57, 385-471.

[57] Giles, C. H, D' Silva, A. P., & Easton, I. A. (1974). A general treatment and classification of the solute adsorption isotherm. II. Experimental interpretation. *J. Colloid Interf. Sci.*, 47, 766-778.

[58] Giles, C. H, Smith, D, & Huitson, A. (1974a). A general treatment and classification of the solute adsorption isotherm. I. Theoretical. *J. Colloid Interf. Sci.*, 47, 755-765.

[59] Ravi Kumar, M. N. A review of chitin and chitosan applications. 2000, *React. Funct. Polym.*, 46, 1-27.

[60] Saha, T. K, Karmaker, S, Ichikawa, H, & Fukumori, Y. (2005). Mechanisms and kinetics of trisodium 2-hydroxy-1, 1'-azonaphthalene-3,4',6-trisulfonate adsorption onto chitosan. *J. Colloid Interf. Sci.*, 286, 433-439.

[61] Chiou, M. S, & Li, H. Y. (2002). Equilibrium and kinetic modeling of adsorption of reactive dye on cross-linked chitosan beads. *J. Hazard. Mater.*, 93, 233-248.

[62] Rudzinski, W, & Plazinski, W. (2008). Kinetics of solute adsorption at solid/aqueous interfaces: searching for the theoretical background of the modified pseudo-first-order kinetic equation. *Langmuir*, 24, 5393-5399.

[63] Liu, Y, & Shen, L. (2008). From Langmuir kinetics to first- and second-order rate equations for adsorption. *Langmuir*, 24, 11625-11630.

[64] Konstandopoulos, A. G, Vlachos, N, Kostoglou, M, & Patrianakos, G. (2007). Proceedings of the 6th International Conference of Multiphase Flow; Leipzig,.

[65] Hitchcock, P. W, & Smith, D. W. (1998). Implications of non-equilibrium sorption on the interception-sorption trench remediation strategy. *Geoderma*, 84, 109-120.

[66] Mckay, G. (1985). The adsorption of dyestuffs from aqueous solutions using activated carbon: An external mass transfer and homogeneous surface diffusion model. *AICHE J.*, 31, 335-339.

[67] Crank, J. (1975). The Mathematics of Diffusion. 2nd ed.; London: Oxford University Press;.

[68] Lazaridis, N. K, Karapantsios, T. D, & Georgantas, G. (2003). Kinetic analysis for removal of a reactive dye from aqueous solution on to hydrotalcite by adsorption. *Water Res.*, 37, 3023-3033.

[69] Larson, A. C, & Tien, C. (1984). Multicomponent liquid phase adsorption in batch, Parts I & II. *Chem. Eng. Commun.*, 27, 339-379.

[70] Coates, J. I, & Glueckauf, E. (1947). Theory of chromatography; experimental separation of two solutes and comparison with theory. *J. Chem. Soc.*, 5, 1308-1314.

[71] Guibal, E. (2004). Interactions of metal ions with chitosan-based sorbents: a review. *Sep. Purif. Technol.*, 38, 43-74.

[72] Smith, M. B, & March, J. (2001). March's Advanced Organic Chemistry. 5th ed.; New York: John Wiley & Sons Inc;.

[73] Guibal, E, Milot, C, & Tobin, J. M. (1998). Metal-anion sorption by chitosan beads: equilibrium a kinetic studies. *Ind. Eng. Chem. Res.*, 37, 1454-1463.

[74] Chu, K. H. (2002). Removal of copper from aqueous solution by chitosan in prawn shell: adsorption equilibrium and kinetics. *J. Hazard. Mater.*, 90, 77-95.

[75] Smith, J. M. (1981). Chemical Engineering Kinetics. New York: McGraw-Hill;.

[76] Brugnerotto, J, Lizardi, J, Goycoolea, F. M, Argüelles-monal, W, Desbrières, J, & Rinaudo, M. (2001). An infrared investigation in relation with chitin and chitosan characterization. *Polymer*, 42, 3569-3580.

[77] Yang, R. T. (1987). Gas separation by adsorption processes. Boston: Butterworths;.

[78] Wang, L, & Wang, A. (2008). Adsorption properties of congo red from aqueous solution onto N,O-carboxymethyl-chitosan. Bioresour. Technol., 99, 1403-1408.

[79] Sakkayawong, N, Thiravetyan, P, & Nakbanpote, W. J. (2005). Adsorption mechanism of synthetic reactive dye wastewater by chitosan. *J. Colloid. Interf. Sci.*, 286, 36-42.

Chapter 8

Textile Dyeing: Environmental Friendly Osage Orange Extract on Protein Fabrics

Heba Mansour

Additional information is available at the end of the chapter

http://dx.doi.org/10.5772/54410

1. Introduction

Dyeing was known as early as in the Indus Valley period (2600-1900 BC); this knowledge has been substantiated by findings of colored garments of cloth and traces of madder dye in the ruins of the Indus Valley Civilization at Mohenjodaro and Harappa. Natural dyes, dyestuff and dyeing are as old as textiles themselves. Man has always been interested in colors; the art of dyeing has a long past and many of the dyes go back into prehistory. It was practiced during the Bronze Age in Europe. The earliest written record of the use of natural dyes was found in China dated 2600 BC. [1-4] Primitive dyeing techniques included sticking plants to fabric or rubbing crushed pigments into cloth. The methods became more sophisticated with time and techniques using natural dyes from crushed fruits, berries and other plants, which were boiled into the fabric and which gave light and water fastness (resistance), were developed. After the accidental synthesis of mauveine by Perkin in Germany in 1856 and its subsequent commercialization, coal-tar dyes began to compete with natural dyes. With advances in chemical techniques, the manufacture of synthetic dyes became possible, leading to greater production efficiency in terms of quality, quantity and the potential to produce low-cost raw materials. As a result, natural dyes were progressively replaced by synthetic dyes, whereas over 80% of which is constituted of the aromatic azo type [5]· However, researches have shown that synthetic dyes are suspected to release harmful chemicals that are allergic, carcinogenic and detrimental to human health. In addition, textile industries produce huge amounts of polluted effluents that are normally discharged to surface water bodies and ground water aquifers. These wastes cause many damages to the ecological system of the receiving surface water, creating a lot of disturbance to the ground water resources [6-9]. Therefore, in 1996, ironically Germany became the first country to ban certain azo dyes [10].

The world-wide demand for fibers and safety dyes is increasing probably according to the greater awareness of the general consumers in the USA, Europe and Japan towards the highly pollutant procedures affecting fiber and textile coloration when using synthetic dyes which act as a sources of skin cancer, disorders and allergic contact dermatitis [11]. Therefore, interest in returning back to natural dyes as synthetic dyes substitute has increased considerably on account of their high compatibility with environment, relatively low toxicity and allergic effects, as well as availability of various natural coloring sources such as from plants, insects, minerals and fungi [12]

Natural dyes can be obtained from plants, animals and minerals, producing different colors like red, yellow, blue, black, brown and a combination of these. For technical application of natural dyes, a number of requirements have to be fulfilled;

Most problems are derived from technical demands, for example:

- adaptation of traditional dyeing processes on modern equipment [13]
- supply of dye-houses with the required amount of plant material [13]
- standardization of extraction and dyeing of the plant material [14]
- selection of plant material and processes that yield products with acceptable fastness properties [14]

From an industrial point of view it would be easier to resort to extracts despite there is at present no definite answer to this prospective solution. The simplest extract would be a watery one although not all the dye pigments are water-soluble. Use of organic solvents might give rise to extracts which are not completely water-soluble [15], provided that the solvent chosen guarantees a series of properties as follows:

- Its extraction capacity is extremely high for practically all the natural pigments present in the raw materials of interest.
- Its boiling temperature and latent heat of vaporization is quite low to allow its separation at low temperatures with minimum energy consumption;
- Its reactivity with colors and pigments is insignificant to avoid any loss in the color quality [15].

Natural dyes are substantive and require a mordant to fix to the fabric, and prevent the color from either fading with exposure to light or washing out. These compounds aid a chemical reaction between the dye and the fiber, so that the dye is absorbed. Traditionally, mordants were found in nature. Wood ash or stale urine may have been used as an alkali mordant, and acids could be found in acidic fruits or rhubarb leaves.

There are three types of mordant: i) Metallic mordants; Metal salts of aluminium, chromium, iron, copper and tin, ii) Tannins; Myrobalan and sumach, iii) Oil mordants; mainly used in dyeing Turkey red color from madder by forming a complex with alum.

In order to obtain high color yield, different shades and good fastness properties, metallic salt mordants are normally employed [16]. The application of mordants for dye fixation was carried out by three methods; i- Pre-mordanting; dipping the fabric in the mordant solution before dye-ing, ii- Simultaneous mordanting; addition of mordant in the dye bath during the dyeing, and iii- post mordanting; dipping the fabric in the mordant solution after dyeing [17].

Osage orange (Maclura pomifera) is a tree in the Moraceae family [18]. The common name is derived from its fruit, which resembles the shape of an orange, and from the fact that its hardwood was used by the Osage Indian tribe to make bows. It is native to Southern Okla-homa and Northern Texas, and is planted throughout the United States. Several compounds have been isolated and identified in various parts of this tree namely, isoflavonoids from the fruit, flavonols and xanthones from the heartwood and stem bark, and flavanones and xan-thones from the root bark [19] It contains lectins, triterpenes, xanthones and flavone-type compounds such as Scandenone and auriculasin as shown in Figure 1 [20] Two predomi-nant isoflavones, pomiferin and osajin, are derived from the simple isoflavone genistein by prenyl substitutions [21] as shown in Figure 2.

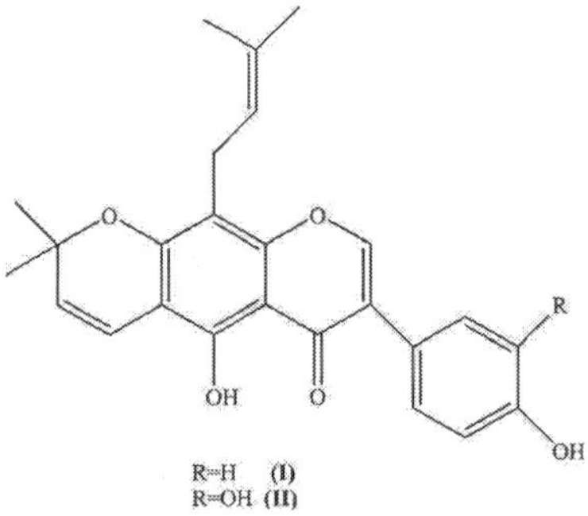

R=H   (I)
R=OH (II)

Figure 1. Chemical structures of scandenone (I) and auriculasin (II)

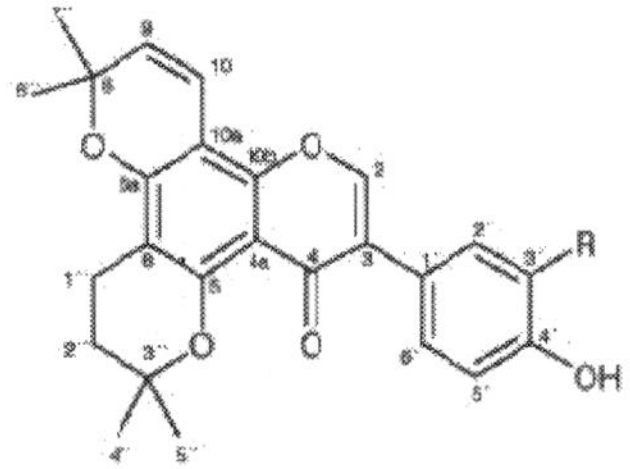

i R= H (Iso-Osajin)
ii R= OH (Iso-Pomiferin)

Figure 2. Chemical structures of Osajin (i) and pomiferin (ii)

The compounds, osajin, iso-osajin, pomiferin, and iso-pomifcrin have been characterized and their chemical structures determined. They are isoflavones with the following structures as shown in Figure 3. [19, 21].

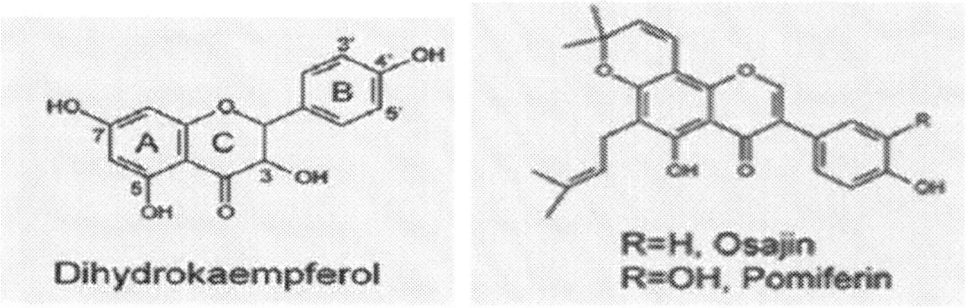

Figure 3. Chemical Structure of iso-pomifcrin

As well as Osage orange was applied as an eco-friendly dye acting as one of the environmental problems solutions. New concepts in the cleaner production are being evaluated to solve the high water and energy consumption in textile industries.

The use of ultrasonic as a renewable source of energy in textile dyeing has been increased due to many advantages associated with it [22-25]. Ultrasonic energy represents a promising technique for assisting silk treatment, dyeing, and mordanting processes in comparison with the conventional heating technique. Sonic energy succeeded in accelerating the rate of, dyeing, and mordanting at lower temperatures rather than the conventional heating technique

Therefore, the present investigation was aimed at identifying the most appropriate leaching solvent for Osage orange pigments to produce an optimum concentrated extract used for dyeing protein fibers; silk and wool fabrics. This has been carried out ultrasonically in comparison with the classical thermal method, using water, in addition to the co-solvents of water-acetone, and water-ethanol mixtures at different concentrations, temperature and time intervals. The optimum condition of the efficiency of ultrasonic assisted dyeing and mordanting methods of Osage orange extraction on the quality of the dyed protein based materials were determined

2. Experimental

2.1. Materials

Degummed and bleached plain Habotain silk fabric, and 100% mill scoured wool fabric purchased from Sherazad Com. New Zealand, were further washed with a solution containing 0.5 g/L of sodium carbonate and 2 g/L of non-ionic detergent (Nonidet ® P 40 Substitute purchased from *Sigma- Aldrich NZ. Ltd.*), keeping the material to liquor ratio at 1:50, for 30 min. at 40-45°C. The scoured materials were thoroughly washed and dried at ambient temperature. [23]

An analytical grade alum and the commercially cream of tartar were used as mordantas.

2.2. Pigment leaching and estimation of extraction yields

To select the best solvent for Osage orange, distilled water and other co-solvents, such as water-acetone, and water- ethanol mixtures all of analytical grade, were tested at concentrations of 10% v/v. 2 g of Osage orange powder (Hands Ashford NZ, LTD, ChristChurch, NZ) was suspended in 20 cm^3 of solvent, and in thermostatic as well ultrasonic baths at 60 °C, for 120 min. Once water-acetone co-solvent, and ultrasonic assisted extraction were chosen as the preferable technique of extraction, 10 % w/v Osage orange powder, dissolved in (2.5- 25) % v/v acetone, at (25- 60) °C, for 30- 120 min, were carried out to determine the standardization method of extraction. [24]

2.3. Dyeing and mordanting

1 % Osage orange extract was filtered and used as a dyeing bath. Silk and wool samples were added to the extract, and the dyeing parameters were studied ultrasonically keeping the material to liquor ratio at L: R of 1:30, for time intervals varied between 30- 120 min, at temperatures from 30-60°C. In terms of the pH used for dyeing; the pH values ranging from (3-11) were carried out to control the dye uptake.

On studying the mordanting methods, the optimum concentration of 8% w/v Osage orange extract was carried out. Stock solutions of 50 gm/ l alum and 25 g/l mixture of each of alum and cream of tartar were prepared. Two different methods of mordanting were used: (1) pre-mordanting method: the samples were first mordanted and then dyed without intermediate washing; and (2) post-mordanting method: the samples were first dyed and then mordanted. Ultrasonic assisted mordanting was carried out in comparison with the conventional heating method at 50-60°C, for 90 min, at pH 5. samples were rinsed, washed with 0.5 g/L sodium carbonate and 2 g/L of non-ionic detergent at 40-45°C for 30 min, keeping the material to liquor ratio at 1:50. Finally washed with water, and dried at ambient temperature. [23, 26, 27]

2.4. Measurements

Dyestuff content of the dyed fabrics was determined according Kubelka–Munk equation [28] using *Cary 100 UV-Vis Spectrophotometer*

$$f(R) = \frac{(1-R)^2}{2R} = \frac{k}{s}$$

R is the absolute reflectance of the sampled layer, K is the molar absorption coefficient and s is the scattering coefficient.

After which the samples were tested for color fastness to light and washing according to AATCC107-1997 [29] The CIE-Lab values of the dyeings were measured and the cylindrical co-ordinates of color were determined after exposure to arc lamp irradiation for 1, 2, 4, 6, 24, 48, and 72 hrs. The colors are given in an internationally commission (CIE L*a*b*) coordinates, L* corresponding to brightness, a* to red–green coordinate (positive sign = red, negative sign = green) and b* to yellow–blue coordinate (positive sign = yellow, negative sign = blue). [30-32].

$$\Delta E^*_{ab} = \{(\Delta L^*)^2 + (\Delta a^*)^2 + (\Delta b^*)^2\}^{1/2}$$

3. Results and discussion

3.1. Extraction parameters

Extraction was carried out by water, water-acetone, and water-ethanol mixtures at 25-60 ᵒC for 30-120 min. The ultrasonic efficiency had been determined simultaneously with extraction parameters, and compared with the conventional heating method. The yield coefficients of co-solvents were definitively greater than water, with much higher values in case of ultrasonic. Water-acetone mixture was found to be the most selective co- solvent followed by water-ethanol. As shown in Figures [4, 5] water-acetone mixture released over 32% of the total dye absorbency, exhibiting 21% of the total color strength when dyeing the woolen sample. Water-ethanol extracted 27% dye and exhibited 19% color strength, while water extracted less than 21% dye, and exhibited 16% color strength. This was relative to [10, 6, and 4] % of absorbance and [18, 15, and 11] % color strength respectively with the co-ordinate solvents when using the conventional heating method.

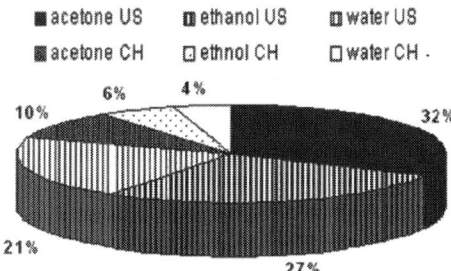

Figure 4. Effect of solvents on the absorbency of Osage orange powder using the conventional [CH] and ultrasonic [US] assisted extraction

Figures 6, 7 and 8 showed the absorption, and color strength values of Osage orange powder extracted by acetone at different concentrations of (2.5 – 25) % v/v at temperatures from 25-60° C, for time intervals varied between 30-120 min.

The maximum values were achieved with 20 % v/v acetone at 60° C, for 60 min. The extraction parameters affected the color strength and are influenced by the properties of solvents such as, the dipole moment, dielectric constant, and refractive index values.

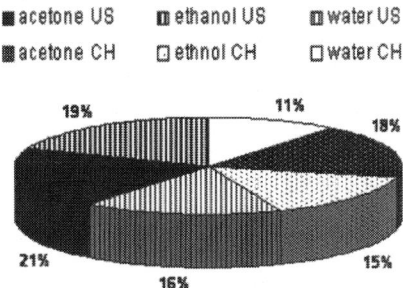

Figure 5. Effect of solvents on the color strength of Osage orange dyed wool using the conventional [CH] and ultrasonic [US] assisted extraction

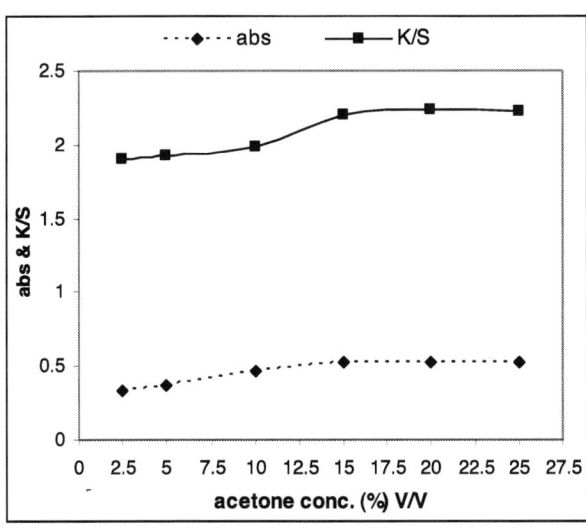

Figure 6. Effect of acetone concentration on the absorbency and color strength of Osage orange dyed wool using ultrasonic assisted extraction

The solvent polarity can change the position of the absorption or emission band of molecules by solvating a solute molecule or any other molecular species introduced into the solvent matrix [33]. By the way, dye molecules are complex organic molecules which might carry charge centers and are thus prone to absorption changes in various media [33, 34]

Acetone acts as the non hydrogen-bond donating solvents (also called as non-HBD type of solvents), while water and ethanol are the hydrogen-bond donating solvents (also called as

HBD type solvents) [33]. The absorbency values of Osage orange in these solvents are given in Figure 4, 5. It can noted from this figure that the absorption maximum of the extract is affected by the solvent type, thus the change in values can be noted as a probe for various types of interactions between the solute and the solvent.

Water and ethanol are considered as polar protic solvents, their polarity stems from the bond dipole of the O-H bond, whereas the large difference in the electro- negativities of the oxygen and hydrogen atom, combined with the small size of the hydrogen atom, warrant separating the Osage orange molecules that contain the OH groups from those polar compounds that do not. On the other hand acetone considered as dipolar aprotic solvent, containing a large multiple bond between carbon and either oxygen or nitrogen e.g. C-O double bond. [33-35].

Although water has the highest dielectric constant among ethanol and acetone solvents, its extraction demonstrated the lowest value of absorbency. This might due to the formation of strong hydrogen bond between the dyes extract and water molecules [33-35].

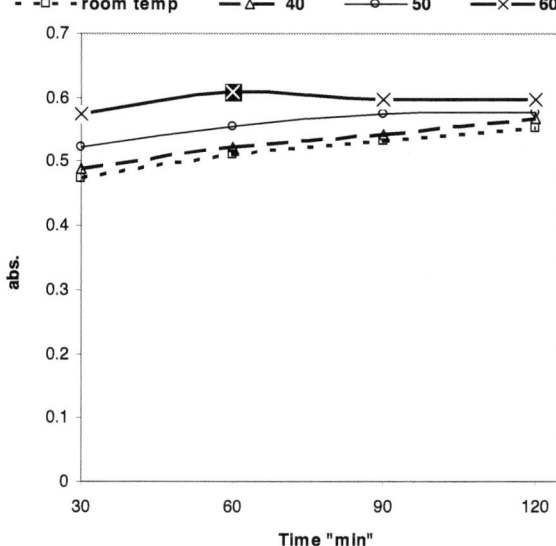

Figure 7. Effect of ultrasonic assisted extraction time on the absorbency of Osage orange at different temperatures

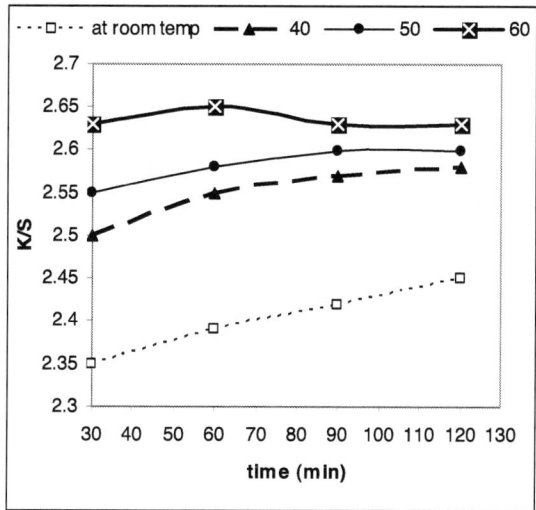

Figure 8. Effect of ultrasonic assisted extraction time on the color strength of Osage orange dyed wool at different temperatures

The dye absorbance is also influenced by the presence of co-solvents. Water-acetone mixture exhibited the highest value of absorbance, followed by the second water-ethanol mixture. In case water-acetone, the salvation of extract is non-HBD type of solvent mainly occurs through charge-dipole type of interaction, whereas in HBD type of solvent, the interaction also occurs by hydrogen bonding besides the usual ion-dipole interaction. In this situation, the methyl groups of acetone are responsible for the solvation of the dye extract. Thus, decreasing the amount of non-HBD acetone solvent "concentration" increasing the amount of HBD solvent (water) shall break these interactions with the dye molecule, thereby decreasing the value of absorbance. Water-ethanol mixtures belong to HBD type of solvents, whereas the dye cation is preferential-ly solvated by the alcoholic component in all mole fractions in aqueous mixtures with ethanol. It is well known that water makes strong hydrogen-bonded nets in the water-rich region, which are not easily disrupted by the co-solvent [33, 34]. This can explain the strong preferential salva-tion by the alcoholic component in this region since water preferentially interacts with itself rather than with the dye. In the alcohol-rich region, the alcohol molecules are freer to interact with the water and with the dye, since their nets formed by hydrogen bonds are weaker than in water. In this situation, the alcohol molecules can, to a greater or lesser extent, interact with wa-ter through hydrogen bonding [33-35]

Wool fiber is considered as relatively easy fiber to dye, the ease with which the polymer sys-tem of wool will take in dye molecules is due to polarity of its polymer and its amorphous nature. The polarity will readily attract any polar Osage orange molecules and draw them into the polymer system. The studies of wool dyeing process have been in two distinct theo-ries (The Gilbert- Rideal's and Donnan theories). The Gilbert and Rideal theory based on

Langmuir's theories of surface adsorption [36], in which the activity coefficient of Osage orange extract ions adsorbed into the wool phase are reduced due to specific binding with sites on wool, which is the formation of ion pairs. This theory proposed that dyeing process is an anion exchange process, in which the Osage orange extract molecules displace smaller anions, depending on four steps: a) diffusion to fiber surface, b) transfer across that surface, c) diffusion within to appropriate "sites" and d) binding at those sites. On the other hand, according to the Donnan equilibrium theory, the Osage orange extract was considered to partition between the external solution and internal solution phase in the wool. The later phase is believed to contain a high concentration of fixed ionic groups, and hence solute molecules have reduced activity co-efficient in that phase due to coulombic interaction between the anionic groups (OH) in fact: O⁻ of Osage orange extract and the protonated amino groups of wool. [36]

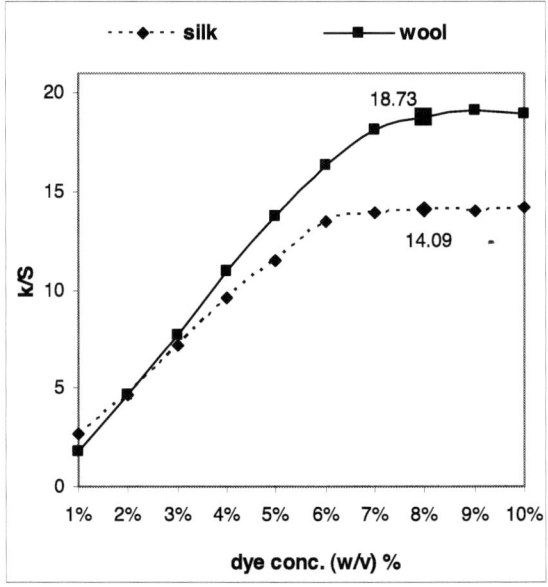

Figure 9. Effect of Osage orange concentration (w/v)%, extracted ultrasonically in 20 % water/ acetone co-solvent at 60 °C for 90 min, on the color strength of silk and wool samples.

Higher color depth was expected from an increase in the extract concentration and the use of high concentrations of mordant [37]

To study the possibility of forming concentrated extracts, different amounts of Osage orange bark powder (1-10) g were extracted per the optimized 20 % water/ acetone co-solvent. The dyeing process was carried out on silk and wool samples at a liquor ratio of L: R 1:20, for 60 min. at 60 °C. It was noted that, the use of more concentrated extracts resulted in somewhat an increase in color depth; whereas the maximum color depth was achieved with 8% w/v powder on both fabrics as shown in Figure 9. The relative high K/S values for dyeings can be explained with the high amount of bark extracted for this series of dyeing experiments.

3.2. Dyeing parameters

Dyeing temperature and time are important parameters influencing the quality of the dyed silk and wool samples. It is well known that dyeing at high temperature for a long time tends to decrease the fabric strength. [36-38] Therefore, it was proffered to dye the samples ultrasonically at temperatures ranging from 30 to 60 °C, relative to the dyeing time that was studied from 30 to 120 mins.

As shown in Figures 10 and 11, it is clear that the standard parameters of dyeing temperature and time were achieved after 90 mins, at 40–50 °C, and 50-60 °C in case of silk and wool respectively, where the color strength increases with the increase in dyeing temperature.

Generally, the increase in dye-uptake can be explained by the fiber swelling which enhanced the dye diffusion. [37] The effect of dyeing time was conducted to reveal the effect of power ultrasonic on the de-aggregation of dye molecules in the dye bath. It was denoted that the color strength obtained increased as the time increased. The decline in the dye-ability may be attributed to the hydrolytic decomposition of the extract molecules under the influence of sonic energy during prolonged dyeing. [38]

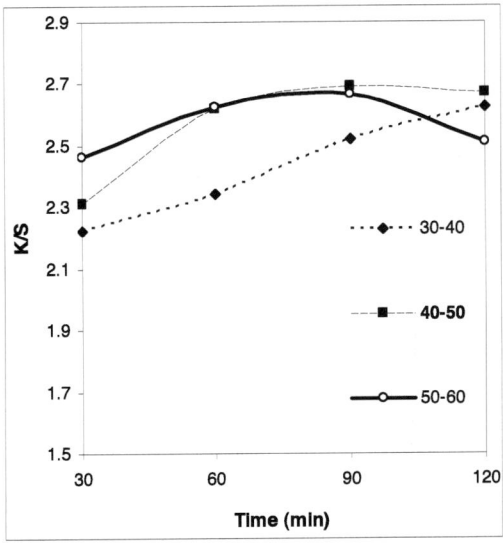

Figure 10. Effect of dye bath temperature at different time intervals on the color strength of silk samples dyed ultrasonically with 1% (w/v) Osage orange extract.

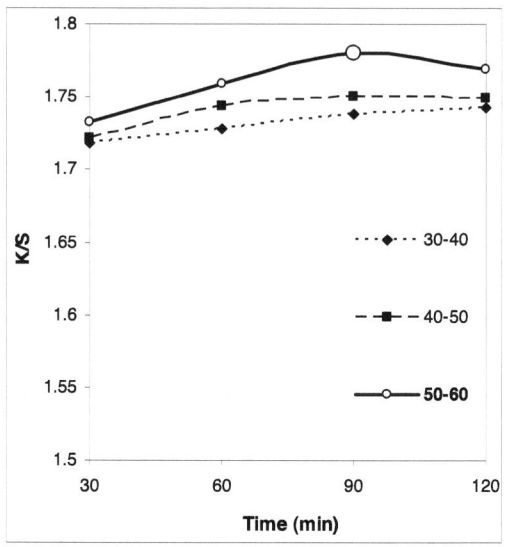

Figure 11. Effect of dye bath temperature at different time intervals on the color strength of wool samples dyed ultra-sonically with 1% (w/v) Osage orange extract

As shown in Figures 12 and 13 the pH values of the dye bath, have a considerable effect on the dyeability of silk and wool fabrics with Osage orange extract under ultrasonic. As the pH increases the dyeability, decreases. The effect of dye bath pH can be attributed to the correlation between dye structure and the protein based materials.

Since the used dye is a water-soluble dye containing hydroxyl groups, it would interact ionic-ally with the protonated terminal amino groups of silk and wool fibers at acidic pH via ion exchange reaction.

The anion of the dye has a complex character, and when it is bound on fiber, further kinds of interactions take place together with ionic forces. This ionic attraction would increase the dye-ability of the fiber as clearly observed in Figures 12 and 13. At pH greater than 5, the ionic interaction between the hydroxyl anion of the dye and the protein fibers decreases due to the decreasing number of protonated terminal amino groups of silk and wool and thus lowering their dye-ability. It is to be mentioned that the lower dye-ability may be attributed to the enhanced desorption of the dye as its ionic bond is getting decreased [36].

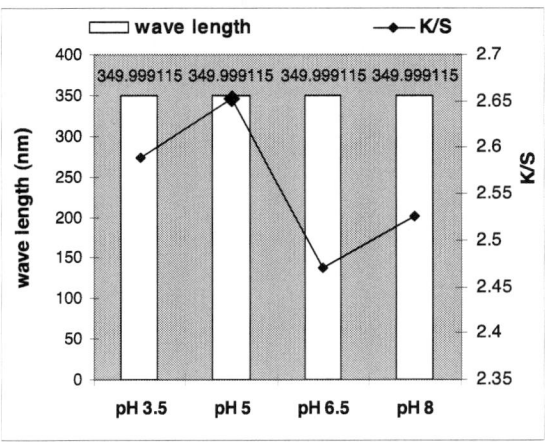

Figure 12. Effect of the dye bath pH on the color strength and wave length of silk samples dyed ultrasonically with 1% Osage orange at 40-50°C for 90 min.

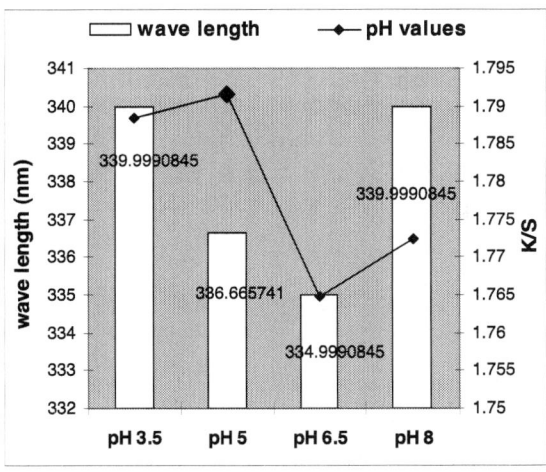

Figure 13. Effect of the dye bath pH on the color strength and wave length of woolen samples dyed ultrasonically with 1% Osage orange at 50-60 °C for 90 min.

3.3. Mordanting methods & colour properties

As shown in Figures 14 and 15, ultrasonic [US] assisted mordanting method possesses a re-markable improvement in the color strength, in comparison with the classical thermal method [CH]. The obtained dyeings are governed by the descending dyeing sequence, and can be ranked as follows: pre-mordanting with a mixture of alum and cream of tartar followed by dyeing > post mordanting with alum > post mordanting with a mixture of alum and cream of tartar > premordanting with alum > unmordanted samples.

Silk and wool fabrics are highly receptive to mordants due to their amphoteric nature; they can absorb acids and bases with equal effectiveness. Mordants during natural dyeing, exhibits fast color due to their complex formation with the dye and fiber [27, 38]

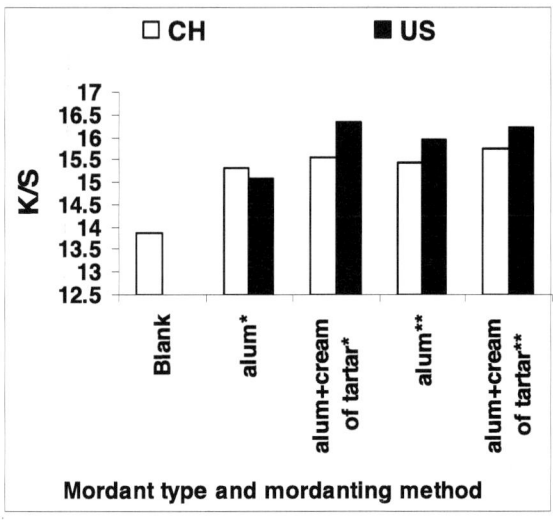

Figure 14. Effect of mordant and mordanting method on the color strength of silk dyeings

It is clear that: (i) pre-mordanting with the nominated mordant brings about a significant enhancement in the K/S values of the obtained dyeings. (ii) the extent of improvement is governed by the physical and chemical states of the dye and degree of fixation, (iii) premordanting followed by dyeing gives dyeings with better fastness properties than those dyed without mordant and mordanting after dyeing (iv) the improvement in the dyeings color strength and fastness, reflects higher extent of dye adsorption, interaction and bridging with the pre-mordanted substrate via different conjugated bonds [27]

The low color strength in post-mordanting condition is due to the accumulation of the metal dye complex in form of clusters. [39]. The high aluminum content might provide useful eco-friendly chelating with Osage orange molecules presented in the extract that might resist their hydrolysis by water. [39]

The commercial cream of tartar (Potassiun Bitartrate) contains a small percentage of calcium tartrate is frequently employed as a mordant for wool. [40]. In this study it was recommended to apply cream of tartar with alum as preferable mordant to get good strong colors as discussed previously. It helps to soften fibers when alum is used, and can also help brighten the yellow color with good levelness. [40]

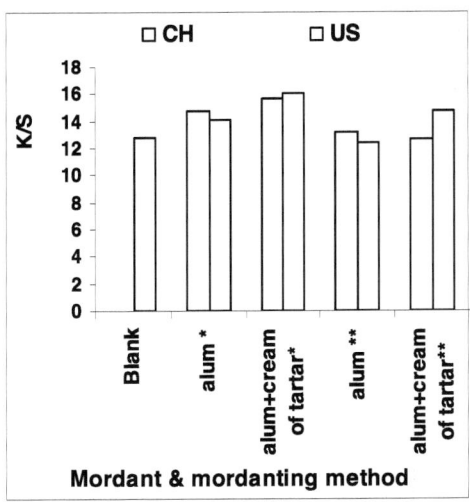

Figure 15. Effect of mordant and mordanting method on the color strength of wool dyeings

Sonic energy succeeded in accelerating the rate of mordanting at lower temperatures rather than the conventional heating technique. The exhibited improvement was generally attributed to the acoustic cavitation, which is the formation of gas-filled micro-bubbles or cavities in a liquid media, producing implosive collapse, which often forming fast-moving liquid jets, where large increases in temperature and pressure are generated. The micro-jets increase the diffusion of solute inside the intermediate spaces of silk and wool fabrics facilitate the de-aggregation of Osage orange molecules in the dye bath and thus increase the dye diffusion rate, and penetration through the fibers. [22-24, 41]

A variety of color hues were obtained with respect to the mordant. It was observed from the color fastness data that the extracted dye from *Osage orange* furnished different color hues with very good affinity for silk and wool fabrics in presence of alum and cream of tartar mordant as illustrated in Tables 1 and 2. The color intensity reached its highest value when the fabrics treated with a mixture of alum and cream of tartar. The brightness of the shades on the dyed samples might be due to the better absorption of Osage orange extract and the easy metal complex formation of mordant with the fibers. Data represented good to very good fastness to washing, because the mordant lead to the formation of a complex dye which aggregates the dye molecules into a large particles insoluble in water. Control samples exhibited poor fastness to washing due to the weak dye- fiber bond, and the ionization of the (OH) groups of the dye during washing under the alkaline condition. [42-45]

Sampls	L*	a*	b*	Wash fastness	Light fastness
Blank US	54.12	4.71	49.2	3	5
alum* CH	50.95	3.57	43.37	3-4	5-6
alum* US	55.05	3.56	51.67	4	6
Alum + cream of tartar * CH	50.57	3.81	37.31	4	6-7
Alum + cream of tartar * US	50.29	6.43	44.8	4	6-7
alum** CH	54.45	4.83	48.08	3-4	6
alum** US	54.99	5.40	52.12	3-4	6
Alum + cream of tartar ** CH	53.01	3.96	38.21	3-4	6
Alum + cream of tartar ** US	52.75	5.52	47.49	3-4	6

Table 1. The CIELab values and fastness properties of Osage orange dyed silk

Sample	L*	a*	b*	Wash fastness	Light fastness
Blank US	68.97	- 0.02	55.19	3	5
alum* CH	65.07	- 0.59	60.08	3-4	5-6
alum* US	67.46	- 2.93	62.40	4	6
Alum + cream of tartar * CH	65.43	- 0.14	56.89	4	6-7
Alum + cream of tartar * US	64.11	- 1.14	52.60	4	6-7
alum** CH	64.61	0.68	62.40	3-4	6
alum** US	64.71	1.56	58.98	3-4	6
Alum + cream of tartar ** CH	64.54	0.40	60.72	3-4	6
Alum + cream of tartar ** US	65.03	1.23	55.29	3-4	6

Table 2. The CIELab values and fastness properties of Osage orange dyed wool

In the light fastness test, mordanted colored samples exhibited better light fastness relative to the control ones. This may be due to: i) the aluminum metal in alum mordant protects, both by the steric and electronic effects of the weak point in the dye structure from attack by means of the reactive species during photochemical reaction, in addition ii) the aluminum metal in alum mordant promotes aggregation of the dye. By the way, the poor light fastness is due to the inherent susceptibility of the dye chromophore to the photochemical degradation. [46-51]

The colorimetric data indicated the depth and natural tone of the control and mordanted dyed samples. The L^* values were found to be lower using alum and cream of tartar as mordant corresponding to deeper shades. The L^* values were found to be higher in case of unmordanted dyed samples corresponding to lighter shades. Similarly, by using alum as mordant the L^* values were also higher corresponding to lighter shades. The higher values of a^* and b^* indicated the brightness, representing the redness and yellowness hues respectively. As a result, alum and cream of tartar might be effectively used as mordant for Osage orange extract.

The light fading of the dyed samples was recorded in terms of the color difference (ΔE) as shown in Figures 16-20. It was denoted that sonic energy assisted alum and cream of tartar mordanting method, exhibited a lower degree of fading in comparison with the conventional heating and the application alum mordant in absence of cream of tartar, whereas the premordanting method appears to be preferred having a great efficiency in lowering the degree of fading in comparison with the post mordanting

The dye physical state is generally important than the chemical structure in determining the color fastness on fibers. (31). It was recognized that the light fastness of many dyed systems has been found to increase with the increase dye concentration applied to the substrate [50].

The fiber swelling was increased with ultrasonic technique due to the sonic energy which i) improves the diffusion and penetration of the dye and mordant molecules inside the pores of the fabric, and ii) the fast breaking-down of the dye molecules, which became much more smaller in size and thus fully dispersed with much higher amount in the dyed samples relative to the samples subjected to the conventional heating method, resulted in, the lower degree of fading in case of ultrasonic assisted dyeing and mordantinting processes. [22-25, 41]

Mixture of alum and cream of tartar mordant renders the dye more bonded and more aggregated onto fibers, therefore the surface area of the dye accessible to light is reduced, and thereby the dye fades at lower degree with nearly constant rate of fading.

Aluminum ions apparently produce metal chelates with improved the overall fastness properties. This either could be evidence of the aggregation of dye molecules within the fiber or perhaps of the formation of dye-metal chelates that forms grater stability of the dye molecules when co-ordinated with the complex aluminum metal atom that might form quite large aggregates giving the highest light fast with the lowest fading degree. [23, 51], resulted in, the low and nearly constant fading rate in case of the mordanted samples.

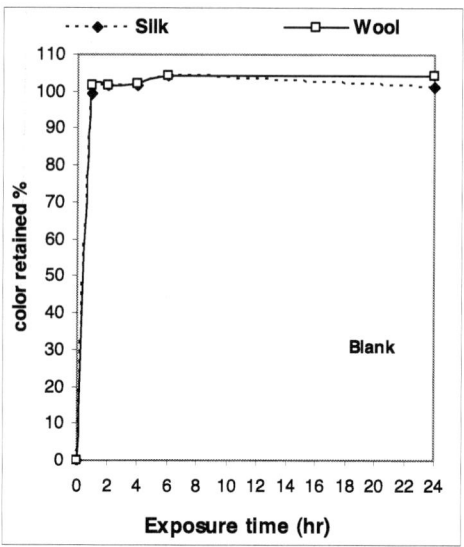

Figure 16. Effect of Xenon arc lamp exposure time on the color retained of silk and wool blank dyed samples

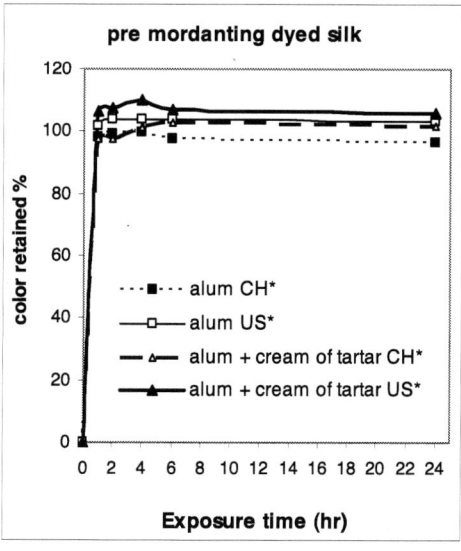

Figure 17. Effect of Xenon arc lamp exposure time on the color retained of the premordanted dyed silk samples

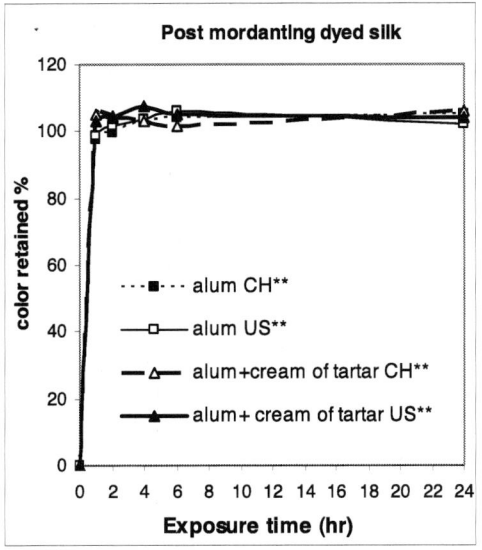

Figure 18. Effect of Xenon arc lamp exposure time on the color retained of the post mordanted dyed silk samples

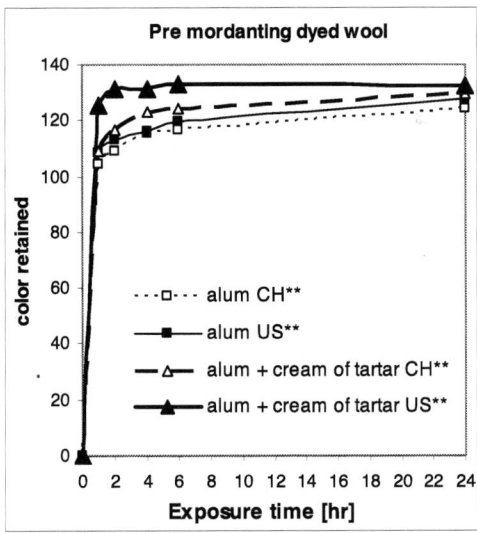

Figure 19. Effect of Xenon arc lamp exposure time on the color retained of the pre-mordanted dyed wool samples

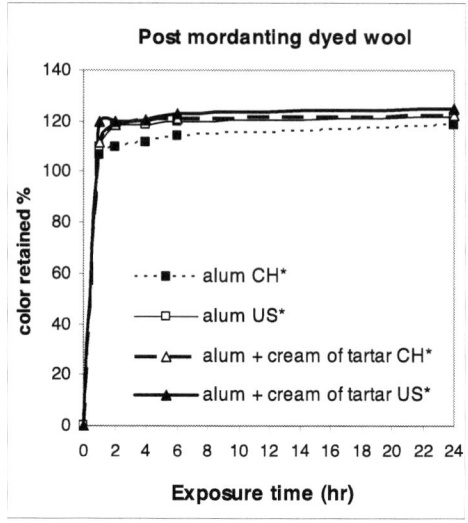

Figure 20. Effect of Xenon arc lamp exposure time on the color retained of the post mordanted dyed wool samples

4. Conclusions

With the demand for more environmentally friendly methods and increasing productivi-ty, the newer ultrasonic energy in assisting the extraction of Osage orange natural dye have been developed, over the conventional heating extraction methods, possessing a shorter extraction times and much higher dye absorbance and color strength at lower temperature. Water-acetone co-solvent, and ultrasonic have been found to be the suitable alternatives to the conventional water heating method. The maximum color yield of dye is dependent on solvent polarity. Solvation of dye molecules probably occurs via dipole-dipole interactions in non-hydrogen- bond donating solvents, whereas in hydrogen-bond donating solvents the phenomenon is more hydrogen bonding in nature. The dye uptake depends on (The Gilbert- Rideal's and Donnan theories) depending on the coulombic in-teraction between the anionic groups (OH) in fact: O⁻ of Osage orange extract and the protonated amino groups of wool fibers.

This research demonstrated the standardization dyeing parameters of Osage orange natural yellow extract on protein based fabrics; silk and wool. The color strength increases with the increase in dyeing temperature and time due to the fiber swelling which enhanced the dye diffusion. The effect of dyeing time was conducted to reveal the effect of power ultrasonic on the de-aggregation of dye molecules in the dye bath. The decline in the dye-ability may be attributed to the hydrolytic decomposition of the extract molecules under the influence of sonic energy during prolonged dyeing. Osage orange is a water-soluble dye containing hy-droxyl groups that interacts ionic-ally with the protonated terminal amino groups of silk and wool fibers at acidic pH 5 via ion exchange reaction. The lower dye-ability at pH greater than 5 may be attributed to the enhanced desorption of the dye as its ionic bond is getting decreased. Improvement in the dyeing color strength and fastness properties reflects the higher extent of dye adsorption, interaction and bridging with the pre-mordanted dyed samples via different conjugated bonds with the mixture of alum and cream of tartar mor-dant. The low color strength in post-mordanting method is due to the accumulation of the aluminum metal dye complex in form of clusters. Sonic energy succeeded in accelerating the rate of mordanting at lower temperatures rather than the conventional heating technique due to the acoustic cavitation which increases the diffusion of solute inside the intermediate spaces of silk and wool fabrics, facilitating the de-aggregation of Osage orange molecules in the dye bath and thus increases the dye diffusion rate, and its penetration through the fibers. A variety of hues were obtained with respect to the mordant. It was observed from the color fastness data that the extracted dye from *Osage orange* furnished different color hues with very good affinity for silk and wool fabrics in presence of a mixture of alum and cream of tartar mordant. Mordanted dyed samples exhibit better wash and light fastness, with a low-est degree of photo fading relative to the control ones.

Acknowledgements

Special thanks for the school of chemical and physical science, at Victoria university of Well-ington in NZ for their financial support. Also many thanks for King abdul-Aziz University, Faculty of Arts and Designs, Fashion Design department for their kindly scientific support

Author details

Heba Mansour[1,2]

1 King Abdul- Aziz University, Faculty of Arts and Design, Fashion Design, Saudi Arabia

2 Helwan University, Faculty of Applied Arts, Textile printing dyeing and Finishing Depart-ment,

References

[1] Gulrajani M L. Introduction to Natural Dyes, Indian Institute of Technology. New Delhi: India;.1992

[2] Gulrajani ML. Natural Dyes- Part 1, Present status of Natural Dyes. Colourage 1999;. 46(7): 25-27.

[3] Gulrajani M L.Present status of natural dyes. Indian Journal Fiber Textile Research. 2001;. 26: 191–201

[4] Siva R. Status of natural dyes and dye-yielding plants in India. Current Science 2007; 92 (7): 1-9

[5] Hameed BH & El-Khaiary MI. Journal of Hazardous Materials. 2008; 154(1-3): 639-648

[6] Georgiou D, Melidis PAivasidis A & Gimouho-poulos K. Dyes and pigments 2002; 52 (2): 69-78.

[7] Tsui LS,.Roy WR &.Gole MA. Coloration Technology 2003; 119 (1): 14-8.

[8] Araceli R, Juan G, Gabriel O & Maria M. Journal of Hazardous Material 2009;172(2-3): 1311-1320.

[9] Uddin MDT, Akhtarul Islam S, Mahmoud Rukanuzzaman MD. Journal of Hazardous Material 2009; 164 (1): 53-60

[10] Singh R V. Healthy hues. Down to Earth 2002; 11: 25–31.

[11] Tsui LS,.Roy WR &.Gole MA. Coloration Technology 2003; 119 (1): 14-8.

[12] Araceli R, Juan G, Gabriel O & Maria M. Journal of Hazardous Material 2009;172(2-3): 1311-1320.

[13] Uddin MDT, Akhtarul Islam S, Mahmoud Rukanuzzaman MD. Journal of Hazardous Material 2009; 164 (1): 53-60.

[14] Borland VS. Natural resources: animal and vegetable fibers for the 21st century American. Textile. Industry 2000; 29: 66–70

[15] Patel B H, Agarwal B J, & Patel H M. Novel padding technique for dyeing Babool dye on cotton. Colourage 2003; January: 21–26.

[16] Deo H T & Desai B K. Dyeing of cotton and jute with tea as a natural dye. Coloration Technology 1999; 115(7):224- 227.

[17] Aydin A H & Zeki T. 'The sorption behaviors between natural dyes and wool fiber. International Journal of Chemistry 2003; (2): 85-91.

[18] Cox P & Leslie P. Osage orange in: Texas Trees, A Friendly Guide. 1999, Cox, P. and Leslie, P. Eds., Corona Publishing Company, San Antonio, TX :147–149.

[19] Martins Teixeiraa D, Teixeira da Costa C. Novel methods to extract flavanones and xanthones from the root bark of Maclura pomifera. Journal of Chromatography A 2005; 1062: 175–181.

[20] Esra Kupeli, Ilkay Orhan, Gulnur Toker, & Erdem Yesilada. Anti-inflammatory and antinociceptive potential of Maclura pomifera (Rafin.) Schneider fruit extracts and its major isoflavonoids, scandenone and auriculasin. Journal of Ethno pharmacology 2006; 107 (2): 169–174.

[21] Tian Li, Jack Blount W & Richard Dixon A. Phenylpropanoid glycosyltransferases from osage orange (Maclura pomifera) fruit. FEBS Letters 2006; 580: 6915–6920.

[22] Padma Vankar S, Shanker R, Mahanta D & Tiwari SC Ecofriendly sonicator dyeing of cotton with Rubia cordifolia Linn. Using biomordant. Dyes and Pigments. 2006; 76 (1): 207-212.

[23] Mansour HF & Heffernan S. Environmental aspects on dyeing silk fabric with sticta coronata lichen using ultrasonic energy and mild mordants. Clean Technologies and Environmental Policy. 2010; DOI: 10.1007/s10098-010-0296-2.

[24] Mansour HF. Environment and energy efficient dyeing of woollen fabric with sticta coronata. Clean Technologies and Environmental Policy. 2009; DOI: 10.1007/s10098-009-0267-7

[25] Vajnhandl S, Majcen Le Marechal A. Ultrasound in textile dyeing and the decolouration and mineralisation of textile dyes. Dyes and Pigments 2005; 65 (2):89-101

[26] De Santis D. & Moresi M. Production of alizarin extracts from Rubia tinctorum and assessment of their dyeing properties. Industrial Crops and Products 2007; 26 (2): 151–162.

[27] Onal A. Extraction of Dyestuff from Onion (Allium Cepa L.) and its application in the dyeing of wool, Feathered, Leather and Cotton. Turkish Journal of Chemistry 1996; 20: 201.

[28] Kubelka P. New contribution to the optics of intensity light-scattering materials-Part I. The Journal of Optical Society of America. 1948; 38 (5): 448-451.

[29] AATCC. Technical Manual 2000; 75, Research Triangle Park: AATCC.

[30] Yoshizumi K & Crews PC. Characteristics of fading of wool cloth dyed with selected natural dyestuffs on the basis of solar radiant energy. Dyes and Pigments 2003; 58 (3): 197–204

[31] Crews PC. The fading rates of some natural dyes. Studies in Conservation 1987; 32 (2): 65-72.

[32] Giles CH. &.McKay RB. The light fastness of Dyes. Textile Research Journal 1963; 33 (7): 528-77.

[33] Muhammad R A, Ahmed SA, and Muhammad K. Solvent effect on the spectral properties of Neutral Red. Chemistry Central Journal 2008; 2:.2-19

[34] Oliveira CS, Bronco KP, Baptista MS & Indig GL. Solvent and concentration effects on the visible spectra of tri-para-dialkylamino-substituted triarylmethane dyes in liquid solutions. Spectochim Acta, Part A 2002; 58 {13)2971-2982.

[35] Bevilaqua T, Goncalves TF, Venturini CG & Machado VG. Solute-solvent and solvent-solvent interactions in the preferential solvation of 4-[4 (dimethylamino)styr-

yl]-1-methylpyridinium iodide in 24 binary solvent mixtures. Spectochim Acta, Part A 2006; 65 (3): 535-542.

[36] Bruce RL, Broadwood NV & King DG. Kinetics of Wool Dyeing with Acid Dyes. Textile Research Journal. 2000; 70 (6): 525-531.

[37] Bechtold T, Amalid MA. & Rita Mussak, A M. Reuse of ash-tree (Fraxinus excelsior L.) bark as natural dyes for textile dyeing: process conditions and process stability. Coloration Technology 2007; 123 (4): 271–279.

[38] Kongkachuichaya P,Shitangkoonb,A, & Chinwongamorna N. Studies on Dyeing of Silk Yarn with Lac Dye:Effects of Mordants and Dyeing Conditions. Science Asia 2002; 28: 161-166.

[39] Mansour HF. Ultrasonic Efficiency on the Photo Fading Of Madder Dyed Silk Using Egg Albumen and Aluminium Ions Chelating. 2009 6th International Textile Conference, NRC, Egypt.

[40] Maiwa 2008. http://www.maiwa.com/stores/supply/mordants.html

[41] Kamel MM, El-Shishtawy RM, Youssef BM &. Mashaly H. Ultrasonic assisted dyeing. IV. Dyeing of cationised cotton with lac natural dye. Dyes and Pigments 2007; 73 (3): 279-284.

[42] Bechtold T, Turcanu A, Ganglberger E., & Geissler S.Natural dyes in modern textile dyehouses — how to combine experiences of two centuries to meet the demands of the future?. Journal of Cleaner Production 2003; 11(5): 499-509.

[43] Walter Gardner M. Notes on mordants applicable in wool dyeing. Journal of the Society of Dyers and Colourists 2008; 6 (2): 37 – 40.

[44] Gulrajani ML. Natural Dyes- Part 1, Present status of Natural Dyes. Colourage 1999; 46 (7): 25-27.

[45] Nadiger GS, Kaushal Sharma P, & Probhaker J. Screening of Natural Dyes. Colourag 2004; 51, annual :130-137.

[46] Feng XX, Zhang LL, Chen JY, & Zhang JC. New insights into solar UV protective properties of natural dyes. Journal of Cleaner Production 2007; 15 (4): 366–372.

[47] Jjaz A, Bhatti AN, Shahid A, Asghar Jamal M, Safda M & Abbas M. Influence of gamma radiation on the colour strength and fastness properties of fabric using turmeric(Curcuma longa L.) asnaturaldye. Radiation Physics and Chemistry 2010; 79 (5): 622–625.

[48] Gupta D, Gulrajani M L, & Kumari S. Light fastness of naturally occurring anthraquinone dyes on nylon. Coloration Technology 2004; 120 (5): 205-212.

[49] Clementi C, Nowik W., Romani A, Cibin F & Favaro G. A spectrometric and chromatographic approach to the study of ageing of madder (Rubia tinctorum L.) dyestuff on wool. Analytica Chimica Acta. 2007; 596 (1): 46-54

[50] Cristea D & Vilarem G. Improving light fastness of natural dyes on cotton yarn. Dyes and Pigments; 70: 238–245.

[51] Mansour HF. Ultrasonic Efficiency on the Photo Fading Of Madder Dyed Silk Using Egg Albumen and Aluminium Ions Chelating. 2009 6th International Textile Conference, NRC, Egypt.

Chapter 9

Physichochemical and Low Stress Mechanical Properties of Silk Fabrics Degummed by Enzymes

Styliani Kalantzi, Diomi Mamma and Dimitris Kekos

Additional information is available at the end of the chapter

http://dx.doi.org/10.5772/53730

1. Introduction

1.1. The silk road

Sericulture or silk production has a long and colorful history unknown to most people. For centuries the West knew very little about silk and the people who made it. For more than two thousand years the Chinese kept the secret of silk altogether to themselves. It was the most zealously guarded secret in history. According to Chinese tradition, the history of silk begins in the 27th century BCE. Its use was confined to China until the Silk Road opened at some point during the latter half of the first millennium BCE.

The writings of Confucius and Chinese tradition recount that in the 27th century BCE a silk worm's cocoon fell into the tea cup of the empress Leizu. Wishing to extract it from her drink, the young girl of fourteen began to unroll the thread of the cocoon. She then had the idea to weave it. Having observed the life of the silk worm on the recommendation of her husband, the Yellow Emperor, she began to instruct her entourage the art of raising silk worms, sericulture. From this point on, the girl became the goddess of silk in Chinese mythology [1].

Though silk was exported to foreign countries in great amounts, sericulture remained a secret that the Chinese guarded carefully. Consequently, other peoples invented wildly varying accounts of the source of the incredible fabric. In classical antiquity, most Romans, great admirers of the cloth, were convinced that the Chinese took the fabric from tree leaves. This belief was affirmed by Seneca the Younger in his Phaedra and by Virgil in his Georgics. Notably, Pliny the Elder knew better. Speaking of the bombyx or silk moth, he wrote in his Natural History "They weave webs, like spiders, that become a luxurious clothing material for women, called silk".

The earliest evidence of silk was found at the sites of Yangshao culture in Xia County, Shanxi, where a silk cocoon was found cut in half by a sharp knife, dating back to between 4000 and 3000 BCE. The species was identified as bombyx mori, the domesticated silkworm. Fragments of primitive loom can also be seen from the sites of Hemudu culture in Yuyao, Zhejiang, dated to about 4000 BCE. Scraps of silk were found in a Liangzhu culture site at Qianshanyang in Huzhou, Zhejiang, dating back to 2700 BCE. Other fragments have been recovered from royal tombs in the Shang Dynasty (c. 1600 - c. 1046 BCE).

During the later epoch, the Chinese lost their secret to the Koreans, the Japanese, and later the Indians, as they discovered how to make silk. Allusions to the fabric in the Old Testament show that it was known in western Asia in biblical times. Scholars believe that starting in the 2nd century BCE the Chinese established a commercial network aimed at exporting silk to the West. Silk was used, for example, by the Persian court and its king, Darius III, when Alexander the Great conquered the empire. Even though silk spread rapidly across Eurasia, with the possible exception of Japan its production remained exclusively Chinese for three millennia.

According to FAO estimates, the world raw silk production for the year 2010 was 164971 tonnes [2]. Approximately 98% of the world's production is in Asia and especially in Eastern Asia (Figure 1). China is the leader in raw silk production followed by India (Table 1).

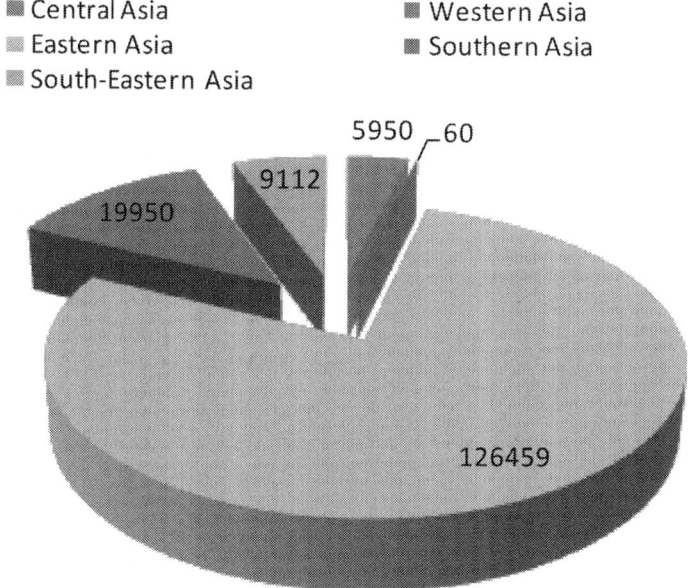

Figure 1. Production of raw silk across Asia (FAO, 2010)

Country	Production (tonnes)	Country	Production (tonnes)
China	126001	Japan	105
India	19000	Afghanistan	50
Viet Nam	7367	Kyrgyzstan	50
Turkmenistan	4500	Turkey	50
Thailand	1600	Cambodia	25
Brazil	1300	Italy	12
Uzbekistan	1200	Lebanon	10
Iran (Islamic Republic of)	900	Bulgaria	5
Democratic People's Republic of Korea	350	Greece	5
Tajikistan	200	Egypt	3
Indonesia	120		

Table 1. Raw silk producing countries [2].

1.2. Types of silk

Silk-producing insects have been classified on the basis of morphological clues, such as fol-licular imprints on the chorine egg, arrangement of tubercular setae on the larvae, and kar-yotyping data [3-4]. Classification based on phenotypic attributes is sometimes misleading because morphological features may vary with the environment [5]. Molecular marker-based analysis has been developed to distinguish genetic diversity among silkworm species [6-9]. Most commercially exploited silk moths belong to either the family Bombycidae or Saturniidae, in the order lepidoptera. Silkworms can be divided in three groups: (a) univol-tine breed (one generation per year) which is usually found in Europe where due to the cold climate the eggs are dormant in winter and they are hatched in spring, (b) bivoltine breed (two generations per year) usually found in Japan China and Korea, where the climate is suitable for developing two life cycle per year and (c) multivoltine breed (up to eight gener-ations per year) usually found in tropical zone.

The finest quality raw silk and the highest fiber production are obtained from the commonly domesticated silkworm, *Bombyx mori*, which feeds on the leaves of the mulberry plant, *Morus* spp. Other than the domesticated *B. mori*, silk fiber production is reported from the wild non-mulberry saturniid variety of silkworms. Saturniid silks are of three types: tasar, muga, and eri (Table 2).

Common Name	Scientific Name	Origin	Primary Food Plant(s)
Mulberry Silkworm	*Bombyx mori*	China	Morus indica, M. alba M.multicaulis, M.bombycis
Tropical Tasar Silkworm	*Antheraea mylitta*	India	Shorea robusta, Terminalia tomentosa T. arjuna
Oak Tasar Silkworm	*Antheraea proylei*	India	Quercus incana, Q. serrate Q. himalayana, Q. leuco tricophora Q. semicarpifolia, Q. grifithi
Oak Tasar Silkworm	*Antheraea frithi*	India	Q. dealdata
Oak Tasar Silkworm	*Antheraea compta*	India	Q. dealdata
Oak Tasar Silkworm	*Antheraea pernyi*	China	Q. dendata
Oak Tasar Silkworm	*Antheraea yamamai*	Japan	Q. acutissima
Muga Silkworm	*Antheraea assama*	India	Litsea polyantha, L. citrate Machilus bombycine
Eri Silkworm	*Philosamia ricini*	India	Ricinus communis, Manihot utilisma Evodia fragrance

Table 2. Commercially exploited sericigenous insects of the world and their food plants [10].

The tasar silkworms are of two categories-Indian tropical tasar, *Antheraea mylitta*, which feeds on the leaves of *Terminalia arjuna*, *Terminalia tomantosa*, and *Shorea robusta*, and the Chinese temperate oak tasar, *Antheraea pernyi*, which feeds on the leaves of *Quercus* spp. and *Philosamia* spp. Indian tropical tasar (Tussah) is copperish colour, coarse silk mainly used for furnishings and interiors. It is less lustrous than mulberry silk, but has its own feel and appeal. Oak tasar is a finer variety of tasar silk [11].

Muga silk is produced by the multivoltine silkworm, *Antheraea assamensis* (also called *A. assama*), which feeds mainly on *Machilus* spp (Table 2). Muga is a golden yellow colour silk. Muga culture is specific to the state of Assam (India) and an integral part of the tradition and culture of that state. The muga silk, a high value product is used in products like sarees, mekhalas, chaddars, etc. [10]. Eri silk is produced by *Philosamia* spp. (*Samia* spp.), whose primary host plant is the castor (*Ricinus* spp.) (Table 2) [11]. The luster and regularity of *B. mori* silk makes it superior to the silk produced by the non mulberry saturniid silkworms, although non-mulberry silk fibers are also used commercially due to their higher tensile strength and larger cocoon sizes. Spider also produced silk fibers that are strong and fine, but have not been utilized in the textile industries [11].

1.3. Structure of the silk fibre

The cocoons of the mulberry silkworm *B. mori* are composed of two major types of proteins: fibroins and sericin. Fibroin, the 'core' protein constitutes over 70% of the cocoon and is a hydrophobic glycoprotein [12] secreted from the posterior part of the silk gland (PSG) [13].

The fibroin, rich in glycine (43.7%), alanine (28.8%) and serine (11.9%), is composed of a heavy chain (~325 kDa), a light chain (~25 kDa) and a glycoprotein, P25, with molar ration of 6:6:1. The heavy and light chains are linked by a disulfide bond. P25 associates with disulfide-linked heavy and light chains primarily by non-covalently hydrophobic interactions, and plays an important role in maintaining integrity of the complex [14]. The light chain has a non-repetitive sequence and plays only a marginal role in the fiber. The heavy chain contains very long stretches of Gly-X repeats (with residue X being Ala in 64%, Ser in 22%, Tyr in 10%, Val in 3%, and Thr in 1.3%) that consist of 12 repetitive domains (R01–R12) separated by short linkers. It is an antiparallel, hydrogen bonded β-sheet and yields the X-ray diffracting structure called the "crystalline" component of silk fibroin [15]. Silk is a typical representative of β-sheet. Each domain consists of sub-domain hexapeptides including: GAGAGS, GAGAGY, GAGAGA or GAGTGA (G is glycine, A is alanine, S is serine and Y is tyrosine) [16]. In contrast, the 151 residues of the N-terminal, 50 residues of the C-terminal, and the 42-43 residues separating the 12 domains are non-repetitive and "amorphous" [17]. Silk fibroin can exist as three structural morphologies termed silk I, II, and III where silk I is a water soluble form and silk II is an insoluble form consisting of extended β-sheets. The silk III structure is helical and is observed at the air-water interface. In the silk II form, the 12 repetitive domains form anti-parallel b-sheets stabilized by hydrogen bonding [18]. Due to the highly oriented and crystalline structure of Silk II, silk fibroin fiber is hydrophobic and has impressive mechanical properties. When controllably spun, its mechanical property may be nearly as impressive as spider dragline silk [19].

Sericins, the 'glue' proteins constitute 20–30% of the cocoon, and are hot water-soluble glycoproteins that hold the fibers (fibroin) together to form the environmentally stable fibroin–sericin composite cocoon structure [20-22]. Sericin, secreted in the mid-region of the silk gland, comprises different polypeptides ranging in weight from 24 to 400 kDa depending on gene coding and post-translational modifications and are characterized by unusually high serine content (40%) along with significant amounts of glycine (16%), [23-24]. Three major fractions of sericin have been isolated from the cocoon, with molecular weights 150, 250, and 400 kDa [24]. Sericin remains in a partially unfolded state, with 35% β-sheet and 63% random coil, and with no α-helical content [18].

The amino acid compositions of fibroin and sericin have been published, with somewhat differences from paper to paper for some specific amino acid contents [16, 25-26].

The chemical composition of raw silk obtained from the silk worm *Bombyx mori* is presented on Table 3. Silk is produced in several countries and the fibres from different regions contain different amounts of sericin which exhibits diverse chemical and physical properties [27].

Component	
Fibroin	70-80
Sericin	20-30
Wax	0.4-0.8
Carbohydrate	1.2-1.6
Inogranic matter	0.7
Pigments	0.2

Table 3. Composition of raw silk from the silk worm *Bombyx mori* [28-29].

1.4. Silk degumming

Silk processing from cocoons to the finished clothing articles consists of a series of steps which include: reeling, weaving, degumming, dyeing or printing, and finishing. Degumming is a key process during which sericin is totally removed and silk fibres gain the typical shiny aspect, soft handle, and elegant drape highly appreciated by the consumers. In addition, the existence of sericin prevents the penetration of dye liquor and other solutions during wet processing of silk. Also, it is the main cause of adverse problems with biocompatibility and hypersensitivity to silk [17]. Furthermore, to prepare pure silk fibroin solution for silk-based biomaterials, separation of silk fibroin fiber from the sericin glue, is a critical step, since (a) residual sericin causes inflammatory responses and (b) non-degummed fibers are resistant to solubilization [30].

The industrial process takes advantage of the different chemical and physical properties of the two silk components, fibroin and sericin. While the former is water-insoluble owing to its highly oriented and crystalline fibrous structure, the latter is readily solubilized by boiling aqueous solutions containing soap [27], alkali [31], synthetic detergents [32]. However, the higher temperature (95°C) and an alkaline pH (8-9) in the presence of harsh chemicals in the treatment bath impose a markedly unnatural environment on the silk, and thus cause partial degradation of fibroin. Fibre degradation often appears as loss of aesthetic and physical properties, such as dull appearance, surface fibrillation, poor handle, drop of tensile strength, as well as uneven dyestuff absorption during subsequent dyeing and printing [33]. More importantly, the large consumption of water and energy contribute to environmental pollution. These costs have fueled interest in developing a new, effective degumming method which minimizes these adverse effects (Table 4).

Degumming method	References
Traditional (alkali, soap, synthetic detergents)	[27, 31-32]
Microwave irradiation	[42]
Plasma method	[41]
Organic acids	[34-37]
Ultrasound method	[43-44]
Enzymatic method	[33, 46-52]

Table 4. Degumming methods of raw silk

Many researches have been performed on degumming and finishing of silk fiber using acid agents for enhancing the physical properties of silk [34-36]. It has been pointed out that the action of organic acids is generally milder and less aggressive than the action of alkali. Khan et al. [37] investigated silk degumming using citric acid. The surface morphology of silk fiber degummed with citric acid was very smooth and fine, showed perfect degumming (almost complete removal of sericin) like traditional soap-alkali method and the tensile strength of silk fiber was increased after degumming. Tartaric and succinic acids demonstrate efficient sericin removal while retaining the intrinsic properties of the fiber. Freddi et al. [34] studied on the degumming of silk fabrics with tartaric acid and showed the excellent performances of tartaric acid, both in terms of silk sericin removal efficiency and of intrinsic physico-mechanical characteristics of silk fibers. The degummed silk fabric with tartaric acid exhibited a good luster and a 'scroopier' handle in compared with soap degummed fabric. They also demonstrated that dyeability with acid dyes and comfort properties (such as wicking, wettability, water retention and permeability) are also enhanced and concluded that the acid degumming process shows potential for possible industrial application. However, there is a tendency toward a gradual decrease of tenacity and elongation values with increasing tartaric acid in the bath.

Low-temperature plasma treatment has been well studied at research level in textile in the last years due to its rapid, water and chemical free process, as well as resource conservation, though it is not yet really used and well established in industry [38-40]. Long et al., [41] reported that degumming efficiency and properties of silk fabric after low-pressure argon plasma treatment were comparable to the conventional wet-chemical treatment process. Unfortunately, plasma methods result in a notable etching effect from physical bombardments and chemical reactions by excited plasma species on sericin layers.

Microwave irradiation and ultrasound are techniques that have been investigated for their performance as degumming agents by several researchers. Microwave treatment of silk resulted in increased weight loss followed by a decrease in strength of the filaments, whereas the elongation increases. This can be explained by the fact that sericin is acting as an adhesive and working as a coating and wrapping material around the fibroin [42].

Ultrasound has been widely used in chemistry and the dyeing, finishing, and cleaning industries because of its obvious advantages in particle treatment, including dispersion and agglomeration effects. Ultrasonic method combined also with natural soaps (olive oil, turpentine and daphne soaps) or proteolytic enzymes (alcalase and savinase) enables an effective clearance in the degumming process, it facilitates the removal of the substances existing on the raw silk like dirt and sericin and yields positive results in terms of weight loss, whiteness degree and mechanical properties [43-44].

The increasing awareness of legislators and citizens for the ecological sustainability of industrial processes has recently stimulated the interest of scientists and technologists for the application of biotechnology to textile processing [45]. In recent years, various studies have dealt with the removal of sericin by using proteolytic enzymes since they can operate under mild conditions and low temperatures which save energy in comparison to the traditional method. Enzymes act selectively and can attack only specific parts of sericin to cause proteolytic degradation. So the pattern of soluble sericin peptides obtained by degumming silk changes as a function of the kinds of enzyme used, attributing to the different target cleavage of the enzymes.

Several acidic, neutral, and alkaline proteases have been used on silk yarn as degumming agents. Alkaline proteases performed better than acidic and neutral ones in terms of complete and uniform sericin removal, retention of tensile properties, and improvement of surface smoothness, handle, and lustre of silk [46-48]. Enzyme degummed silk fabric displayed a higher degree of surface whiteness, but higher shear and bending rigidity, lower fullness, and softness of handle than soap and alkali degummed fabric, owing to residual sericin remaining at the cross over points between warp and weft yarns [49]. Freddi et al. [33] applied acidic, neutral, and alkaline proteases to silk degumming and found that alkaline and neutral proteases performed better than acidic proteases in terms of complete sericin removal. After complete sericin removal with proteolytic methods, the quality of appearance and retention of tensile properties is expected to be superior to those silks degummed through traditional methods due to less chemical and physical stress applied to the silk during enzymatic processing. Nakpathom et al., [50] degummed Thai *Bombyx mori* silk fibers with papain enzyme and alkaline/soap and reported that the former exhibited less tensile strength drop and gave higher color depth after natural lac dyeing, especially when degumming occurred at room temperature condition. Alcalase, savinase, (two commercial proteolytic preparations) and their mixtures also proved to be feasible for degumming applications [51]. Gulrajani et al., [52] degummed silk with the combination of protease and lipase enzymes, and obtained efficient de-waxing and degumming effects, while maintaining favorable wettability of silk fibers.

Silk degumming is a high resource consuming process as far as water and energy are concerned. Moreover, it is ecologically questionable for the high environmental impact of effluents. The development of an effective degumming process based on enzymes as active agents would entail savings in terms of water, energy, chemicals, and effluent treatment. This could be made possible by the milder treatment conditions, the recycling of processing water, the recovery of valuable by-products such as sericin peptides, and the lower environmental impact of effluents [33]. However, the limitations of higher cost of enzymes compared to chemicals and the necessary continuous use of enzymes may limit the development of industrial processes using proteolytic degumming methods [41, 49].

1.5. Potential applications of sericin

Sericin is at present an unutilized by-product of the textile industry and the discarded degumming wastewater also ultimately leads to environmental contamination due to the high oxygen demand for its degradation by microbes [53]. It is estimated that out of the 1 million tons (fresh weight) of cocoon production worldwide, or about 400 000 tons of dry cocoon, approximately 50 000 tons of sericin could be recovered from the waste solution [54]. If sericin was recovered, perhaps it could be used as a 'value added' product for many sericin-derived products and purposes [55] and this would also be beneficial in terms of the economy and the environment.

Limitations on devising specific applications are caused by its ability to exist in many forms that depend on its method of extraction and purification, etc. Each specific application requires a particular form so it will be necessary to devise and understand how to prepare consistent products suitable for each. Non-textile applications of sericin range from cosmetics to biomedical products, which include its use in anticancer drugs, anticoagulants, and cell culture additives, for its antioxidant properties [11, 56]. Furthemore, its ability to form crosslink or blends with other polymers to produce more effective films that can be used for new drug delivery methods with reduced immunogenicity and increased drug stability or even new food packaging materials worth further investigation [56].

2. Materials & methods

2.1. Silk fabric

The treatments were carried out on a 100% raw silk fabric (crêpe). Construction parameters are listed on Table 5.

	Wrap yarn	Weft yarn
Number of ends (cm⁻¹)	46	110
Fabric weight (g·m⁻²)	89.27	

Table 5. Construction properties of silk fabric

2.2. Enzymes

Esperase® 8.0L and Lipolase® Ultra 50T were kindly provided from Novo Nordisk Co. (Bagsvaerd, Denmark). Papain was purchased from Sigma. All other chemicals were laboratory-grade (analytical reagents, Sigma).

2.3. Enzyme activities

The proteolytic activity was assayed spectrophotometrically with azocasein as a substrate [57]. One unit of activity was defined as the amount of enzyme required to produce a 0.1 increase in absorbance at 440 nm under the assay conditions (50°C and pH 8.0).

Lipase activity was determined against p-nitrophenyl- propionate (pNPP) at pH 7.0 and 25°C [58]. The release of p-nitrophenol was monitored spectrophotometrically at 410 nm with the aid of a microplate reader (Molecular Devices Corporation, Sunnuvale, USA). The reaction was initiated by adding 10 μL of properly diluted enzyme to 190 μL of substrate solution (0.4 mM). Control reactions with inactivated lipase were used to correct nonenzymatic pNPP hydrolysis. One unit of activity was defined as the amount of enzyme which released 1 μmol of product per minute under the conditions described.

2.4. Soap degumming (conventional method)

Silk fabrics were soaked overnight in a solution of 5 $g \cdot l^{-1}$ Marseille soap at pH 9.5 at a liquor-to-fabric ratio 40:1. Next day the silk fabrics were degummed in a boiled alkaline solution containing 10 $g \cdot l^{-1}$ Marseille soap and 1 $g \cdot l^{-1}$ sodium carbonate for 2 h at a liquor-to-fabric ratio 40:1 and pH 9.5. Degummed silk fabrics were first rinsed at 50°C with 1 $ml \cdot l^{-1}$ ammonia and consequently two times at 40°C with 1 $ml \cdot l^{-1}$ ammonia. Finally the fabrics were rinsed with cool water.

2.5. Enzymatic degumming

A fabric sample of approximate weight 3.0 g was immersed in Tris-HCl buffer (50 mM) at pH 8.0. A non-ionic wetting agent (Sadopane SF 0.1% w/v) and the appropriate amount of enzyme(s) supplemented the solution. The liquor-to-fabric ratio was adjusted to 40:1, and the mixture was incubated at 50°C and 50 rpm. Blank samples were obtained by treating silk with buffer alone, without enzyme. Enzyme dosage and treatment time were changed. Inactivation of the enzyme(s) was carried out in hot distilled water for 10 min. At the end of the treatment, silk fabrics were rinsed with distilled water and dried at room temperature. All degumming tests were performed in duplicate.

2.6. Bleaching of degummed silk fabrics

The enzymatic and conventional degummed silk fabrics were bleached using: 0.4% (o.w.g) bleaching agent (Belphor BH); 0.5% (o.w.g) stabilizator (Sifa FL); 20 $ml \cdot l^{-1}$ H_2O_2; 3 $g \cdot l^{-1}$ Na_2CO_3; liquor-to-fabric ratio (30:1); wetting agent Clariant Sandoclean PC-FL (1 $g \cdot l^{-1}$).

The bleaching agent was mixed with the stabilizator, H_2O_2 and the Na_2CO_3. The mixture was stirred and heated at 40°C. The silk fabrics were soaked in the mixture and stirred until the temperature reached 90°C. The fabrics were left for 100 min. The fabrics were removed and rinsed initially with hot tap water and finally they were immersed in water with 2-3 drops of formic acid.

2.7. Weight loss and degumming efficiency determination

Fabric weight loss was recorded as dried sample weight loss. The drying conditions were 105°C in an air-circulated oven for 1 h. The samples were weighed, after cooling in a desiccator. The following equation (Eq. (1)) was used to calculate the weight loss (wt%):

$$Wt\% = (\frac{W1 - W2}{W1}) \times 100 \tag{1}$$

where, *W1* and *W2* are the weights of the fabric before and after treatment, respectively [59].

The efficiency of the degumming was calculated through a comparison of the enzyme process for silk fabrics with the standard method (degumming using Marseille soap), using the following equation (Eq. (2)):

$$DegumEff = \frac{W_E}{W_{MST}} \tag{2}$$

where, *DegumEff* is the Degumming Efficiency (%), W_E is the percentage of weight loss by the enzyme treatment and W_{MST} is the percentage of weight loss by the Marseille soap treatment [51].

The degumming with Marseilles soap was taken as the standard 100% weight loss.

2.8. Wettability: Drop test

Wettability of the fabric was measured, by means of the "drop test" before and after the degumming process. The dried samples at room temperature were tested using AATCC Test Method 39-1980 (evaluation of wettability) [60]. The time period (in sec) between the contact of the water drop with the fabric and the disappearance of the water drop into the fabric was counted as the wetting time. The time of drop disappearance was averaged from measurements in different points of the fabric sample. Wetting times equal or less than 1 sec were considered as indication of adequate absorbency of the fabrics [61]. All measurements were performed in triplicate.

2.9. Whiteness

The whiteness index (Berger degree) of the fabrics was determined using a reflectance measuring Datacolor apparatus at standard illuminant D65 (LAV/Spec. Excl.,d/8, D65/10o) [62].

2.10. Crystallinity index (CrI)

An X-ray diffractometer (Siemens D5000), was used in order to determine the Crystallinity Index, using copper Ka radiation. The angles scanned were 10–30° at 0.01°/s. The Crystallinity Index was determined according to the empirical method of Segal et al. [63] applying Eq. (3)

$$CrI = (\frac{I_{002} - I_{am}}{I_{002}}) \times 100 \qquad (3)$$

where I_{002} is the peak intensity from the lattice plane, and I_{am} is the peak intensity of amorphous phases.

Triplicate sets of data were used to establish the relative error associated with the X-ray diffraction method [63].

2.11. The Kawabata evaluation system

Basic mechanical properties namely tensile, bending and shearing were measured by the KES-FB system under high sensitivity conditions. The temperature was $20 \pm 0.5°C$, and relative humidity is $65\% \pm 5$. The properties measured were shear rigidity (G, gf·cm·degree^{-1}), bending rigidity per unit length (B, gf·cm^2·cm^{-1}) and extensibility (EMT, %) at 500 gf·cm^{-1}. All measurements were performed in triplicate.

3. Results and discussion

3.1. Characteristics of the enzymes used for degumming

The commercial enzyme preparations used for degumming of the silk fabric are listed in Table 6. Novozymes launched Lipolase® in 1988, the first commercial lipase developed for the detergent Industry. Lipolase® was the first lipase produced by recombinant DNA technology. This lipase, originating from *Thermomyces lanuginosus*, formerly *Humicola lanuginosa*, was expressed in *Aspergillus oryzae*. This enzyme is widely used in detergent formulations to remove fat-containing stains and it also has a broad range of substrate specificity. Furthermore is stable in proteolytic wash solutions. Novozymes launched two variants of Lipolase® issued from rational protein design: Lipolase® Ultra and LipoPrime™. These variants were also expressed in *A. oryzae* [64]. The enzyme preparation used in the present study (Lipolase® Ultra 50T) exhibited optimal pH and temperature of 9.0 and 50°C respectively. Furthermore, it was stable at temperatures 30-50°C and pH values of 7.0-10.0 (the remaining activity was measured after incubation for 24 h at the above mentioned temperatures and the pH values).

Esperase® 8.0L is also a product of Novozymes. It is a bacterial serine type alkaline protease produced by *Bacillus* sp. Esperase® is characterized by excellent perfomance at elevated temperature and pH. It exhibited optimal pH and temperature of 10.0 and 70°C respectively. The enzyme preparation was stable (>90% of its original activity) at pH values of 7.0 and 8.0 and temperatures of 30-50°C (the remaining activity was measured after incubation for 24 h at the above mentioned temperatures and the pH values).

Papain is a cysteine protease, isolated from Papaya (*Carica papaya*) Latex. Papain consists of a single polypeptide chain with three disulfide bridges and a sulfhydryl group necessary for activity of the enzyme. The pH optimum of papain was found 8.0 while temperature optimum at 50°C. Papain was stable at pH values of 7.0 and 8.0 and temperatures of 30-50°C (the remaining activity was measured after incubation for 24 h at the above mentioned temperatures and the pH values).

Enzyme	Origin	Characteristics	Activity[a]	pH[a]	T (oC)[a]
Papain	*Carica papaya*	Cysteine protease	1.15 U·mg⁻¹ dry matter	7.0-9.0 (8.0)	30-50 (50)
Esperase° 8.0L	*Bacillus* sp.	Serine-type protease	19.3 U·ml⁻¹	7.0-8.0 (10.0)	30-50 (70)
Lipolase° Ultra 50T	Aspergillus oryzae (genetically modified)	Lipase	6.1 U·ml⁻¹	7.0-9.0 (9.0)	30-50 (50)

[a] Optimum pH and temperature values in parentheses.

Table 6. Enzymes used for silk degumming

3.2. Protease treatment of silk fabrics: effect of enzyme dosage

Raw silk fabrics were degummed with Papain and Esperase® 8.0L using different enzyme loadings (expressed as Units of protease per g of silk fabric, $U \cdot g^{-1}_{fabric}$) and the physicochemical (weight loss, crystallinity index, whiteness, whiteness after bleaching) and low-stress mechanical properties (Kawabata evaluation system) of the preterated fabrics were assessed. Untreated and conventionally degummed (using Marseille soap) silk fabrics were used as controls.

3.2.1. Weight loss

The weight loss (or degumming loss) represents a quantitative evaluation of the degumming efficiency after standard or enzymatic degumming. The effect of enzyme dosage on the extent of sericin removal was studied by treating silk fabric samples for 60 min with different amounts of the two proteases. The results are depicted in Figure 2 both as weight loss (Fig. 2a) and degumming efficiency (Fig. 2b).

Degumming loss increased linearly as the amount of papain increased, attaining a value of 10.2 % (w/w) (degumming efficiency 50%). Esperase® 8.0L was more efficient in sericin removal as judged by the weight loss (Fig. 2a). Maximum value (20.1%, w/w, degumming efficiency nearly 99%) was obtained at enzyme loading of 75 $U \cdot g^{-1}_{fabric}$. Degumming loss of silk fabrics treated with Marseille soap attained a value of 20.4% (w/w).

Without enzymes, the degumming loss was negligible, owing to the low treatment temperature. In fact, it is well known that sericin can be removed by using water alone, but high temperature is needed to attain complete degumming (110-120°C, under pressure).

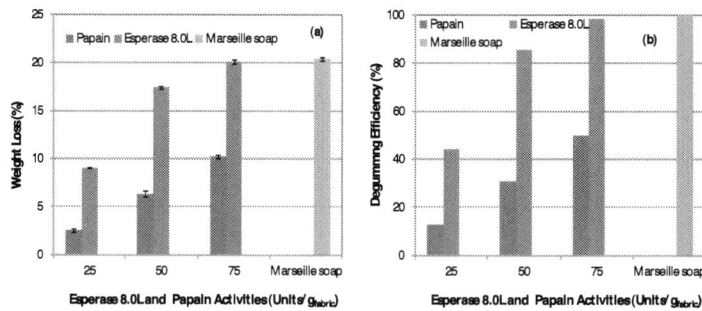

Figure 2. a) weight loss of silk fabrics treated with different types of proteases at enzyme dosages ranged from 25 to 74 U$\cdot$g^{-1}$_{fabric}$ and (b) degumming efficiency of the enzymatic process

The amount of sericin in raw silk measured in terms of weight loss varies between 17and 38%. Some of the good Chinese and Japanese varieties show about 17 to 17.5%, while yellow Italian silk has about 23% sericin and some Thai varieties have as high as 38% [27].

3.2.2. Wettability and crystallinity

Wettability was a function of both enzyme dosage and type of protease. Silk fabrics treated with Esperase® 8.0L at 75 U$\cdot$g^{-1}$_{fabric}$ showed adequate water absorbency (<1 sec). On the other hand silk fabrics treated with papain showed higher wetting times namely more hydrophobic fabrics. Although the implementation of higher papain activities formed silk fabrics with lower wetting times (Table 7) these values were higher than 1 sec, which is the maximum wetting time required for efficient dyeing and finishing. The lowest wetting time achieved using papain was 3.65 sec at the highest papain activity (75 U$\cdot$g^{-1}$_{fabric}$).

Treatment	Wetting time (sec)	Crystallinity Index (%)
Papain (U$\cdot$g^{-1} $_{fabric}$)		
25	6.12 ± 0.25	61.73 ± 1.48
50	4.16 ± 0.16	62.27 ± 0.98
75	3.65 ± 0.19	63.90 ± 0.89
Esperase® 8.0L (U$\cdot$g^{-1} $_{fabric}$)		
25	1.80 ± 0.20	63.90 ± 1.30
50	1.36 ± 0.15	66.61 ± 1.10
75	<1	66.72 ± 1.20
Marseille soap	<1	66.48 ± 1.35
No enzyme	6.85 ± 0.20	63.90 ± 1.27

Table 7. Wetting time and Crystallinity Index of silk fabrics treated conventionally (Marseille soap) or enzymatically.

The microstructure of cotton fabrics was investigated by X-ray diffraction. The results are presented on Table 7. The diffraction curves of all silk fibers exhibit the typical pattern of a silk II crystal with high crystallinity [19].

Silk fabrics treated with Esperase® 8.0L exhibited higher Crystallinity Index values compared to papain treated ones. Furthermore, at higher Esperase® 8.0L the crystallinity of the fabric was superior to that treated conventionally (Marseille soap).

3.2.3. Whiteness and whiteness after bleaching with H_2O_2

Natural coloring matters present in silk are associated mainly with sericin and hence are eliminated during degumming. The natural colouring matter of silk can be roughly divided into yellow, green and brown pigments. However the residual pigments are adsorbed by fibroin and hence silk fabrics made from yellow raw silk after degumming are not white but have a cream colour [65]. Lustre is one of the most important properties of silk, hence the method of degumming is significant.

A slight increase in whiteness was observed after treatment with the two different proteases compared to the fabric treated in the absence of enzyme. Esperase® 8.0L exhibited better results compared to papain. However, whiteness of silk fabrics treated with Marseille soap was superior to that of enzymatically treated fabrics (Figure 3a).

The results are in contrast to those reported by Chopra et al., [49] who demonstrated that enzyme treated samples rate marginally better than the soap-treated samples.

Bleaching of silk is for white and pastel shades only. Degummed silk fabrics present a slightly off-white in colour, because of some sericin, which is stubbornly stuck to the fibrin [65]. After enzymatic and conventional degumming the silk fabrics were bleached with H_2O_2. The results indicated that the whiteness after bleaching of all Esperase® 8.0L treated fabrics were the highest (Figure 3b).

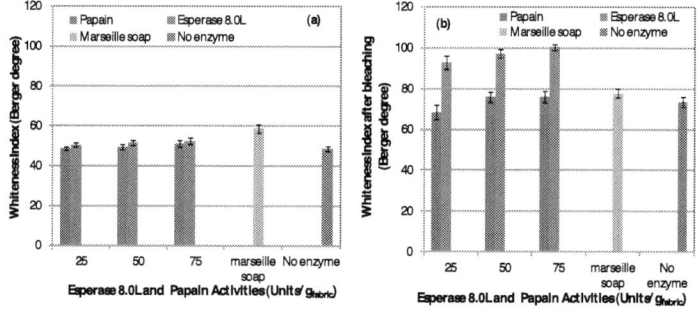

Figure 3. a) Whiteness Index and (b) Whiteness Index after bleaching with H_2O_2 of degummed silk fabrics using Marseille soap (conventional treatment) or papain and Esperase® 8.0L at different loadings.

3.2.4. Bending property

Fabric bending property is apparently a function of the bending property of its constituent yarns. B is bending rigidity, a measure of a fabric's ability to resist bending deformation. In other words, it reflects the difficulty with which a fabric can be deformed by bending. This parameter is particularly critical in the tailoring of lightweight fabrics. The higher the bending rigidity, the higher a fabric's ability to resist when it is bent by external force like what happens during fabric manipulation in spreading and sewing.

Enzymatically treated silk fabrics exhibited higher B values compared to silk fabrics treated only with buffer solution (no enzyme) (Table 8). Bending rigidity was affected by both type of protease and enzyme concentration. Increasing enzyme concentration resulted in higher B values, namely in more stiff fabrics. Silk fabrics treated with papain exhibited 25-116% increase in B values while silk fabrics treated with Esperase® 8.0L showed higher increase namely 44-137% in comparison with the buffer treated silk fabrics. Enzymatically treated silk fabrics are less rigid, softer compared to conventionally degummed fabrics.

Chopra et al., [49] reported that enzyme treated silk fabrics exhibited increased bending rigidity compared to soap treated ones. In the present study, even though an increase in bending rigidity was observed through the different experimental conditions, fabric treated with Marseille soap exhibited the highest value of this property.

3.2.5. Shear property

Shear rigidity G provides a measure of the resistance to rotational movement of the warp and weft threads within a fabric when subjected to low levels of shear deformation. The lower the value of G, the more readily the fabric will conform to three-dimensional curvatures.

Enzymatically treated silk fabrics showed lower G values compared to untreated and conventionally treated silk fabrics. Shear rigidity (G) of biotreated fabrics, for a given protease, decreased, when enzyme loading was increased (Table 8). The silk fabrics treated with papain showed 5-63% decrease in G values for the different enzyme activities, while the fabrics treated with Esperase® 8.0L showed a higher decrease in G values which ranged between 65-75%. Lower G values, of the enzymatically treated silk fabrics is "translated" in softer fabrics with better drape (Table 8).

3.2.6. Tensile property

The tensile behaviour of fabric is closely related to the inter-fiber friction effect, the ease of crimp removal and the load-extension properties of the yarns themselves. Through the KES apparatus the EMT parameter was determined which reflects fabric's extensibility, a measure of a fabric's ability to be stretched under tensile load. The larger the EMT, the more extensible the fabric.

Type of treatment	Bending rigidity (B) (gf·cm²·cm⁻¹)			Shear stiffness (G) (gf·cm⁻¹·degree⁻¹)		
	warp	weft	mean	warp	weft	mean
No enzyme	0.0205±0.0005	0.0117±0.0001	0.0161	1.80±0.02	1.60±0.01	1.70
P 25 U·g⁻¹ fabric	0.0256±0.0007	0.0147±0.0003	0.0202	1.70±0.03	1.53±0.02	1.62
P 50 U·g⁻¹ fabric	0.0356±0.0009	0.0185±0.0004	0.0271	1.30±0.01	1.25±0.01	1.27
P 75 U·g⁻¹ fabric	0.0463±0.0006	0.0234±0.0002	0.0349	0.75±0.01	0.50±0.01	0.63
E 25 U·g⁻¹ fabric	0.0337±0.0004	0.0126±0.0003	0.0232	0.85± 0.01	0.35±0.01	0.60
E 50 U·g⁻¹ fabric	0.0321±0.0006	0.0283±0.0002	0.0302	0.55±0.01	0.35±0.01	0.45
E 75 U·g⁻¹ fabric	0.0479±0.0007	0.0283±0.0005	0.0381	0.50±0.01	0.35±0.01	0.43
Marseille soap	0.0537±0.0006	0.0264±0.0003	0.0401	1.80±0.03	1.60±0.01	1.70

P: Papain, E: Esperase® 8.0L

Table 8. Comparison of micromechanical properties of silk fabrics treated for 60 min with various levels of papain and Esperase® 8.0L, with reference and conventionally degummed materials

Tensile property increased when enzyme concentration increased for both protease used (Table 9). Treatment with Esperase® 8.0L and particularly with high concentrations resulted in silk fabrics more elastic compared to those treated with Marseille soap (Table 9).

Type of treatment	Tensile (%)		
	warp	weft	mean
No enzyme	5.15 ± 0.2	1.63 ± 0.07	3.39
P 25 U·g⁻¹ fabric	4.09 ± 0.3	1.24 ± 0.04	2.67
P 50 U·g⁻¹ fabric	3.97 ± 0.2	0.94 ± 0.02	2.53
P 75 U·g⁻¹ fabric	4.53 ± 0.3	1.33 ± 0.04	2.93
E 25 U·g⁻¹ fabric	4.51 ± 0.3	1.05 ± 0.04	2.78
E 50 U·g⁻¹ fabric	5.37 ± 0.1	1.49 ± 0.02	3.43
E 75 U·g⁻¹ fabric	6.32 ± 0.3	1.15 ± 0.01	3.74
Marseille soap	4.29 ± 0.2	1.42 ± 0.07	2.86

P: Papain, E: Esperase® 8.0L

Table 9. Comparison of micromechanical properties of silk fabrics treated for 60 min with various levels of papain and Esperase® 8.0L, with reference and conventionally degummed materials

3.3. Synergistic action of protease-lipase on silk degumming: Effect of treatment time

Proteases and lipases are normally used in combination for degumming and removing others impurities such as waxes, fats, mineral salts and pigments [49]. Waxes and fats as well as colorants and mineral components occur exclusively in silk gum layer (sericin) [66]. The combined effect of enzyme activity (expressed as Units of protease per g of silk fabric) and treatment time on physicochemical and low-stress mechanical properties of cotton fabric was investigated. Raw silk fabrics were treated with two different proteases namely papain and Esperase® 8.0L at an enzyme activity of 50 and 75 U·g⁻¹fabric combined with a lipase (Lipolase® Ultra 50T) at a constant activity of 50 U·g⁻¹fabric ic for 30, 60, 90 min. Furthermore, buffer treated and conventional degummed silk fabrics were used as controls.

3.3.1. Weight loss

The results are depicted in Figure 4 both as weight loss (Figure 4a) and degumming efficiency (Figure 4b). Without enzymes, the weight loss was negligible.

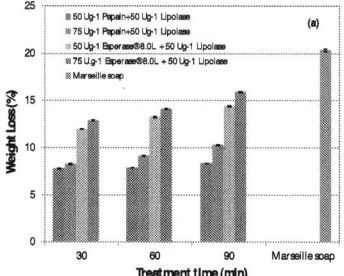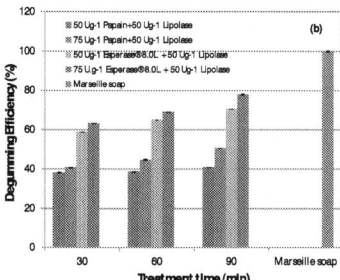

Figure 4. a) weight loss of silk fabrics treated with mixtures of proteases/lipase for 30, 60 and 90 min and (b) degumming efficiency of the enzymatic process

The degree of fabric's degradation was found to be affected by both enzyme concentration and treatment time. Increasing enzyme concentration and treatment time resulted in increased weight loss values. The combination of Esperase® 8.0L with Lipolase® Ultra 50T was found superior to that of papain. The highest weight loss (15.9%, w/w) was achieved when fabrics were treated with 75 U·g⁻¹fabric Esperase® 8.0L + 50 U·g⁻¹fabric Lipolase® Ultra 50T for 90min, while weight loss during conventional degumming was higher (20.4 %, w/w).

Degumming efficiency did not exceed 78% (Figure 4b). The proteases and lipase combination did not improve the results obtained when the fabrics degummed only with proteases.

3.3.2. Wettability and crystallinity index

Wettability seemed to depend more on the enzymes used for degumming and less on treatment time. (Table 10). The most effective combination is that of Esperase® 8.0L with Lipolase® Ultra 50T. This combination at the highest enzyme loading used, exhibited adequate water absorbency (<1 sec) after 60 min of treatment. The mixture of papain with Lipolase® Ultra 50T did not improve the wettability of the silk fabrics compared to those treated only with papain, even at higher treatment times.

Crystallinity index increased with treatment time for all enzyme combination tested (Table 10). The combination of Esperase® 8.0L with lipase was more effective compared to that of papain. However, the addition of lipase did not seem to improve Crystallinity Index in any condition tested, compared to the action of single protease.

Treatment	Wetting time (sec)	Crystallinity Index (%)
50 U g⁻¹ fabric Papain + 50 U g⁻¹ fabric Lipolase® Ultra 50T		
30 min	5.90 ± 0.18	62.80 ± 1.30
60 min	5.80 ± 0.12	63.20 ± 1.24
90 min	5.60 ± 0.10	63.32 ± 1.31
75 U g⁻¹ fabric Papain + 50 U g⁻¹ fabric Lipolase® Ultra 50T		
30 min	4.72 ± 0.20	62.90 ± 1.36
60 min	3.60 ± 0.15	63.20 ± 1.40
90 min	3.30 ± 0.30	63.32 ± 1.26
50 U g⁻¹ fabric Esperase® 8.0L + 50 U g⁻¹ fabric Lipolase® Ultra 50T		
30 min	2.20 ± 0.20	64.80 ± 1.60
60 min	2.00 ± 0.19	65.00 ± 1.80
90 min	1.50 ± 0.18	65.10 ± 1.10
75 U g⁻¹ fabric Esperase® 8.0L + 50 U g⁻¹ fabric Lipolase® Ultra 50T		
30 min	1.47 ± 0.09	65.20 ± 1.50
60 min	<1	65.40 ± 1.40
90 min	<1	66.00 ± 1.10
Marseille soap	<1	66.48 ± 1.35
No enzyme	6.85 ± 0.20	63.90 ± 1.27

Table 10. Wetting time and Crystallinity Index of silk fabrics treated conventionally (Marseille soap) or enzymatically.

3.3.3. Whiteness and whiteness after bleaching with H₂O₂

The synergistic action of protease-lipase improved the whiteness of the silk fabrics (Figure 5a). Increasing enzyme concentration and treatment time resulted in increased whiteness values. The highest whiteness values for each combination were observed at highest enzyme loading and after 90 min of treatment. Those values were 55.2 and 59.8 Berger degree for papain+ Lipolase® Ultra 50T and Esperase® 8.0L+ Lipolase® Ultra 50T, respectively. It should be noted that the conventional degumming resulted in a lower whiteness value (58.8 Berger degree) (Figure 5a). The addition of lipase improved the whiteness of the silk fabrics compared to results obtained by the use of protease only. For example the whiteness of the silk fabrics treated with 75 $U \cdot g^{-1}_{fabric}$ Esperase® 8.0L for 60 min was found 52.2 Berger degree, while the corresponding value of the fabrics treated at the same conditions adding 50 $U \cdot g^{-1}_{fabric}$ Lipolase® Ultra 50T was 57.3 Berger degree.

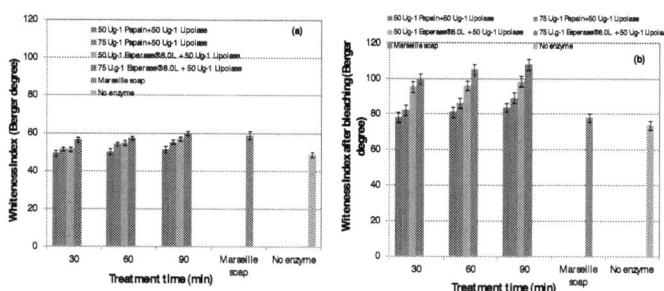

Figure 5. a) Whiteness Index and (b) Whiteness Index after bleaching with H_2O_2 of degummed silk fabrics using Marseille soap (conventional treatment) or mixture of papain and Esperase⁺ 8.0L with Lipolase⁺ Ultra 50T at different loadings and treatment times.

The synergistic action of protease-lipase improved the bleaching effect of the silk fabrics (Figure 5b). Increasing enzyme concentration and treatment time resulted in increased whiteness values after bleaching. The whiteness after bleaching of the silk fabrics treated with papain and Lipolase® Ultra 50T ranged between 78.0-89.0 Berger degree, while the corresponding values for the fabrics treated only with papain were found between 68.4-76.0 Berger degree. Higher whiteness values were observed for the fabrics treated with the mixture of Esperase® 8.0L with Lipolase® Ultra 50T. The whiteness ranged between 95.23-108 Berger degree. The silk fabrics treated with Esperase® 8.0L + Lipolase® Ultra 50T exhibited 40% increase in whiteness in comparison with those treated with the conventional method with Marseille soap, while the corresponding increase in whiteness for the fabrics treated with papain and Lipolase® Ultra 50T was 15%.

3.3.4. Bending property

Enzymatically and soap treated silk fabrics exhibited higher B values compared to untreated silk fabrics. Bending rigidity of the enzymatically treated silk fabrics was affected by enzyme concentration and treatment time. Increasing enzyme concentration and treatment time resulted in lower B values, namely in less rigid fabrics compared to conventionally treated ones (Table 11).

One interesting aspect is that addition of Lipolase® Ultra 50T caused further decrease in B values compared to those obtained when silk fabrics were treated only with protease. For example, silk fabrics treated with 75 $U \cdot g^{-1}_{fabric}$ Esperase® 8.0L exhibited a B value of 0.0381 ($gf \cdot cm^2 \cdot cm^{-1}$) after 60 min (Table 8). Addition of 50 $U \cdot g^{-1}{}_{fabric}$ Lipolase® Ultra 50T resulted in 43% decrease in B value (0.0291, $gf \cdot cm^2 \cdot cm^{-1}$). The same pattern was observed for all conditions tested.

Type of treatment	Bending rigidity (B) ($gf \cdot cm^2 \cdot cm^{-1}$)			Shear stiffness (G) ($gf \cdot cm^{-1} \cdot degree^{-1}$)		
	warp	weft	*mean*	warp	weft	*mean*
No enzyme	0.0205±0.0005	0.0117±0.0001	*0.0161*	1.80±0.02	1.60±0.01	*1.70*
50 Ug⁻¹ Papain + 50 Ug⁻¹ Lipolase* Ultra 50T						
30 min	0.0300±0.0003	0.0200±0.0004	*0.0250*	1.50±0.02	1.34±0.01	*1.42*
60 min	0.0250±0.0005	0.0170±0.0002	*0.0210*	1.44±0.03	1.30±0.03	*1.37*
90 min	0.0200±0.0002	0.0160±0.0002	*0.0180*	1.28±0.02	1.22±0.02	*1.25*
75 Ug⁻¹ Papain + 50 Ug⁻¹ Lipolase* Ultra 50T						
30 min	0.0220±0.0002	0.0180±0.0001	*0.0200*	1.34±0.02	1.30±0.01	*1.32*
60 min	0.0170±0.0003	0.0140±0.0002	*0.0155*	1.29±0.03	1.25±0.01	*1.27*
90 min	0.0160±0.0002	0.0130±0.0003	*0.0145*	1.26±0.01	1.23±0.02	*1.25*
50 Ug⁻¹ Esperase* 8.0L + 50 Ug⁻¹ Lipolase* Ultra 50T						
30 min	0.0424±0.0009	0.0233±0.0002	*0.0329*	1.85±0.04	1.50±0.05	*1.68*
60 min	0.0341±0.0005	0.0200±0.0001	*0.0271*	1.80±0.04	1.45±0.05	*1.63*
90 min	0.0330±0.0007	0.0187±0.0003	*0.0259*	1.60±0.06	1.40±0.05	*1.50*
75 Ug⁻¹ Esperase* 8.0L + 50 Ug⁻¹ Lipolase* Ultra 50T						
30 min	0.0491±0.0005	0.0231±0.0002	*0.0361*	1.90±0.06	1.20±0.04	*1.55*
60 min	0.0242±0.0003	0.0195±0.0005	*0.0219*	1.58±0.03	0.90±0.03	*1.24*
90 min	0.0214±0.0007	0.0170±0.0003	*0.0192*	1.53±0.02	0.10±0.02	*0.81*
Marseille soap	0.0537±0.0006	0.0264±0.0003	*0.0401*	1.80±0.03	1.60±0.01	*1.70*

Table 11. Comparison of micromechanical properties of silk fabrics treated at varying durations with various levels of Papain and Esperase* 8.0L combined with constant level of Lipolase* Ultra 50T, with reference and conventionally degummed materials

3.3.5. Shear property

Combined action of proteases with Lipolase resulted in silk fabrics with lower G values compared to untreated and conventional treated silk fabrics (Table 11). Increasing protease concentration and treatment time resulted in silk fabrics less rigid with better drape (lower G values) compared to those treated with Marseille soap.

Shear rigidity was also affected by the addition of Lipolase® Ultra 50T. For example, silk fabrics treated with 75 $U \cdot g^{-1}{}_{fabric}$ Esperase® 8.0L exhibited a G value of 0.43 ($gf \cdot cm^{-1} \cdot degree^{-1}$) after 60 min (Table 8). Addition of 50 U/g Lipolase® Ultra 50T resulted in 65% increase in G value (1.24, $gf \cdot cm^{-1} \cdot degree^{-1}$). The same pattern was observed for all conditions tested.

3.3.6. Tensile property

The elasticity, (tensile strength) of the silk fabrics treated with papain and Lipolase® Ultra 50T was higher than the elasticity of the untreated and soap treated silk fabrics and seemed to be dependent on protease dosage and treatment time (Table 12). Increasing papain dosage and treatment time resulted in an increase elasticity of the enzymatically treated of silk fabrics. At highest papain dosage and for treatment times 60 min and more, the elasticity was superior to that of Marseille soap treated silk.

Opposite trend was observed for the fabrics treated with Esperase® 8.0L + Lipolase® Ultra 50T combination. Increasing protease dosage and treatment time the silk fabrics became more anelastic (Table 12).

Type of treatment	Tensile (%)		
	warp	**weft**	*mean*
No enzyme	5.15 ± 0.2	1.63 ± 0.07	*3.39*
50 U·g⁻¹ fabric Papain + 50 U·g⁻¹ fabric Lipolase® Ultra 50T			
30 min	3.30 ± 0.10	2.70 ± 0.20	*3.00*
60 min	3.60 ± 0.10	3.00 ± 0.10	*3.30*
90 min	3.90 ± 0.20	3.10 ± 0.10	*3.50*
75 U·g⁻¹ fabric Papain + 50 U·g⁻¹ fabric Lipolase® Ultra 50T			
30 min	4.10 ± 0.20	3.30 ± 0.20	*3.70*
60 min	4.20 ± 0.20	3.40 ± 0.30	*3.80*
90 min	4.40 ± 0.40	3.50 ± 0.20	*3.95*
50 U·g⁻¹ fabric Esperase® 8.0L + 50 U·g⁻¹ fabric Lipolase® Ultra 50T			
30 min	4.85 ± 0.10	1.74 ± 0.05	*3.30*
60 min	4.75 ± 0.20	1.65 ± 0.02	*3.20*
90 min	4.60 ± 0.20	1.50 ± 0.03	*3.05*
75 U·g⁻¹ fabric Esperase® 8.0L + U·g⁻¹ fabric Lipolase® Ultra 50T			
30 min	4.96 ± 0.10	1.87 ± 0.09	*3.42*
60 min	4.62 ± 0.40	1.55 ± 0.05	*3.09*
90 min	4.51 ± 0.30	1.36 ± 0.04	*2.94*
Marseille soap	4.29 ± 0.20	1.42 ± 0.02	*2.86*

Table 12. Comparison of micromechanical properties of silk fabrics treated for different times with various levels of papain and Esperase® 8.0L combined with constant level of Lipolase® Ultra 50T, with reference and conventionally degummed materials

Addition of Lipolase® Ultra 50T in papain degumming mixture produces more elastic fabrics compared to those obtained by single papain treatement. For example silk fabrics treated with papain at 75 U·g⁻¹$_{fabric}$ fabric for 60 min exhibited a tensile strength value of 2.93%. Addition of 50 U·g⁻¹$_{fabric}$ Lipolase® Ultra 50T increases this value at 3.80%. On the other hand the use of Esperase® 8.0L + Lipolase® Ultra 50T mixture causes a decrease in the fabric's elasticity.

4. Conclusions

Enzymatic degumming of silk fabric using proteolytic enzymes (papain, Esperase® 8.0L) and mixtures of thereof with a lipolytic (Lipolase® Ultra 50T) enzyme under mild conditions was investigated. The results obtained are encouraging in comparison with those of the conventional method of degumming (Marseille soap).

Silk fabrics treated with Esperase® 8.0L (an alkaline protease) at high concentration (75 U/g fabric) for 60 min exhibited degumming efficiency nearly 99%, which indicate almost complete removal of sericin from the surface of the fabric. Esperase® 8.0L treatment resulted in fabrics with adequate wettability, higher CrI and whitenees after bleaching compared to those treated conventionally (Marseille soap).

On the other hand papain was not so effective in the degumming process. This could probably be attributed to the different substrate specificity of the proteases, that is, the chemical structure of the target cleavage site. The silk fabrics treated with both proteolytic enzymes exhibited low bending and shear rigidity and higher elasticity compared to Marseille soap treated silk fabrics. This means that the silk fabrics after enzymatic degumming were softer, less rigid, with better drape and higher elasticity compared to those treated with Marseille soap.

Since the silk of *Bombyx mori* apart of the proteins fibroin and sericin, also contains fats, wax etc., the combined effect of proteolytic enzymes with a lipolytic one was investigated. Combined action of protease with lipase (Lipolase® Ultra 50T) resulted in lower degumming efficiency compared (under the same conditions) with protease treatment only, but generated silk fabrics with significant improvement in whiteness after bleaching. As far as the properties measured in the Kawabata evaluation system the combination of proteolytic with lipolytic enzymes resulted in silk fabrics with extremely low bending rigidity, reduced shear stiffness and with higher elasticity compared to Marseille soap treated. This means that the fabrics were softer, less rigid with better drape compared to conventionally treated. Addition of lipolase caused decrease in bending and increase in shear rigidity and elasticity of the silk fabrics compared to those treated only with protease.

Silk degumming is a high resource consuming process as far as water and energy are concerned. Moreover, it is ecologically questionable for the high environmental impact of effluents. The development of an effective degumming process based on enzymes as active agents would entail savings in terms of water, energy, chemicals, and effluent treatment. This could be made possible by the milder treatment conditions, the recycling of processing water, the recovery of valuable by-products such as sericin peptides, and the lower environmental impact of effluents.

Author details

Styliani Kalantzi, Diomi Mamma and Dimitris Kekos*

*Address all correspondence to: kekos@chemeng.ntua.gr

Biotechnology Technology, School of Chemical Engineering, National Technical University of Athens, Zografou Campus, Athens, Greece

References

[1] The Silk Association of Great Britain. http://www.silk.org.uk/history.php (accessed 30 March 2012)

[2] FAO: Food and Agriculture Organization of the United Nations. www.fao.org (accessed 25 March 2012)

[3] Jolly MS, Sen SK. Patterns of follicular imprints in egg shell-a species-specific character in Antheraea. Bull Entomol 1969;10 32-38.

[4] Jolly MS,Narasimhanna MN,Wartkar VB. Arrangement of tubercular setae in Antheraea species (Lepidoptera: Saturniidae). Entomologist 1970;103 18-23.

[5] Souche YS, Patole M. Sequence analysis of mitochondrial 16S ribosomal RNA gene fragment from seven mosquito species. Journal of Biosciences 2000;25 61-66.

[6] Tan YD, Wan C, Zhu Y, Lu C, Xiang Z, Deng HW. An amplified fragment length polymorphism map of the silkworm. Genetics 2001;157 1227-1284.

[7] Mahendran B, Padhi BK, Ghosh SK, Kundu SC. Genetic variation in ecoraces of tropical tasar silkworm, Antheraea mylitta D. using RFLP technique. Current Science 2005;90 100-103.

[8] Mahendran B, Acharya C, Dash R, Ghosh SK, Kundu SC. Repetitive DNA in tropical tasar silkworm Antheraea mylitta. Gene 2006;370 51-57.

[9] Li A, Zhao Q, Tang S, Zhang Z, Pan S, Shen G. Molecular phylogeny of the domesticated silkworm, Bombyx mori, based on the sequences of mitochondrial cytochrome b genes. Journal of Genetics 2005;84 137-142.

[10] http://www.csb.gov.in/silk-sericulture/silk/ (accessed 30 March 2012)

[11] Kundu SC, Dash BC, Dash R, Kaplan DL. Natural protective glue protein, sericin bioengineered by silkworms: Potential for biomedical and biotechnological applications. Progress in Polymer Science 2008;33 998-1012.

[12] Sinohara H, Asano Y, Fukui A. Glycopeptides from silk fibroin of Bombyx mori. Biochimica et Biophysica Acta 1971;237 273-279.

[13] Prudhomme J-C, Couble P., Garel J-P. and Daillie J. Silk synthesis. In: Kerkut GA, Gilbert LJ. (eds) Comprehensive Insect Physiology, Biochemistry and Pharmacology. Oxford: Pergamon Press; 1985. Vol 10 p571-594.

[14] Inoue S, Tanaka K, Arisaka F, Kimura S, Ohtomo K, Mizuno S. Silk fibroin of Bombyx mori is secreted, assembling a high molecular mass elementary unit consisting of H-chain, L-chain, and P25, with a 6:6:1 molar ratio. Journal of Biological Chemistry 2000;275(51) 40517-40528.

[15] Zhou CZ, Confalonieri F, Jacquet M, Perasso R, Li ZG, Janin J. Silk fibroin: structural implications of a remarkable amino acid sequence. Proteins:Structure, Function and Genetics 2001;44 119-122.

[16] Vepari C, Kaplan DL. Silk as a biomaterial. Progress in Polymer Science 2007;32 991-1007.

[17] Li G, Liu H, Li TD, Wang J. Surface modification and functionalization of silk fibroin fibers/fabric toward high performance applications. Materials Science and Engineering C 2012;32 627-636.

[18] Teramoto H, Kakazu A, Asakura T. Native structure and degradation pattern of silk sericin studied by 13C NMR spectroscopy. Macromolecules 2006;39 6-8.

[19] Shao ZZ, Vollrath F. Surprising strength of silkworm silk. Nature 2002;418 741.

[20] Jin HJ, Kaplan DL. Mechanism of silk processing in insects and spiders. Nature 2003;424 1057-1061.

[21] Sinohara H. Glycopeptides isolated from sericin of the silkworm, Bombyx mori. Comparative Biochemistry and Physiology Part B: Comparative Biochemistry 1979;63(1) 87-91.

[22] Vollrath F, Knight DP. Liquid crystalline spinning of spider silk. Nature 2001;410 541-548.

[23] Sprague KU. Bombyx mori silk proteins: Characterization of large polypeptides. Biochemistry 1975;14 925-931.

[24] Takasu Y, Yamada H, Tsubouchi K. Isolation of three main sericin components from the cocoon of the silkworm, Bombyx mori. Bioscience Biotechnology and Biochemistry 2002;66 2715-2718.

[25] Colomban P., Dinh HM, Riand J, Prinsloo LC, Mauchamp B. Nanomechanics of single silkworm and spider fibres: a Raman and micro-mechanical in situ study of the conformation change with stress. Journal of Raman Spectroscopy 2008;39(12) 1749-1764.

[26] Freddi G, Pessina G, Tsukada M. Swelling and dissolution of silk fibroin (Bombyx mori) in N-methyl morpholine N-oxide. International Journal of Biological Macromolecules 1999;24(2-3) 251-263.

[27] Gulrajani ML. Degumming of silk. Review of Progress in Coloration 1992;22 79-89.

[28] Rajasekhar A, Ravi V, Reddy MN, Rao KRSS. Thermostable Bacterial Protease - A New Way for Quality Silk Production. International Journal of Bio-Science and Bio-Technology 2011;3(2) 43-58.

[29] Uddin K, Hossain S. A comparative study on silk dyeing with acid dye and reactive dye. International Journal of Engineering & Technology IJET-IJENS 2010;10(6) 21-26.

[30] Wray LS, Hu X, Gallego J, Georgakoudi I, Omenetto FG, Schmidt D, Kaplan DL. Effect of processing on silk-based biomaterials: Reproducibility and biocompatibility. Journal of Biomedical Materials Research B: Applied Biomaterials 2011;99B(1) 89-101.

[31] Rajkhowa R, Wang LJ, Kanwar JR, Wang XG. Molecular weight and secondary structure change in eri silk during alkali degumming and powdering. Journal of Applied Polymer Science 2011;119 1339-1347

[32] Svilokos Bianchi A, Colonna GM. Developments in the degumming of silk. Melliand Textilberichte 1992;73 68-75.

[33] Freddi G, Mossotti R, Innocenti R. Degumming of silk fabric with several proteases. Journal of Biotechnology 2003;106(1) 101-112.

[34] Freddi G, Allara G, Candiani G. Degumming of silk with tartaric acid. Journal of the Society of Dyers and Colourists 1996;112 191-195.

[35] Gulrajani ML, Chatterjee A. Degumming of silk with oxalic acid. Indian Journal of Fibre and Textile Research 1992;17 1.

[36] Gulrajani ML, Sethi S, Guptha S. Some studies in degumming of silk with organic acids. Journal of the Society of Dyers and Colourists 1992;108(2) 79-86.

[37] Khan MR, Tsukada M, Gotoh Y, Morikawa H, Freddi G, Shiozaki H. Physical properties and dyeability of silk fibers degummed with citric acid. Bioresource Technology 2010;101 8439-8445.

[38] Vohrer U, Müller H, Oehr C. Glow-discharge treatment for the modification of textiles. Surface and Coatings Technology 1998;98(1-3) 1128-1131.

[39] Borcia G, Anderson CA, Brown NMD. Surface treatment of natural and synthetic textiles using a dielectric barrier discharge. Surface and Coatings Technology 2006;201 3074-3081.

[40] Yip J, Chan K, Sin KM, Lau KS. Low temperature plasma-treated nylon fabrics. Journal of Materials Processing Technology 2002;123(1) 5-12.

[41] Long JJ, Wang HW, Lu TQ, Tang RC, Zhu YW. Application of Low-Pressure Plasma Pretreatment in Silk Fabric Degumming Process. Plasma Chemistry and Plasma Processing 2008;28 701-713.

[42] Mahmoodi NM, Moghimi F, Arami M, Mazaheri F. Silk Degumming Using Microwave Irradiation as an Environmentally Friendly Surface Modification Method. Fibers and Polymers 2010;11(2) 234-240.

[43] Mahmoodi NM, Arami M, Mazaheri F, Rahimi S. Degradation of sericin (degumming) of Persian silk by ultrasound and enzymes as a cleaner and environmentally friendly process. Journal of Cleaner Production 2010;18 146-151.

[44] Yuksek M, Kocak D, Beyit A, Merdan N. Effect of Degumming Performed with Different Type Natural Soaps and Through Ultrasonic Method on the Properties of Silk Fiber. Advances in Environmental Biology 2012;6(2) 801-808.

[45] Gübitz GM, Cavaco-Paulo A. Editorial: Biotechnology in the textile industry – perspectives for the new millennium. Journal of Biotechnology 2001;89 89-90.

[46] Gulrajani ML, Gupta SV, Gupta A, Suri M. Degumming of silk with different protease enzymes. Indian Journal of Fibre and Textile Research 1996;21 270-275.

[47] Gulrajani ML, Sen S, Soria A, Suri M. Efficacy of proteases on degumming of dupion silk. Indian Journal of Fibre and Textile Research 1998;23, 52-58.

[48] Gulrajani ML, Agarwal R, Chand S. Degumming of silk with fungal protease. Indian Journal of Fibre and Textile Research 2000b;25 138-142.

[49] Chopra S, Chattopadhyay R, Gulrajani ML. Low stress mechanical properties of silk fabric degummed by different methods. Journal of the Textile Institute 1996;87 542-553.

[50] Nakpathom M, Somboon B, Narumol N. Papain Enzymatic Degumming of Thai Bombyx mori Silk Fibers. Journal of Microscopy Society of Thailand 2009;23(1): 142-146.

[51] Arami M, Rahimi S, Mivehie L, Mazaheri F, Mahmoodi NM. Degumming of Persian Silk with Mixed Proteolytic Enzymes. Journal of Applied Polymer Science 2007;106 267–275.

[52] Gulrajani ML, Agarwal R, Grover A, Suri M. Degumming of silk with lipase and protease. Indian Journal of Fibre and Textile Research 2000a;25 69-74.

[53] Fabiani C, Pizzichini M, Spadoni M, Zeddita G. Treatment of waste water from silk degumming processes for protein recovery and water reuse. Desalination 1996;105 1-9.

[54] Zhang YQ. Applications of natural silk protein sericin in biomaterials. Biotechnology Advances 2002;20 91-100.

[55] Vaithanomsat P, Kitpreechavanich V. Sericin separation from silk degumming wastewater. Separation and Purification Technology 2008;59 129-133.

[56] Aramwit P, Siritientong T, Srichana T. Potential applications of silk sericin, a natural protein from textile industry by-products. Waste Management and Research 2012;30(3) 217-224.

[57] Kocabiyik S, Erdem B. Intracellular alkaline proteases produced by thermoacido-philes: detection of protease heterogeneity by gelatin zymography and polymerase chain reaction (PCR). Bioresource Technology 2002;84 29-33.

[58] Bastida A, Sabuquillo P, Armisen, P, Fernández-Lafuente R, Huguet J, Guisán JM. A single step purification, immobilization, and hyperactivation of lipases via interfacial adsorption on strongly hydrophobic supports. Biotechnology and Bioengineering 1998;58(5) 486-493.

[59] Kalantzi S, Mamma D, Christakopoulos P, Kekos D. Effect of pectate lyase bioscour-ing on physical, chemical and low-stress mechanical properties of cotton fabrics. Bio-resource Technology 2008;99(17) 8185-8192.

[60] AATCC Technical Manual. AATCC: Test Method 39 1980;55˙286

[61] Li Y, Hardin IR. Enzymatic scouring of cotton: surfactants, agitation, and selection of enzymes. Textile Chemist and Colorist 1998;30(9) 23-29.

[62] Tzanov T, Calafell M, Guenitz, GM, Cavaco-Paulo A. Bio-preparation of cotton fab-rics. Enzyme and Microbial Technology 2001;29 357-362.

[63] Segal L, Creely JJ, Martin AE, Conrad C.M. An empirical method for estimation the degree of crystallinity of native cellulose using the X-ray diffractometer. Textile Re-search Journal 1959;29(10) 786-794.

[64] Houde A, Kademi A, Leblanc D. Lipases and their industrial applications. Applied Biochemistry and Biotechnology 2004;118 155-170.

[65] Sonthisombat A, Speakman PT. Silk: Queen of Fibres -the Concise Story 2004 1-28. (available at http://www.en.rmutt.ac.th/prd/Journal/Silk_with_figuresnew.pdf)

[66] Zahn H, Krasowski A. Silk. In: Ullmann's Encyclopedia of Industrial Chemistry 2002